Knocking in Gasoline Engines

Michael Günther · Marc Sens
Editors

Knocking in Gasoline Engines

5th International Conference, December 12–13, 2017, Berlin, Germany

Springer

Editors
Michael Günther
IAV GmbH
Stollberg
Germany

Marc Sens
IAV GmbH
Berlin
Germany

ISBN 978-3-319-69759-8 ISBN 978-3-319-69760-4 (eBook)
https://doi.org/10.1007/978-3-319-69760-4

Library of Congress Control Number: 2017959600

Printed on acid-free paper

This Springer imprint is published by Springer Nature
The registered company is Springer International Publishing AG
The registered company address is: Gewerbestrasse 11, 6330 Cham, Switzerland

Preface

Dear colleagues,

If we want to continue to offer our customers vehicle drive systems that unite key characteristics such as low fuel consumption, impressive driving dynamics, flexibility in use, and high comfort combined with the sustainable use of resources and the lowest exhaust emissions, then we're going to have to ramp up our development budgets. Current market trends, which are seeing a decline in registrations of vehicles with diesel engines, are increasing the pressure to develop high-efficiency gasoline engines. As a bridging technology, the gasoline engine will be a key determinant of our car fleet's consumption in the coming years. Further increases in efficiency, however, will be limited by the well-established phenomenon of gasoline engine knock.

This conference thus addressed one of the most important issues in the ongoing development of the gasoline engine.

The majority of today's high-efficiency gasoline engines utilize downsizing or combine high mean pressures with Miller cycles and increased geometric compression, known as "rightsizing." When designing engines for high compression ratios—an essential requirement for efficiency increases—the knock limit is the defining criterion. In addition, the rise in specific torque at very low speeds as the basis of low fuel consumption, assisted by downsizing and downspeeding, has long been associated with the preignition phenomenon.

Ideal efficiencies are achieved by charge dilution combined with very high compression. This, however, causes "extreme knock," potentially leading to catastrophic events in the mid- to high-speed range. Improving charge-diluted concepts demands a greater focus of research in this field.

The introduction of RDE legislation this year will further grow the requirements for combustion process development, as residual gas scavenging in the low-end torque range and enrichment to reduce knock and exhaust temperatures will be legally limited. There is still, however, the need to reach the center of heat release with the highest possible compression. Preventing damage to high-efficiency gasoline engines demands a deep understanding of the phenomenology and precise detection and control engineering, resulting in new approaches to improved thermodynamics, more exact control algorithms, and optimized applications.

Together with its partners, IAV presented a wide range of research findings.

We hope that this conference gave all participants an opportunity to gain new insights, impressions, and ideas for their future development work.

We would like to thank our partners at UNIVERSAL Kongress & Event Marketing GmbH, and in particular Mr. Kniehase, for their proactive support in preparing and running the conference, our IAV colleague Lars Gamasin, and Springer Verlag.

Special thanks go to the international speakers, without whose active contribution this conference could not have been such a success.

The conference explored important ways to improve engine thermodynamics and control technology of interest to combustion process specialists, design engineers, and application experts. In the process, it presented solutions of relevance to advanced engineering, volume, and application specialists. IAV and its expert colleagues around the world are working hard on solutions that prevent or reliably control knock and other irregular combustion events such as preignition. These solutions create the conditions to further increase the efficiency of the gasoline engine and thus meet ambitious CO2 emission reduction targets.

We would be pleased to welcome you again as a participant or exhibitor to this IAV conference.

December 2017

Michael Günther
Marc Sens

Contents

Introduction

Proposal of Knock Mitigation Method Through Enhancement of Local Heat Transfer 3
Katsuya Matsura, Yoshihisa Sato, Kazuo Yoshida and Hiroshi Sono

Approaches to Meeting Fluctuating Natural Gas Quality in Large Bore Engine Applications 17
Jan Zelenka, Claudio Hoff, Martin Kirsten and Andreas Wimmer

Pre-ignition

Investigating the Cause of Initial Pre-ignition - A New Approach 37
Ina Volz, Jürgen Pfeil, Thomas Koch and Frank Altenschmidt

Optical Diagnostic Tools for Detection and Evaluation of Glow Ignitions 55
Arndt Döhler and Peter Schaffner

Study of HSPI/LSPI from Spark Plugs on Turbocharged Gasoline Engines 71
Tomohiro Iwatsuka, Masaru Kano, Kiyoteru Mori, Waldemar Werner and Stefan Schulte

Simulation

Fundamental Mechanism Analysis on the Underlying Processes of LSPI Using Experimental and Modeling Approaches 89
Masaharu Kassai, Taisuke Shiraishi and Toru Noda

Chemical Analysis of Potential Initiating Fluid for Low-Speed Pre-ignition 112
Terence F. Alger and Thomas E. Briggs, Jr.

Simulation of the Effects of Spark Timing and External EGR on Gasoline Combustion Under Knock-Limited Operation at High Speed and Load 121
Michal Pasternak, Corinna Netzer, Fabian Mauss, Michael Fischer, Marc Sens and Michael Riess

Development of a Model for Predicting the Knock Boundary in Consideration of Cooled Exhaust Gas Recirculation at Full Load . . . 143
Alexander Fandakov, Michael Grill, Michael Bargende, Max Mally, Marco Günther, Stefan Pischinger, Liming Cai, Heinz Pitsch, Karl Alexander Heufer, Ajoy Ramalingam, Heiko Minwegen and André Casal Kulzer

Optical Measurement Technique

Visualization of Fuel Wall Wetting, Oil Dilution by Fuel, and Oil Transport Mechanisms in an Optically Accessible Engine by LIF Imaging 189
Stefan Wigger, Torben Müller, Hans-Jürgen Füßer and Sebastian Kaiser

Irregular Combustion Events in RDE Test Situations Diagnostics – Analysis – Improvements 203
Ernst Winklhofer, Alois Hirsch and Paul Kapus

Quantitative Optical Measurement Techniques for Mixture Formation and Combustion Process Analysis 214
Thomas Berg, Stefan Seefeldt and Olaf Thiele

Detection and Analysis Methods for Irregular Combustion in SI Engines 225
Ulrich Spicher

Fuel/Lubricating Oil

Impact of Detailed Fuel Chemistry on Knocking Behaviour in Engines 245
Roger F. Cracknell, Arjun Prakash, Kieran P. Somers and Chongming Wang

Knock Detection

Model Based Knock Detection 257
Matthias Biehl and Michael Meister

Trustworthy Estimation and Control of Engine Knocking Level for Transient Operation 267
Maxime Jean, Thomas Leroy and Fabien Vidal-Naquet

Potential of Series-Compatible In-Cylinder Pressure Sensors for Gasoline Engines Using the Example of Ignition Angle Control 279
Benedikt van Booven, Harry Schüle, Thorben Walder and Hermann Rottengruber

Water Injection

Water Injection for Gasoline Engines - Quo Vadis? 299
Ingo Hermann, Claus Glahn, Matthias Kluin, Martin Paroll and Werner Gumprich

Gasoline Water Direct Injection (GWDI) as a Key Feature for Future Gasoline Engines 322
Christoph Heinrich, Heinrich Dörksen, Andreas Esch and Kevin Krämer

Potential of Direct Water Injection to Reduce Knocking and Increase the Efficiency of Gasoline Engines 338
Matthias Hunger, Tobias Böcking, Ulrich Walther, Michael Günther, Normann Freisinger and Günter Karl

Combustion Process

Suppressing Knocking by Using CleanEGR – Better Fuel Economy and Lower Raw Emissions Simultaneously 363
Michael Fischer, Michael Günther, Carsten Berger, Ralf Troeger, Michal Pasternak and Fabian Mauss

Introduction

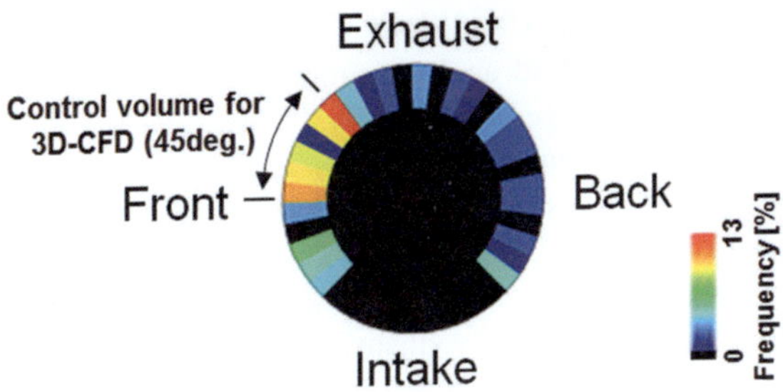

Fig. 3. Knock occurrence frequency at each position

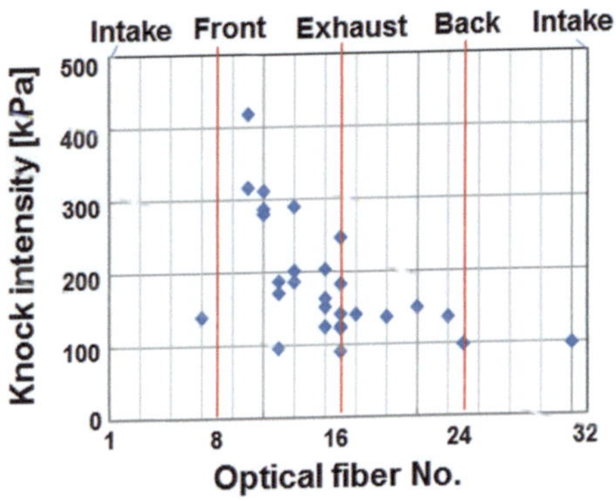

Fig. 4. Relationship between auto-ignition onset position and knock intensity (*Ne*: 2000 rpm, *A/F*: 14.7, *Pin*: 90 kPa, θig: −5 deg. ATDC)

3 Enlarging Knock Limit by Enhancement of Local Heat Transfer

3.1 Evaluation of Near-Wall Flow at Knock Onset Position

The objective of this study was to enlarge the knock limit by appropriately strengthening the flow at the knock onset position to locally promote heat transfer, so it was important to understand the flow at this position. When evaluating the near-wall flow in experiments, installation of a measurement probe results in evaluation of an altered flow. Optical measurement such as PIV is limited to the region that the incident laser beam can reach, and there are concerns of reduced measurement accuracy due to light source reflection from the wall surface. Therefore, this study used in-cylinder CFD to evaluate the flow near the knock onset and investigate piston shapes that manipulate the flow.

3.1.1 Calculation Model and Conditions

Calculation used the thermal fluid analysis program Converge (Convergent Science, Inc.), a RANS (Reynolds Averaged Navier-Stokes) based RNG κ-ε model was used as the turbulence model, and a wall function was applied near the wall surface. Flow calculations were performed using the geometric shape (Fig. 5) and specifications (Table 1) of the test engine used to measure the knock onset in the previous section. Table 2 shows the calculation conditions. The calculation mesh size of the calculation region extending for 10 mm from the wall surface toward the center of the combustion chamber was fixed to 0.5 mm for the entire calculation period. The mesh size in other regions was set to 0.5, 1.0, or 2.0 mm in accordance with the change in volume of the combustion chamber. However, the mesh size around the intake and exhaust valves and valve seats was set to 0.25 mm during the gas exchange stroke in order to converge the calculations.

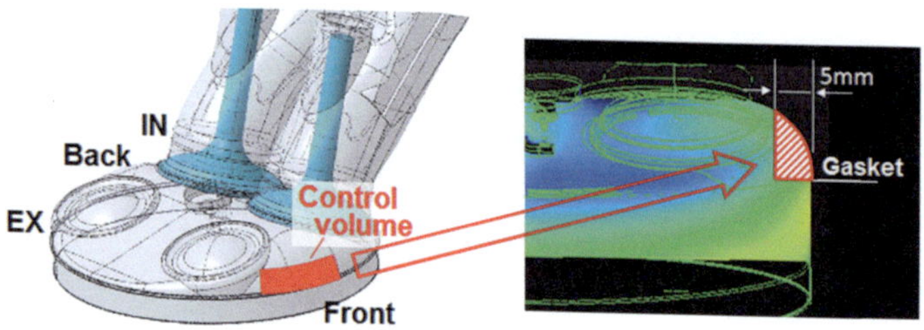

Fig. 5. Control volume for evaluating in-cylinder flow on 3D-CFD

Table 2. Calculation conditions

Engine speed	2000 rpm
Gas	Air (N_2: 79%, O_2: 21%)
Turbulent Prandtl number	0.9
Intake pressure	90 kPa (averaged)
Cylinder head temperature	353 K
Liner temperature	353 K
Piston temperature	353 K

In addition, a one-dimensional calculation code (Gamma Technologies GT-Power) was used to set the boundary conditions, with the measured cylinder pressure and intake and exhaust pulsations applied to a one-dimensional engine model. This enabled to calculate the pressure pulsation at the intake manifold upstream end and the exhaust manifold downstream end positions of the three-dimensional calculation model, and these values were input as the boundary conditions for three-dimensional calculations.

3.1.2 Design of Piston Shape

In the engine used in the tests, the air fuel mixture led from the intake manifold into the cylinder flows down along the sleeve on the exhaust side. When this flow reaches the piston top surface, it splits into left and right components that form a tumble flow, and a component that flows along the sleeve is also formed. That is to say, the exhaust side of the piston end is the onset point where the air fuel mixture led into the cylinder forms a complex flow field after reaching the piston top surface, and modifying the shape at this position can be expected to manipulate the flow. Figure 6 shows the piston top surface shapes evaluated by CFD. The objective was to use a groove (hereafter, guide) provided around the top surface to lead the flow reaching the exhaust side end of the piston top surface to the knock onset position.

Fig. 6. Piston top surface shapes to manipulate the in-cylinder flow

In order to extract the effects of the guide shape, calculations were performed for the three specifications of a reference shape with a flat piston top surface (FLAT), a guide width of 2.5 mm and depth of 5.0 mm (W2.5/D5.0), and a guide width of 5.0 mm and depth of 10.0 mm (W5.0/D10.0).

3.1.3 Evaluation of In-Cylinder Flow

Except during cold start, heat transfer from the combustion chamber wall surface to the air fuel mixture starts during the intake stroke and continues until the middle of the compression stroke. Piston compression then increases the temperature of the air fuel mixture, and when the temperature exceeds that of the wall surface, the heat exchange reverses and heat is transferred from the air fuel mixture to the wall surface. However, it is known that the tumble flow breaks down and intense turbulence is formed near the top dead center (TDC) [7], and this turbulence is presumed to strongly influence heat transfer around the TDC. In addition, end gas auto-ignition occurs near the wall surface, so the evaluation region was the interrogation volume shown in Fig. 5 (see Fig. 3 for the position). The average values of the flow velocity and turbulence intensity within the interrogation volume were calculated, and the change in the in-cylinder flow due to the guide was investigated.

3.1.4 Calculation Results

Figure 7(a) shows the averaged flow velocity and Fig. 7(b) shows the averaged turbulence intensity inside the interrogation volume. For all specifications, both the flow velocity and turbulence intensity decreased until the middle of the compression stroke, and then both increased from around −90 deg. ATDC. The averaged flow velocity peaked at around −45 deg. ATDC, and the averaged turbulence intensity then peaked

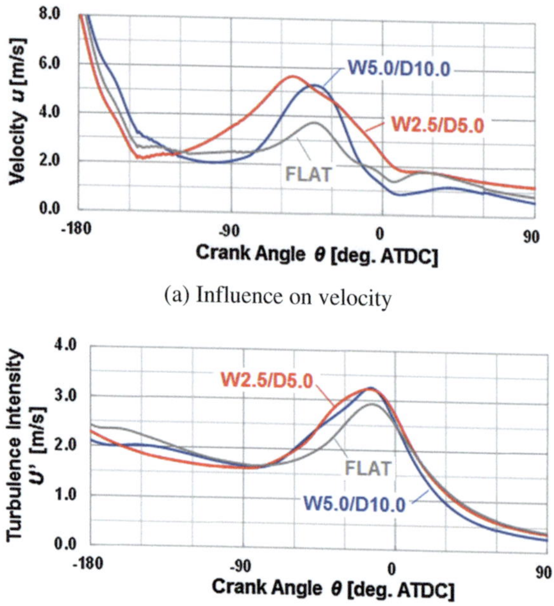

(a) Influence on velocity

(b) Influence on turbulence intensity

Fig. 7. Effect of piston top shape on velocity (upper) and turbulence intensity (lower) in knock onset region (*Ne*: 2000 rpm, Gas: Air, Motoring)

at around −15 deg. ATDC. The reason for this is that as the cylinder volume decreased, the bulk flow velocity increased, so the bulk flow split at the compression TDC and intense turbulence was formed by the breakdown of tumble flow.

Examination of the change in flow due to the difference in piston shape shows that compared to the FLAT shape, providing a guide increases the flow velocity which contributes to increasing the turbulence intensity. Figure 8 compares the center section of the combustion chamber in the piston pin direction and the velocity distribution at the section 2.5 mm from the wall surface in the direction perpendicular to that center section for the FLAT and W2.5/D5.0 guide specifications. The bulk flow reaching the piston top surface splits to the left and right at −90 deg. ATDC and then rises along the sleeve surface for both piston specifications. However, for the specification with the guide, the air flowing inside the guide begins to rise along the sleeve surface from around the arrows in the figure, and the high flow velocity has a wider distribution compared to that of the FLAT shape. The −40 deg. ATDC and −20 deg. ATDC timing charts confirm that this flow is maintained even when the piston moves to the TDC.

3.2 Heat Transfer at Knock Onset Position

This section describes the instantaneous heat flux measurement performed to examine heat transfer between the end gas and the wall surface during the intake and compression strokes at the position where knock onset frequently was high. In addition, the influence of the near-wall flow obtained by CFD on heat transfer was also investigated.

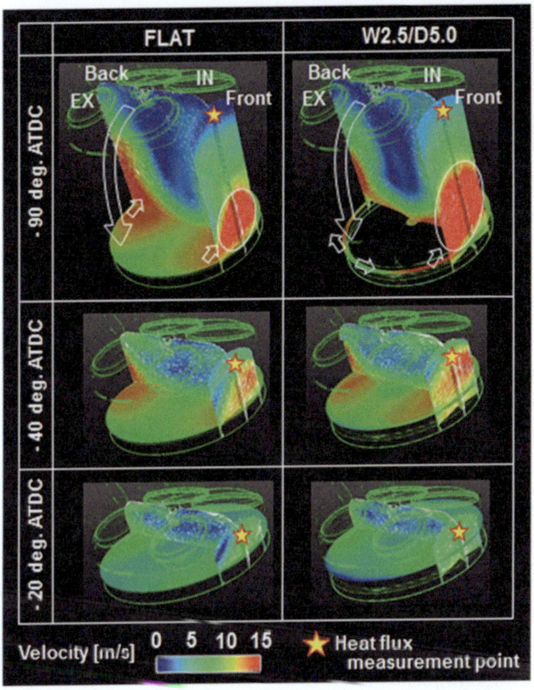

Fig. 8. Comparison of in-cylinder air flow behavior between piston without guide (left) and piston with guide (right)

3.2.1 Experimental Apparatus and Procedure

Figure 9 shows the photograph of the heat flux sensor. The heat flux sensor (Medtherm TCS-K-10702A) has K type (Chromel/Alumel) thermocouples at the sensor tip (to measure the surface temperature) and at the position 4 mm from the tip (to measure the wall inside temperature), and obtains the instantaneous heat flux by measuring the time history in the temperature difference between these two points over time. When calculating the instantaneous heat flux from the measured values, it was assumed that Fourier's one-dimensional differential equation of heat transfer (1) is satisfied at the measurement position, and the difference Eq. (2) that takes into account the amount of

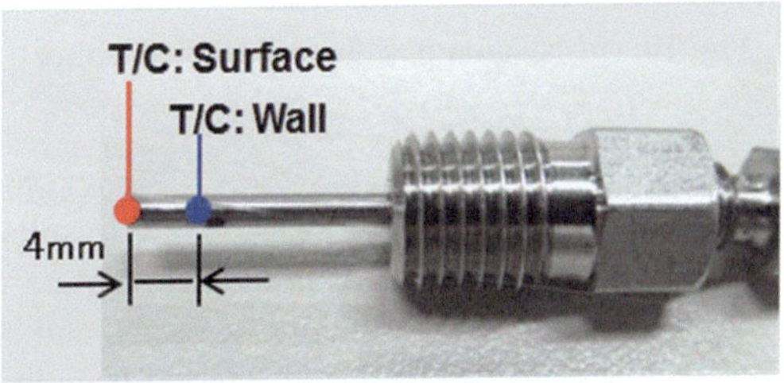

Fig. 9. Photo of heat flux sensor and position of thermo couples (T/C)

heat stored during the time Δt was used between the wall surface and the distance Δx inside the wall [8].

$$\frac{\partial T}{\partial t} = \frac{\lambda}{C\rho}\frac{\partial^2 T}{\partial x^2} \tag{1}$$

$$q(k) = \frac{\lambda}{\Delta x}\{T(1,k) - T(2,k)\} + C\rho\frac{\Delta x}{2}\frac{\{T(1,k+1) - T(1,k)\}}{\Delta t} \tag{2}$$

Here,

$T(i, k)$: Temperature [K] at position i (i = 1, 2, 3, …) and time k (k = 1, 2, 3, …)
λ: Thermal conductivity [W/m* K]
C: Specific heat [J/kg* K]
ρ: Density [kg/m^3]
$q(k)$: Instantaneous heat flux [W/m^2]
Δx: Depth division width (distance between i and i+1) [m]
Δt: Time division width (time difference between k and k+1) [s]

A heat flux sensor was installed on the front side of the test engine cylinder head used to investigate the knock onset as shown in Fig. 10. The test settings were engine speed *Ne*: 2000 rpm, regular gasoline (RON90.1), *A/F*: 14.7, intake air temperature *Tin*: 25 °C, lubricant temperature *Toil*: 80 °C, cooling water temperature *Tw*: 85 °C, and intake pressure *Pin*: 55 kPa. Combustion was assumed to start at the 5% mass fraction burned timing CA5, and the test was performed at the ignition timing where CA5 = −5 deg. ATDC. The heat flux was recorded with a time resolution of 1 deg. crank angle, and was evaluated using the averaged value of 250 cycles.

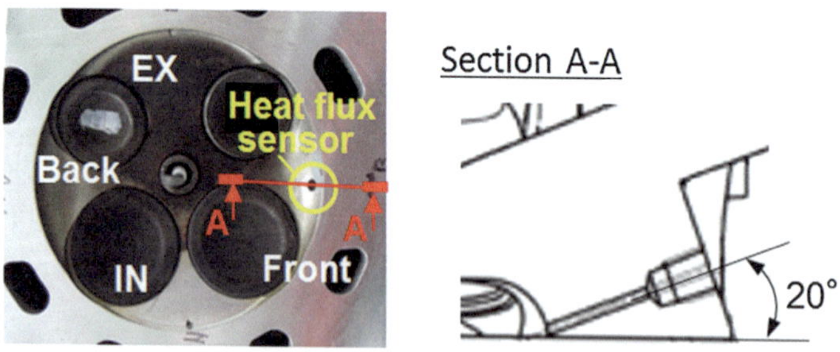

Fig. 10. Installation position of heat flux sensor

3.2.2 Test Results

Figure 11 shows the instantaneous heat flux measurement results, with the cylinder volume *Vcyl* on the horizontal axis. The wall surface temperature was higher than the air fuel mixture led into the cylinder during the intake stroke and continues to be until the middle of the compression stroke, so heat is transferred from the wall surface to the air fuel mixture during this period. That is to say the instantaneous heat flux value is negative. Piston compression then increases the temperature of the air fuel mixture, and when the temperature exceeds that of the wall surface, the heat transfer direction reverses and the instantaneous heat flux value becomes positive. In the compression stroke, the values for all three specifications trace similar histories until around −40 deg. ATDC, but thereafter the pistons with guides exhibit higher heat flux values than the FLAT shape.

The star marks in Fig. 8 show the heat flux measurement points. Here, the flow led into the cylinder splits to the left and right after reaching the piston top surface, and the flow rising along the sleeve wall surface reaches the heat flux measurement position at approximately −40 deg. ATDC. Furthermore, the high flow velocity range is wider for piston specifications with guides compared to FLAT shapes. Therefore, it can be inferred that a difference occurs in the instantaneous heat flux history during the compression stroke.

In addition, the maximum instantaneous heat flux values in Fig. 11 indicate that providing a guide and increasing the guide size increases the maximum value. Figure 12 shows the *P-V* diagram at this time. This tendency appears as a drop in the maximum pressure, and is particularly evident for the W5.0/D10.0 specification. One reason for this is thought to be that, as shown in Fig. 7, the W5.0/D10.0 specification exhibits lower flow velocity and turbulence intensity after the TDC compared to the FLAT specification, so the combustion duration increases, which increases the heat transfer amount from the gas to the wall surface, and reduces the degree of constant volume. As a result, the indicated work for the W5.0/D10.0 specification is 14 kPa lower than the IMEP = 510 kPa of the FLAT specification. In contrast, the indicated work for the W2.5/D5.0 specification is approximately the same as that of the FLAT specification.

The above results confirmed that providing an appropriately shaped guide can locally promote heat transfer at the knock onset position without adversely affecting the

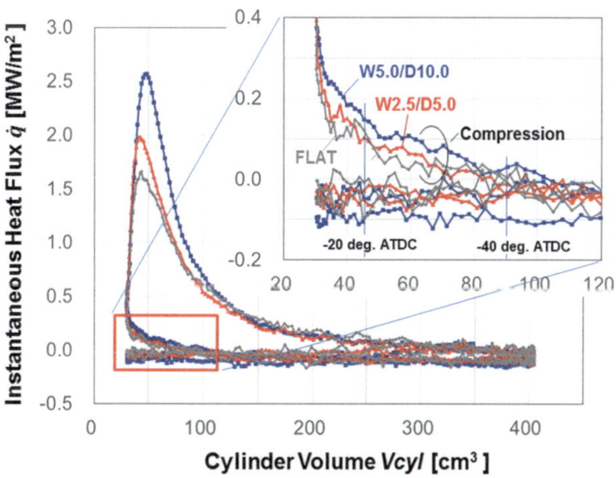

Fig. 11. Comparison of instantaneous heat flux histories between three different piston top shapes (*Ne*: 2000 rpm, *A/F*: 14.7, *Pin*: 55 kPa, *Tin*: 25 °C)

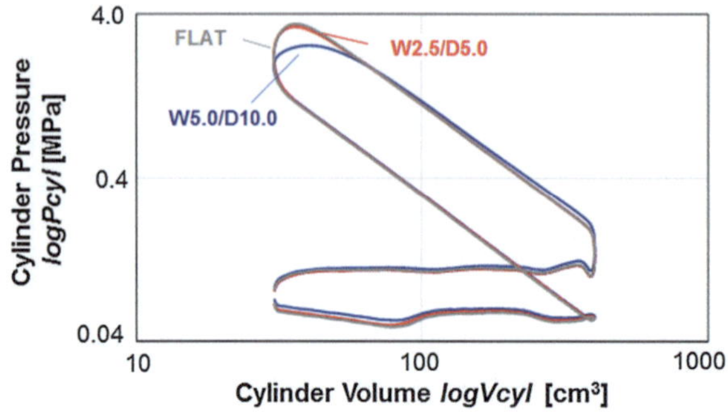

Fig. 12. Comparison of cylinder pressure traces obtained simultaneously during heat flux measurement (Same test conditions as Fig. 11)

degree of constant volume. This means that higher power can be expected due to the avoidance of knock during high load operation.

3.3 Engine Performance on High Load Condition

The instantaneous heat flux measurements in the previous section confirmed that heat transfer from the end gas to the wall surface at the position where knock frequently generates can be promoted by modifying the piston end shape to manipulate the flow. This section describes the engine performance under high load conditions and evaluation of the knock mitigation effects of manipulating the piston shape.

3.3.1 Experimental Conditions and Procedure

This test used the same test engine used to investigate the knock onset direction, and the test settings were engine speed *Ne*: 2000 rpm, regular gasoline (RON90.1), air fuel ratio *A/F*: 14.7, intake air temperature *Tin*: 25 °C, intake air humidity 50%, lubricant temperature *Toil*: 80 °C, cooling water temperature *Tw*: 85 °C, and intake pressure *Pin*: 90 kPa. The engine power and knock limit were investigated to compare the FLAT and W2.5/D5.0 guide specifications. A microphone was installed in the test engine cylinder block position and the knock sound was monitored during the test. Knock sound was audibly recognized when the high frequency components (5 to 12 kHz) of the cylinder pressure exceeded 30 kPa, so 30 kPa was used as the threshold value for determining the presence of knock with the cylinder pressure in this test.

3.3.2 Test Results

Figure 13 shows the IMEP, knock limit (when retarded by 1° from the trace knock ignition timing), and percentage of unburned hydrocarbons and carbon monoxide in the exhaust. Compared to the piston specification without a guide, providing a guide enabled combustion phase advancement by 2.2°, and this advanced combustion phase increased the IMEP by 17.2 kPa. This corresponds to an increase in indicated thermal efficiency of 0.7%. Regarding the exhaust components, there was no apparent change in the unburned hydrocarbon emission rate with or without a guide, but providing a

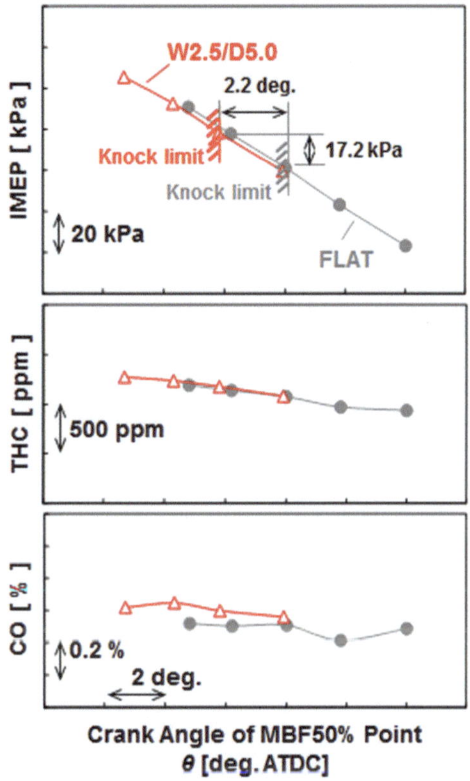

Fig. 13. Comparison of indicated power and unburned component concentration in exhaust gas due to flow guide (*Ne*: 2000 rpm, *A/F*: 14.7, *Pin*: 90 kPa, *Tin*: 25 °C, Spark timing swing)

guide tended to increase the carbon monoxide emission rate. Potential reasons for this are thought to be that strengthening the flow to avoid knock lowered the temperature near the wall toward the end of the combustion process, which froze the carbon monoxide oxidation reaction, or that providing a guide created a locally rich region, which changed the equilibrium concentration, etc. Detailed analysis is planned for the future to reduce carbon monoxide emissions.

The above results show that the piston specifications with a guide slightly increase both the carbon monoxide in the exhaust and the heat transfer from the gas to the wall, so the IMEP tends to drop. However, this drop in IMEP is small, and knock can be avoided with the combustion phase as advanced as described above, so it was confirmed that use of a simple method based on this study concept enabled an engine power increase without major changes to the engine structure. In addition, the knowledge obtained through this study also suggests that if an appropriate flow can be arranged at the knock onset position, knock can also be avoided by other methods without relying on piston shape manipulation.

4 Conclusion

The applicability of strengthening the local in-cylinder flow to promote heat transfer between the end gas and the wall surface was investigated as a simple method of enlarging the knock limit. Engine performance was evaluated using CFD, instantaneous heat flux measurement, and prototype pistons, and the following knowledge was obtained.

- The flow formed in the second half of the compression stroke is closely related to combustion and knock. Manipulation of the piston top shape to lead the flow to the knock onset position can strengthen both the flow velocity and turbulence intensity at that position.
- Instantaneous heat flux measurements confirmed that use of a guide to lead the flow to the knock onset position promotes heat transfer from the end gas to the wall surface.
- Use of a simple method based on this study concept enabled to enlarge the knock limit and increase engine power without major changes to the engine structure.

References

1. Rudolf, K., Rüdiger, S., Hermann, M., Michael, F., Alfons, L., Stefan, V.: Neuer Ottomotor mit Direkteinspritzung und Doppelaufladung von Volkswagen. MTZ **66**(11), 844–856 (2005)
2. Konig, G., Maly, R.R., Bradley, D., Lau, A.K.C., Sheppard, C.G.W.: Role of exothermic centres on knock initiation and knock damage. SAE Technical Paper 902136 (1990)
3. Hiroshi, T., Mitsuo, K.: Mechanisms of strong pressure wave generation in end-gas autoignition during knocking combustion. Combust. Flame **162**, 1944–1956 (2015)
4. Midgley, T., Boyd, T.A.: Industrial engine chemistry, vol. 14, p. 584 (1922)
5. Hiromitsu, A.: Combustion control technologies for GDI engines-Removal of restriction in mixing control, no. 10, pp. 24–32. Mitsubishi Motors Technical Review (1998)
6. Philipp, H., Hirsch, A., Baumgartner, M., Fernitz, G., Beidl, C., Piock, W., Winklhofer, E.: Localization pf knock events in direct injection gasoline engines. SAE Technical Paper, 2001-01-1199 (2001)
7. Kuwahara, K., Watanabe, T., Takamura, J., Omori, S., Kume, T., Ando, H.: Optimization of in-cylinder flow and mixing for a center-spark four-valve engine employing the concept of barrel-stratification. SAE Technical Paper, 940986 (1994)
8. Yoshiteru, E., Toshihiro, O., Mitsunori, I., Kenjiro, E., Nobuhiro, K.: Study on analysis of instantaneous heat flux flowing into the combustion chamber wall of an internal combustion engine. Examination in the case of consideration of heat storage term and the temperature dependency of the thermocouple's thermophysical properties. J. Jpn. Soc. Mech. Eng. B **57**, 2421–2427 (1991)

Approaches to Meeting Fluctuating Natural Gas Quality in Large Bore Engine Applications

Jan Zelenka[1(✉)], Claudio Hoff[1], Martin Kirsten[1], and Andreas Wimmer[1,2]

[1] LEC GmbH (Large Engines Competence Center), Graz, Austria
jan.zelenka@lec.tugraz.at
[2] Institute of Internal Combustion Engines and Thermodynamic, Graz, Austria

Abstract. State-of-the-art large bore gas engines for power generation are traditionally designed for long-term operation at high load. They run at high power density and high engine efficiency, which is guaranteed as long as the engine boundary conditions, e.g., fuel gas quality, stay within certain limits. A growing share of gases from alternative sources such as biomass, hydrogen from power-to-gas technologies or LNG will be found in the pipeline grids of the future. This is further promoted by the harmonization process for gas qualities in Europe. The European standard EN 16726 as well as EASEE Gas Directive set rather wide limits for relative density, calorific value and methane number. For both engine manufacturers and operators, new challenges arise that have to be met by improving operating strategies.

Operating a stationary large engine with gaseous fuels – spark or direct ignited (diesel pilot) – leads to the problem of auto-ignition. The detection of knocking combustion is an important measure for the active prevention of engine damage. Different regimes during the combustion process result in a diversification of the cylinder pressure signal as well as the acceleration sensor signal. This requires different strategies for the distinct identification of knocking combustion. The methods shown are mainly based on the use of the cylinder pressure signal and focus on knock detection, detection of knock onset and evaluation of knock intensity.

If knocking combustion is detected, the engine control system has to react appropriately in order to ensure stable and safe engine operation. Here one can distinguish between approaches that adjust soft parameters of state-of-the-art engines (ignition timing or mixture quality), and approaches that need additional hardware and their functionality to be realized (variable intake valve timing or variable compression ratio). It is shown how these approaches to knock control impact engine performance.

1 Introduction

1.1 Volatility of Gas Quality

Current large bore gas engines are highly efficient, very robust and nearly maintenance-free engines for power generation. In addition to these attributes, such combustion engines can be used to burn a wide range of gases of various origins. Over

M. Günther and M. Sens (eds.), *Knocking in Gasoline Engines*,
https://doi.org/10.1007/978-3-319-69760-4_2

the last few years it has become very common for gas engine manufacturers to offer many different configurations derived from the same engine platform that run on a specific gaseous fuel-composition.

Figure 1 indicates the most important sources of gases and how they are converted into electricity and fed into the grid.

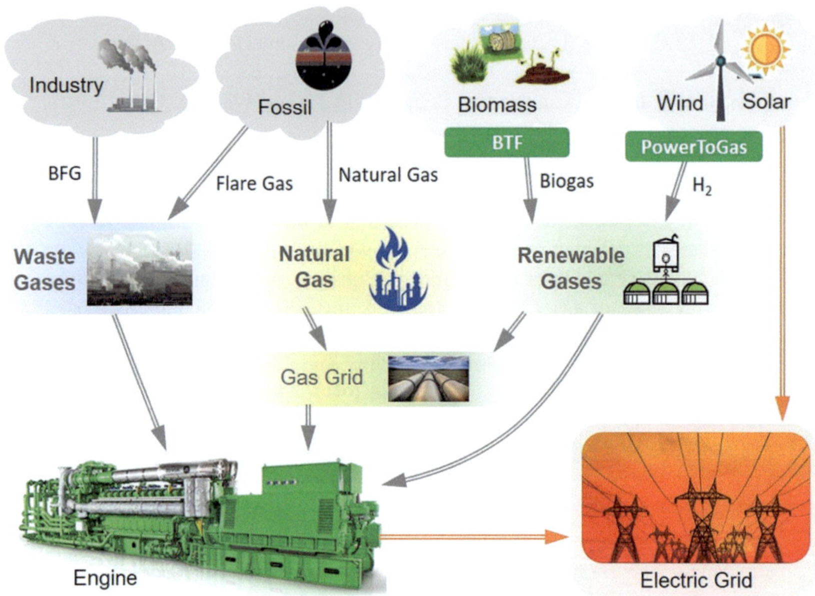

Fig. 1. Power generation with gases from various sources [1]

In general, gas engines have a comparatively low environmental impact, which has been steadily reduced in recent years due to progress in development. If renewable gases are burned, this benefit is even enhanced. As shown above, such gases are created from biomass via gasification processes, which are often referred to by the synonym biomass to fuel (BTF), or via power to gas, where hydrogen is produced from excess electric power from wind parks or solar plants. Renewable gases can either be directly fed into the engine or added to the conventional gas grid. The latter presents a great challenge for stable operation of a gas engine because this is the main driver of volatility in pipeline gas quality. The properties of the gas in the grid change depending on the amount of renewable gases fed into the pipeline system. The great unknown is the exact composition in the grid, hence which quality can be expected at each engine's inlet pipe.

Besides renewable gases, the most pervasive approach is natural gas from fossil resources. Special adaptations enable gas driven large bore engines to use flare gases and waste gases, e.g., blast furnace gas (BFG), from industrial processes.

For sustainable energy production, it has become more and more important to use gases from alternative, renewable resources. To this end, the European Union and the

European Committee for Standardization (CEN) in particular created the standard EN 16726 "Gas infrastructure – Quality of gas – Group H" in 2015. This standard defines requirements for gas quality with the goal of allowing free gas flow in the pipeline grid within the CEN member states [2, 3]. The restrictions from EN 16726 are shown in Fig. 2.

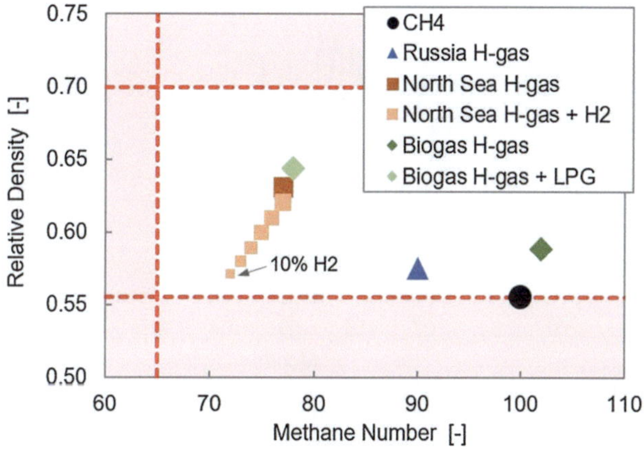

Fig. 2. Real pipeline gases and mixtures of them with limits according to EN 16726

In EN 16726 the member states were able to agree on limitations in methane number and relative density. The methane number has a lower limit of just 65. To calculate the relative density, the gas density is divided by the air density at reference conditions (0 °C and 101 kPa). The resulting value's lower limit is 0.555 and the upper limit is 0.7. As illustrated in the figure, the examples stay within the limits indicated by the dashed red line.

It can be seen that an admixture of hydrogen (e.g., from Power to Gas) is not possible with each starting gas composition. Pure methane, biogas as well as H-gas from Russia are already close to the minimum relative density. Hydrogen can only be admixed to North Sea H-gas due to the latter's high relative density, providing even up to 10% of volumetric share without exceeding the lower limit.

1.2 Consequences for Engine Operation

The most important gas property for combustion in a common gas engine is the methane number. It can be seen as the key parameter to quantifying the knocking tendency of a specific gas composition. A high methane number indicates good resistance of the gaseous fuels to knocking combustion and vice versa. Pure methane, for example, has a methane number of 100, as illustrated by the black dot in Fig. 2.

In addition to changes in environmental boundary conditions such as temperature, ambient pressure and humidity, another great challenge that has to be taken into account is volatile gas composition. Figure 3 provides an example of the variations in methane number in the natural gas grid in southeastern Austria observed over the last two years.

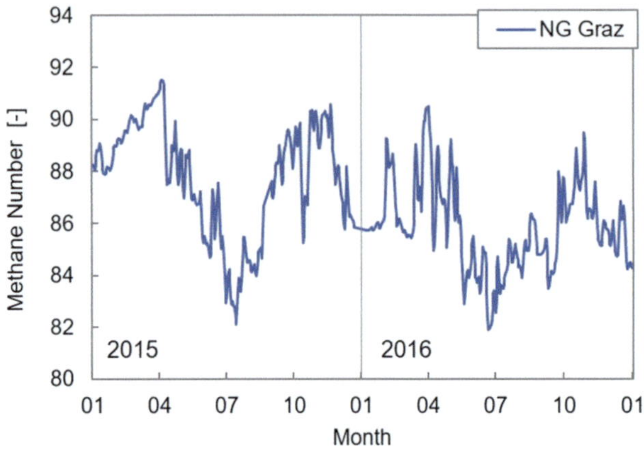

Fig. 3. Fluctuations in methane number of natural gas in Graz, Austria

The seasonal influence on gas quality is obvious; especially during spring and autumn, the methane amount in the gas grid seems to be quite high, whereas in summer, the methane number reaches its annual minimum. Fluctuations in this quantity have to be considered in order to guarantee stable and satisfactory engine performance during the entire year.

Diverse measures can be taken to respond to comparatively low methane numbers; what they have in common is that they should reduce the tendency for knocking combustion. Possible measures include [4]:

- Lowering air or mixture temperatures in the intake manifold
- Employing exhaust gas recirculation
- Cooling the cylinder charge via Miller timing
- Avoiding deposits to keep wall heat transfer constant
- Increasing charge motion to reduce time for pre-reactions
- Raising the engine speed level
- Increasing excess air ratio

Many of these measures attempt to reduce the temperature of the cylinder charge, which also has a positive effect on nitrogen oxide production. For a detailed description of selected measures, see Sect. 3.

2 Methods for Detecting Knocking Combustion

Knock detection aims to separate knocking combustion cycles from non-knocking combustion cycles using calibrated algorithms as well as to determine the precise knock onset.

If severe, knock can result in fatal engine damage within a couple of combustion cycles. [5] Thus to prevent engine damage and ensure knock-free engine operation at any time during commercial engine operation, a knock detection algorithm is mandatory. The knock detection algorithm provides the basis for counteractive measures such as control of engine parameters and/or advanced hardware functionalities. Researchers also use knock detection algorithms to determine the knock limit under certain boundary conditions.

Following, the measured cylinder pressure trace is the basic physical parameter for knock detection because it is superimposed by high frequency pressure oscillations when knock occurs.

Since combustion regimes for spark ignited combustion and diesel pilot ignited combustion differ, they require different knock detection algorithms.

2.1 Spark Ignited Combustion (Gas Engine Concept)

With spark ignited combustion using gaseous fuels, knock detection based on the in-cylinder pressure trace involves identifying the sudden transition to knocking combustion, whereas the in-cylinder pressure trace is superimposed by high frequency pressure oscillations as shown in Fig. 4.

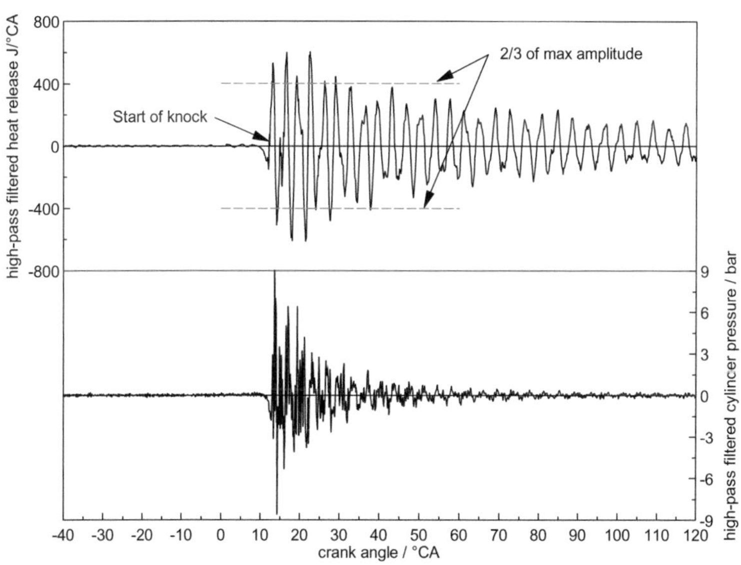

Fig. 4. High-pass filtered heat release and cylinder pressure [9]

The knock detection algorithm is basically a combination of several criteria. To increase sensitivity by differentiation, the apparent heat release rate is calculated from the pressure curve using the first law of thermodynamics. The high frequency part is extracted using a high-pass filter, after which a critical threshold value is determined that is two thirds of the maximum amplitude. Second, the amplitude that first exceeds this threshold value is determined. The zero value before this amplitude is the start of knocking, see Fig. 4.

To ascertain that this is in fact the start of knock, the high frequency pressure oscillations in the area of several degrees crank angle before and after the start of knock are integrated and set in relation to each other. This ratio must also exceed a threshold value. In addition to the start of knock, knock intensity is another parameter used to characterize knock behavior and can be determined from the high-frequency part of the cylinder pressure curve [6–9].

2.2 Diesel Pilot Ignition (Dual Fuel Concept)

The challenge in setting up a knock detection system with diesel pilot ignited combustion systems is the appearance of ringing combustion, or simply ringing. Ringing is a well-known phenomenon induced by diesel pilot combustion. Both ringing and knocking are the result of a sudden conversion of large parts of fresh cylinder charge that leads to pressure oscillations.

Three parameters influence the severity of ringing:

- Rail pressure (p_Rail)
- Start of injection (SOI)
- Amount of diesel injected (φ_Diesel)

Increasing the rail pressure, advancing the SOI and increasing the amount of injected diesel increase the tendency for ringing to occur.

Investigations have also shown that ringing intensity greatly fluctuates in consecutive in-cylinder pressure traces. An example is given in Fig. 5. Cycle 1 exhibits strong ringing whereas cycle 2 has light ringing and cycle 3 has a ringing intensity somewhere in between.

Under extreme conditions, ringing and knocking can naturally occur within the same combustion cycles. As seen in Fig. 5, ringing combustion is present throughout the entire combustion cycle, whereas knocking combustion sets in after TDC – glow ignition is not taken into account.

Therefore, knock detection with diesel/natural gas dual fuel combustion must be capable of distinguishing between ringing and knocking combustion. This can be achieved by calculating the energy of the superimposed oscillations in a time range of the combustion cycle when only ringing combustion occurs and in a time slot during which knocking combustion occurs.

The superimposed oscillations are extracted from the in-cylinder pressure trace using a bandpass filter (p_{bp}). Based on this data, the signal energy (SE) of the signal in the particular time slots during the combustion cycle are calculated according to:

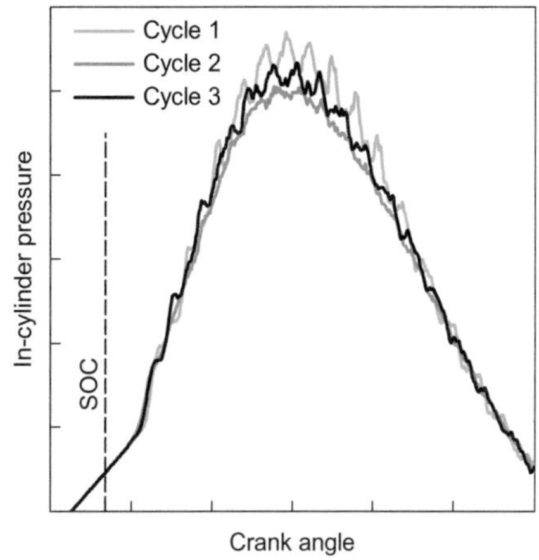

Fig. 5. Consecutive in-cylinder pressure traces with various ringing intensities [9]

$$SE = \int_{\varphi_1}^{\varphi_2} p_{bp}^2 \, d\varphi$$

Both phases are compared to each other and an index is calculated.

To increase the sensitivity of this method, the peak pressure of the bandpass filtered in-cylinder pressure trace is also evaluated in both of these combustion phases according to:

$$p_{max,} = max(p_{bp}|_{\varphi_1}^{\varphi_2})$$

In the same manner as with the signal energy, a second index is calculated with the peak pressure. Both indices are multiplied and the product is a collective knock indicator – the so-called dual fuel knock indicator (DFKI), which is then compared to a threshold value.

In fact, this threshold already shows good agreement, but it can be improved by introducing a second threshold. According to the mathematical definition of the previously described knock detection algorithm, dividing by a figure close to zero may lead to an exceeding of the first threshold value. The remedy for this mathematical misinterpretation was the determination of a minimum amount of energy of the phase where knock is expected – which is the second threshold value. Investigations of a broad variety of non-knocking and knocking combustion cycles in diesel/natural dual fuel operation have shown that this approach is independent of various operation conditions and the definition of just two thresholds is required.

Figure 6 shows the results of a previously conducted subjective separation of the measurement matrix into non-knocking and knocking combustion cycles (gray

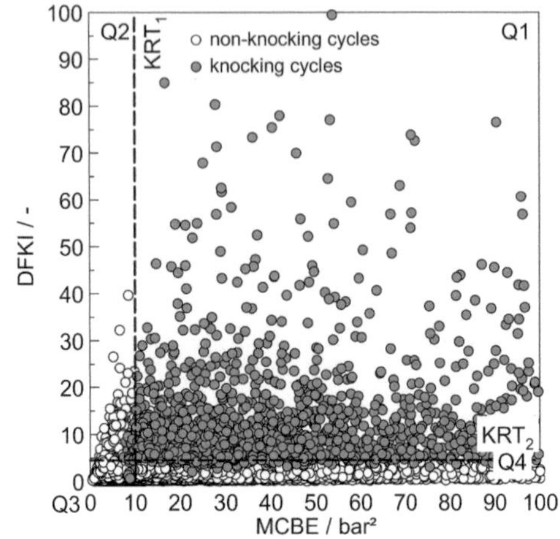

Fig. 6. Overall results of knock detection with diesel/natural gas dual fuel operation

dots = knocking combustion cycles, white dots = non-knocking combustion cycles). It also provides the results of the knock detection algorithm based on the x-axis and y-axis values; the dashed lines indicate the thresholds (knock related threshold, or KRT). These thresholds result from the differentiation between the knocking and non-knocking combustion cycles. The two thresholds divide the measurement matrix into four quadrants (Q1 to Q4).

For a more detailed description of knock detection with diesel/natural gas dual fuel combustion see [9].

Figure 7 provides an example of a series of consecutively measured measurement points where the EAR was varied in equal steps. Each EAR step is a separate

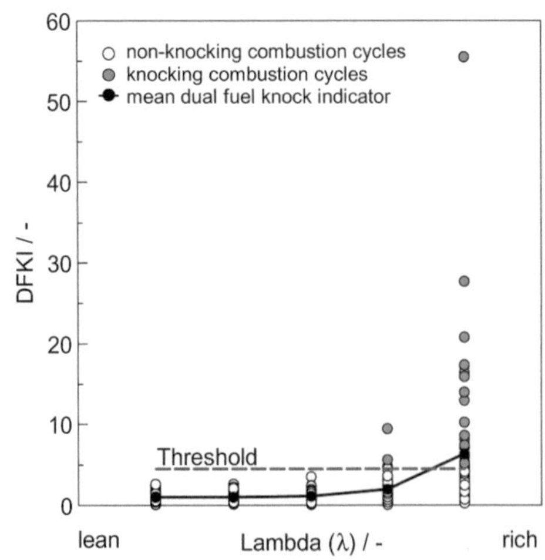

Fig. 7. EAR variation

measurement point that consists of 60 consecutively measured combustion cycles and is indicated by the dots.

This example was chosen because all measurement points exhibited decent ringing whereby knock detection is exclusively dependent on the first threshold value (DFKI). A gray dot once again indicates a knocking combustion cycle whereas a white dot indicates a non-knocking combustion cycle. When the charge is enriched, knocking combustion occurs. The first three measurement points from left to right contain no knocking combustion cycles. The fourth measurement point already contains a few knocking combustion cycles and is considered to indicate light knocking. Enriching the cylinder charge even further increases both knock frequency and knock intensity. With more than 30% knock frequency (31 out of 60 consecutively measured combustion cycles) and a knock intensity of 8 bar, this measurement point is considered to exhibit severe knocking.

3 Approaches to Preventing Knocking in Large Bore Engines

Four different approaches were chosen to react to knocking that occurs during engine operation and ensure stable and safe combustion. These approaches are described in the following chapter and all follow the same principle of cooling down the temperature of the in-cylinder charge, which directly results in a lower temperature in the unburned zone (two thermodynamic zones: burned zone and unburned zone). This can be understood as the common goal of all strategies to prevent knocking.

The investigated approaches can be divided into two categories. The first includes strategies applicable to nearly all common large bore engines and are often already in use. The second category consists of strategies that require additional devices or more advanced hardware, which requires changes or even adaptations in engine design.

All four approaches were investigated on a state-of-the-art multi-cylinder engine (MCE) application based on simulation in GT-Power. The simulations were conducted on a spark ignited lean burn gas engine with a scavenged prechamber. Such engines are equipped with a turbocharger unit followed by a mixture cooler. In our case, we chose a two-stage turbocharged configuration so that the model engine was able to reach an overall power level of more than 4 MW. The engine load is controlled by a standard throttle valve as well as an additional valve that opens a compressor bypass path.

The engine's behavior with reference to its knock tendency was derived from an extensive database of measurement results from a single cylinder research engine at the LEC. The single cylinder was operated under real-life conditions and ran on the same configuration later used for the MCE simulations. From these investigations, a knock limit was determined for this specific engine configuration. Providing examples of six different measurement points, Fig. 8 indicates how the knock criterion defines whether it is a knocking or a non-knocking operation point. As can be seen in the figure, there is good agreement of the points with the linear approach to the knock limit over the entire range of methane numbers.

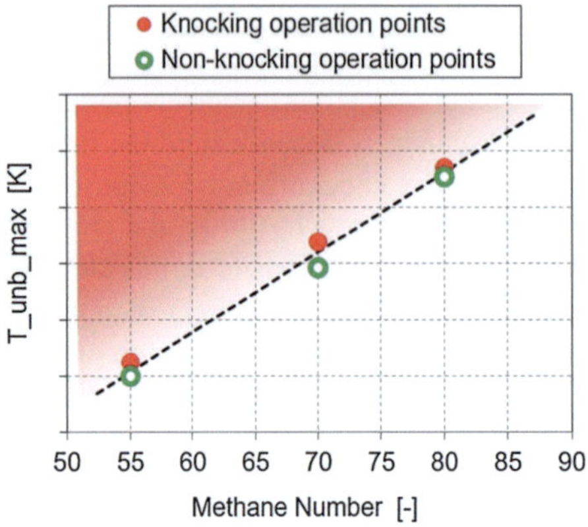

Fig. 8. Definition of knock limit (= dashed line)

Next, the correlation for the maximum temperature of the unburned zone (T_unb_max) was transferred to the MCE model. This made it possible to quantify the knock margin by taking the difference from T_unb_max (from a particular operating point) to the knock limit in Kelvin, subsequently referred to as ΔT_knock. A constant value of 10 K was selected in order to guarantee a standard knock margin for all the anti-knock strategies described in the next chapters.

3.1 Control of Engine Parameters

Adjustment of ignition timing

It was first considered how lowering the methane number affects the combustion process itself without any other intervention, cf. the gray line in Fig. 9. Since an increased number of higher hydrocarbons speeds up combustion, its phasing advances with lower MNs. Consequently, the maximum temperature of the unburned zone rises and ΔT_knock is reduced. While MN goes down, knocking combustion will start; this can be seen at MN 69, where the gray line meets the knock limit. It should be noted that the two operating points below the knock limit (against the red background) cannot actually be reached and only represent theoretical behavior.

In practice, every engine retards combustion itself by adjusting the start of ignition to withstand upcoming knocking. This is accomplished very fast and quite effectively by means of engine control. The blue curve in Fig. 9 illustrates this common strategy. Starting from MN 85 on the right, the knock margin decreases until it reaches the chosen distance of 10 K. To keep this constant, it is necessary to adjust ignition to late timings. For example, when the MN is 70, ignition has to be retarded by 2 °CA. As a result, it is not surprising that engine efficiency becomes worse, although it must be remembered that the gray line is theoretical and cannot be reached under any circumstances. Nevertheless, the gray line will serve as the *baseline* for assessing the engine efficiency potential, called ΔETA_eff, of all further approaches.

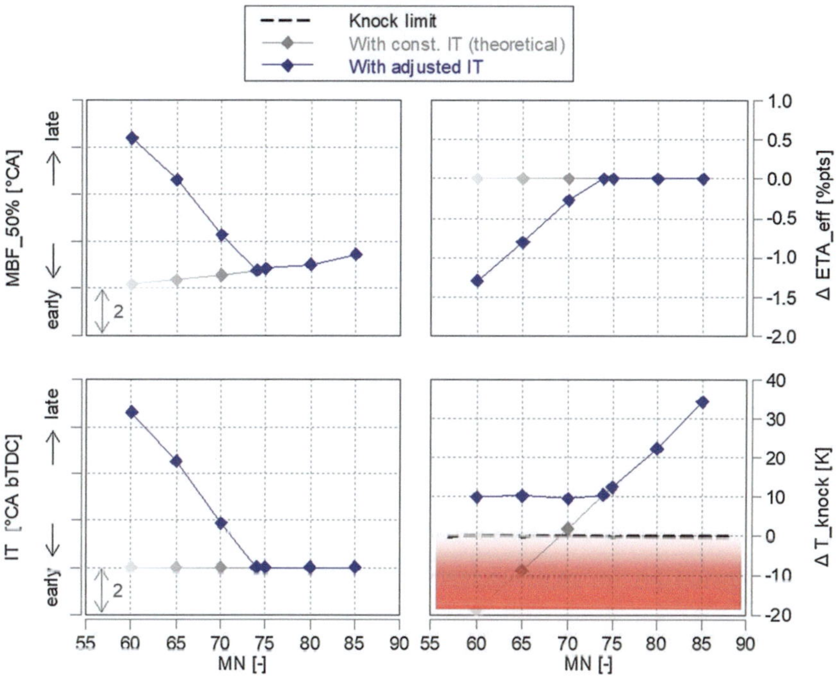

Fig. 9. Reacting to MN change by adjusting ignition timing (strategy 1)

Adjustment of mixture temperature

Another approach that is also standard practice is to cool down the charge's temperature directly by lowering the manifold air temperature (MAT). This measure is not as fast as adjustment of ignition timing because cooling down the entire air flow of a MCE with the attached mixture cooler takes time. Especially with stationary large bore engines, the size of the mixture cooler unit as a quantity for capacity is not restricted by the packaging; thus this strategy can be used on these types of engines.

Figure 10 shows the simulation results using this strategy. Engine control begins to intervene at a MN of 74. As methane number continues to decrease, the charge temperature goes down along the cyan-colored line. To ensure knock-free combustion at MN 67, the MAT has to be decreased by roughly 15 K, which is quite a large amount considering that MAT usually has a value in the range of 45 to 60 °C.

By applying *strategy 2*, combustion phasing remains fairly unaffected, hence the efficiency continues to become better than the *baseline* with lower MNs. Yet in very humid conditions, this particular strategy could lead to condensation problems along the intake air-path.

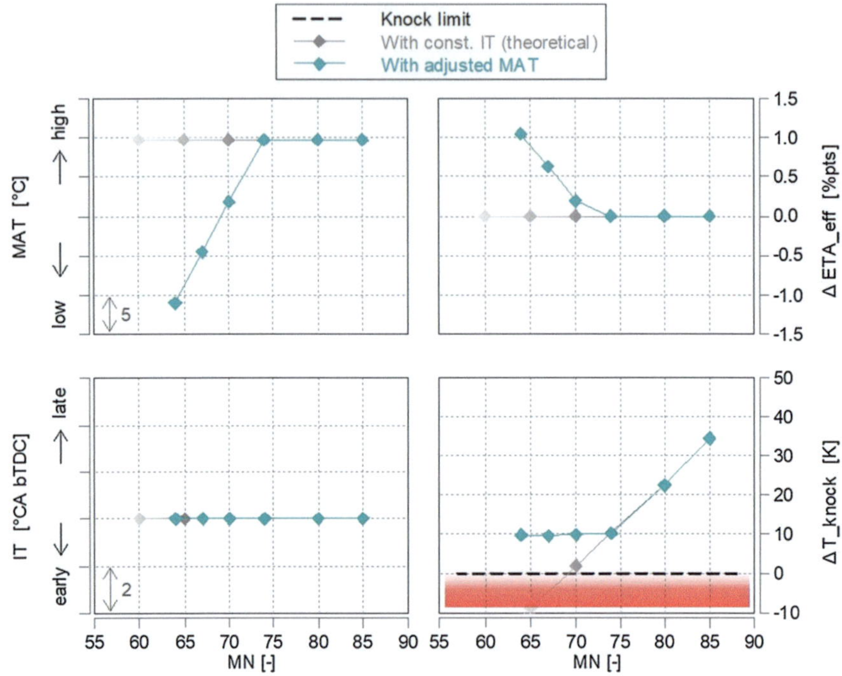

Fig. 10. Reacting to MN change by adjusting manifold air temperature (strategy 2)

3.2 Enhanced Hardware Functionalities

Variable valve timing

With this approach, the engine must be capable of controlling the timing of the valves. In the specific case implemented into the simulation model described in this paper, variability only refers to the intake valves. To keep it simple, only the closing of the intake valve is changed, i.e. the valve's opening ramp remains the same with all different closing angles (IVC) while the opening duration is varied. Such behavior can be obtained using a variable valvetrain, which has already been described in previous publications [10, 11].

The simulations summarized in Fig. 11 were carried out to demonstrate the impact of variable valve timing. At the starting point of MN 85, the efficiency of the IVC controlled model (green curve) as well as the distance to the knock limit is the same as the *baseline*. The shape of the valve lift curves is different, which is the reason for a retarded IVC in relation to standard valve timing at MN 85 [12]. IVC is controlled so that engine load is kept constant. In the *baseline* case, this is achieved using the compressor bypass valve so that the ignition timing can remain unchanged.

If the MN decreases so that combustion advances, IVC occurs later until MN 76, at which point the maximum duration of intake valve opening is reached. From that point on, the controller starts to advance the IVC; the ignition has to be retarded in parallel to

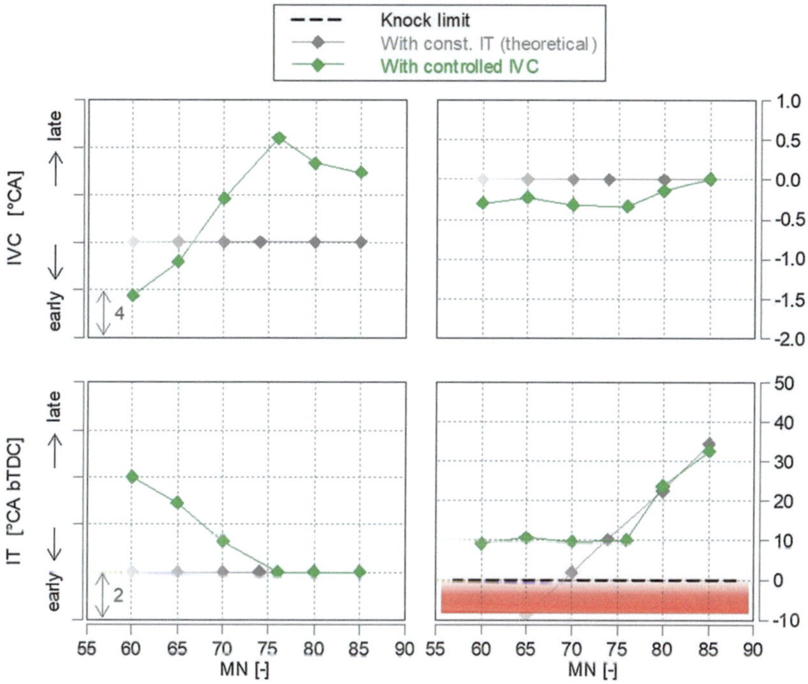

Fig. 11. Reacting to MN change by adjusting intake valve closing angle (strategy 3)

ensure that the engine load remains constant. With the earlier IVC, the temperature of the unburned zone decreases so that ΔT_knock can be kept constant at 10 K. The efficiency stays at nearly constant throughout the entire range below MN 76.

Variable compression

A more complex change in engine design is recommended when variable compression is used. There are several theoretical approaches to how to achieve a variable compression ratio [13] and the first passenger vehicle equipped with this new technology will be launched on the market in 2018 [14].

Our focus was on developing a strategy to prevent knocking, not to increase efficiency. As previously mentioned, we used a MCE simulation model for our investigations and changed the compression ratio (CR) within this model, disregarding how it might be put into practice on a large bore engine.

As explained above for *strategies 1* and *2*, an adjustment must be made only if the MN is lower than 74. In this case, the CR is steadily reduced while ignition timing remains unchanged. This results in worse conditions at ignition timing in terms of pressure and temperature, which in turn leads to a shift in combustion, which lasts longer. This fact is shown in MFB_50% curve in the top right diagram of Fig. 12. Together with the reduced compression ratio the latter is the reason for the drop in effective efficiency, indicated by ΔETA_eff, which steadily rises as the MN decreases.

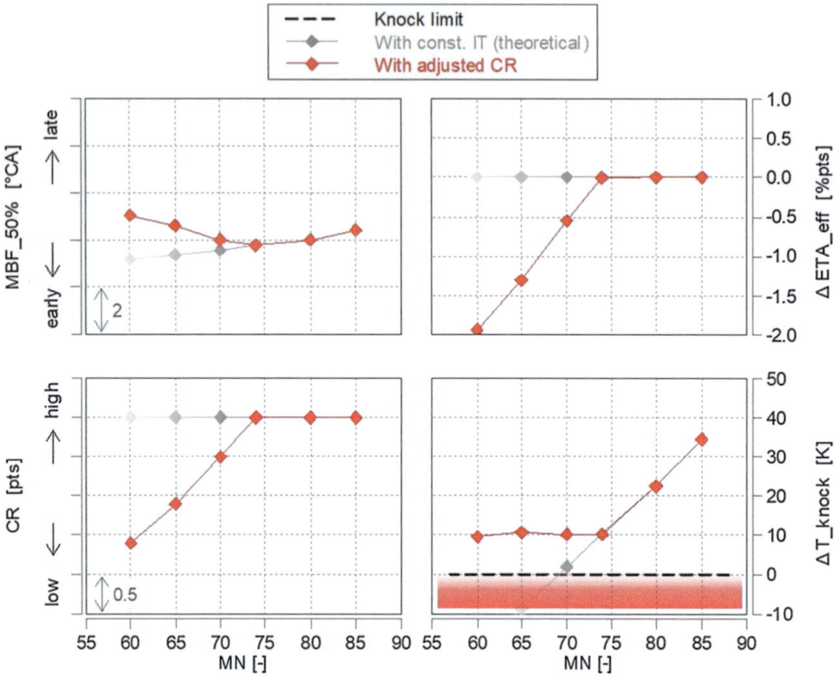

Fig. 12. Reacting to MN change by adjusting compression ratio (strategy 4)

4 Summary

Knock is a combustion anomaly that – in severe cases – can lead to engine failure. Since large bore gas engines have to deal with a broad variety of gas qualities, knock is likely to occur. To prevent this harmful form of combustion, robust knock detection is mandatory. To this end, algorithms are used that detect knocking combustion cycles. It was pointed out that spark ignited combustion and diesel pilot ignited combustion both have different combustion regimes and thus require different strategies for detecting knock.

When knock is detected, countermeasures are initiated. The strategies described in this paper provide an overview of potential approaches to preventing knocking yet make no claim to completeness. It is certain that adjustment of ignition timing (strategy 1) is the most straightforward way, especially in contrast to the enhanced hardware functionalities that already have to be considered during the engine design phase.

Although the focus of the strategies was on preventing knocking, we would like to take a look at the expected efficiencies shown in Fig. 13. Two operating points are compared for each strategy, one at the highest MN of 85 and the other at the lower limit of MN 65 defined in EN 16726 (see Sect. 1.1).

As can be seen in the figure, if the ignition timing is adjusted to operate the engine at enough distance to knocking (ΔT_knock = 10 K), efficiency (ETA_eff) drops by approximately 0.4%pts. Only *strategy 4*, where the compression ratio is reduced by

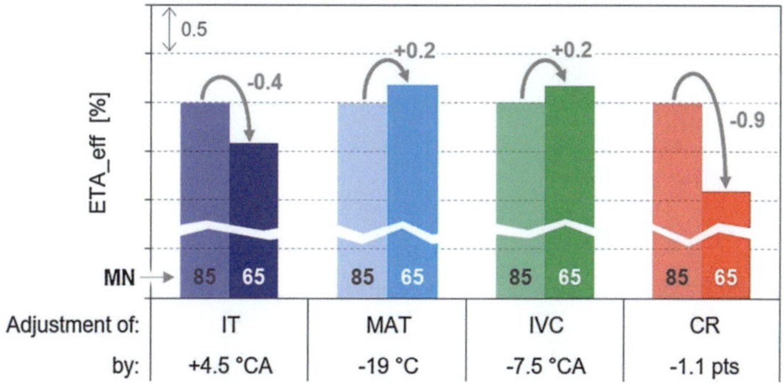

Fig. 13. Direct comparison of the four strategies

almost 1.1 pts, is worse regarding ETA_eff. The two strategies in the middle of Fig. 13 perform the same. But it has to be considered that the mixture temperature must be cooled down by 19 °C, which is quite a large amount, so that this strategy – *strategy 2* – cannot be applied to every engine concept. Consequently, the best efficiency can be achieved using intake valve variability.

Acknowledgements. The authors would like to acknowledge the financial support of the "COMET - Competence Centres for Excellent Technologies Programme" of the Austrian Federal Ministry for Transport, Innovation and Technology (BMVIT), the Austrian Federal Ministry of Science, Research and Economy (BMWFW) and the Provinces of Styria, Tyrol and Vienna for the K1-Centre LEC EvoLET. The COMET Programme is managed by the Austrian Research Promotion Agency (FFG).

Glossary

bp	Index for a bandpass filtered value
BFG	Blast furnace gas
BTF	Biomass to fuel
CR	Geometric compression ratio
DFKI	Dual fuel knock indicator
EAR	Excess air ratio
ETA_eff	Engine break efficiency
IT	Ignition timing
IVC	Intake valve closing angle
KRT	Knock related threshold
MAT	Manifold air temperature

MBF_50%	Crank angle at which 50% of fuel mass is burned
MN	Methane number
NG	Natural gas
P_RAIL	Rail pressure
SE	Signal energy
SOC	Start of combustion
SOI	Start of injection
TDC	Top dead center
T_knock	Temperature of the unburned zone where knocking occurs
T_unb	Temperature of the unburned zone
φ_Diesel	Energetic amount of diesel

References

1. Pirker, G., Wimmer, A.: Sustainable power generation with large gas engines. In: 11th sdewes (2016)
2. European Standard EN 16726: Gas infrastructure – Quality of gases – Group H (2015)
3. Zelenka, J., Kammel, G., Pichikala, K., Tritthart, W.: The quality of gaseous fuels and consequences for gas engines. In: 10. Internationale Energiewirtschaftstagung an der TU Wien (2017)
4. Pischinger, R., Klell, M., Sams, T.: Thermodynamik der Verbrennungskraftmaschine, 3rd edn. Springer, Vienna (2009)
5. Nates, R.J.: Thermal stresses induced by knocking combustion in spark-ignition engines. SAE International (2002). https://doi.org/10.4271/2000-01-1238
6. Worret, R.: Zylinderdruckbasierte Detektion und Simulation der Klopfgrenze mit einem verbesserten thermodynamischen Ansatz. Dissertation, Universität Karlsruhe (2002)
7. Dimitrov, D., Chmela, F., Wimmer, A.: Eine Methode zur Vorausberechnung des Klopfverhaltens von Gasmotoren. In: 4. Dessauer Gasmotoren Konferenz, Dessau (2005)
8. Dimitrov, D., Strasser, C., Chmela, F., Wimmer, A.: Vorhersage des Klopfverhaltens für Groß-Gasmotoren mit Direktzündung oder Vorkammer. In: 2. Tagung Klopfregelung für Ottomotoren – Trends für Serienentwickler, Berlin (2006)
9. Kirsten, M., Pirker, G., Redtenbacher, C., Wimmer, A., et al.: Advanced knock detection for diesel/natural gas engine operation. SAE International (2016). https://doi.org/10.4271/2016-01-0785
10. Christen, C., Codan, E.: Engine control and performance enhancement with variable valve train for gas engines. In: 16th Supercharging Conference, Dresden/Germany (2011)
11. Mathey, C.: Variable valve timing – a necessity for future large diesel and gas engines. Paper 298, CIMAC Congress Bergen (2010)
12. Zelenka, J., Hoff, C., Wimmer, A., Berger, R., Thalhauser, J.: Variable intake valve train to optimize the performance of a large bore gas engine. In: ICEF2016-9358, Proceedings of the ASME 2016 Internal Combustion Engine Fall Technical Conference, Greenville/USA (2016)

13. Roberts, M.: Benefits and challenges of Variable Compression Ratio (VCR). SAE 2003-01-0398, 2003 SAE Word Congress, Detroit/USA (2003)
14. Kiga, S., Moteki, K., Kojima, S.: The world's first production variable compression ratio engine – the new Nissan VC-T engine. In: 38th International Vienna Motor Symposium (2017)

Pre-ignition

Investigating the Cause of Initial Pre-ignition - A New Approach

Ina Volz[1(✉)], Jürgen Pfeil[1], Thomas Koch[1], and Frank Altenschmidt[2]

[1] Karlsruhe Institute of Technology (KIT), Karlsruhe, Germany
ina.volz@kit.edu
[2] Daimler AG, Stuttgart, Germany

Abstract. The irregular combustion phenomenon of low speed pre-ignition (LSPI) is still an important issue in modern internal combustion engine development and research, despite numerous publications looking into different directions such as the influence of injection strategy, scavenging, engine oil and fuel properties, and solid particulate matter as a potential ignition source.

This paper gives an insight into a new approach to investigate the phenomenon of initial pre-ignition from a rather fundamental perspective, triggered by the general research objective: Which conditions in the combustion chamber are necessary for the occurrence of an initial LSPI event?

A major goal of the herein presented path is to reduce the complexity arising from the interaction of diverse processes typically taking place in a gasoline engine. Coming from the main question, which methods to adopt and which new ones to develop in order to achieve this goal, a completely new test bed with a constant volume combustion cell as the centerpiece has been developed.

While maintaining characteristic thermodynamic conditions for a modern boosted gasoline engine, the interaction of the in-cylinder flow and the detachment of liquid drops from the liner as well as a typical fuel wall interaction possibly followed by engine oil dilution are minimized for this baseline study.

At the first step this experimental approach focuses on liquids and their physical and chemical properties as a potential trigger for auto-ignition.

The investigations are meant to result in a better understanding of the thermodynamic conditions leading to LSPI, which, according to the authors, is crucial to overcome the high risk of damage, which hinders the development of downsized gasoline engines with a higher compression ratio and thus better fuel efficiency.

1 Introduction

In general, concerning terms and definitions of abnormal combustion phenomena in gasoline engines, knocking combustion, hot surface ignition and pre-ignition (PI) can be differentiated. High pressure gradients and elevated maximum pressures because of high heat release rates are the characteristics of knocking combustion, which occurs in the remaining, unburnt gas fuel mixture after the ignition has been triggered by the sparkplug. Hot surface ignition starts at hot components or at hot deposits. This can

M. Günther and M. Sens (eds.), *Knocking in Gasoline Engines*,
https://doi.org/10.1007/978-3-319-69760-4_3

happen before or after the pre-defined spark timing and is often followed by knocking combustion. Pre-ignitions cannot always be definitely distinguished from hot surface ignition, whereas they differ regarding the ignition mechanism. In common, pre-ignitions are auto-ignitions, which take place before the ignition by the spark plug. Depending on the time of auto-ignition in the combustion chamber of an internal combustion engine (ICE), this can result in extreme knocking with very high pressure gradients and peak pressures of more than 350 bar. This means, that the risk of damage is very high.

PI in internal combustion engines occur characteristically at engine speeds below 2000 rpm and at comparably high mean indicated pressures. In this range of engine speeds the flow velocity during intake is relatively low, resulting in only little turbulence and thus in long burn durations, which rather promote auto ignition phenomena. PI events are stochastically distributed and unpredictable. The frequency strongly depends on the boundary conditions. Values in the range of 5 to 30 pre-ignitions per 10,000 cycles are not unusual [5]. Pre-ignitions either occur as individual events or often in series. Contemplating PI series, several PI events can occur consecutively or a first "initial" pre-ignition is followed by a normal combustion and then several times alternately a pre-ignition is followed by a normal combustion. The latter is more often the case. Having a closer look at the triggering mechanisms, a distinction must be made between the initial pre-ignition and the so-called subsequent pre-ignition.

Apparently, the mechanism can at least in part be explained for subsequent pre-ignitions. Optical measurements in several publications show that due to the initial pre-ignition and the subsequent knocking, deposits are flaked off and thus a multitude of particles are present in the combustion chamber [2, 19]. Some of these particles can remain in the combustion chamber after the charge exchange and heat up strongly during the next normal combustion. When these particles remain inside the combustion chamber after the ensuing charge exchange process, they can cause an ignition during the next cycle [18]. A sufficient thermal mass and a high temperature of the solid particles are required in order to prevent too much cooling off during the charge exchange [11]. Several other publications name loosening of oil droplets from the cylinder liner and the top land crevice as a precursor to PI events [6, 10, 15, 16].

When contemplating the cause of initial pre-ignitions the situation is complex. Current knowledge suggests that a pure gas phase ignition for example due to inhomogeneities in the combustion chamber of the engine is rather unlikely. This is shown both in the reaction-kinetic calculations as well as in experiments. In the event tree prepared here, the assumption is made that it is a two-phase phenomenon [9].

According to this, liquid or solid particles are involved in the self-ignition. In regard to liquid particles, various studies show a dependency of the pre-ignition frequency on the lubricating oil, in which both base oils, the additive as well as the dilution with fuel are critical. Because solid particles or deposits also depend on the fuels and lubricants, in this regard an indirect dependency of the pre-ignition on the system of lubricating oil/fuel must be noted here. Fundamentally, a differentiation between three main hypotheses can be made.

Considering direct ignition in the surroundings of oil droplets, optical investigations indicate that oil droplets are separated from the liner or the top land area and enter the gas phase. Due to the fact that oil has a higher ignition tendency than fuel, under certain conditions auto-ignition can occur in the surrounding gas phase of such a droplet [1, 11, 17]. However, this is counteracted by the vaporization on the droplet surface, because the enthalpy of vaporization causes local cooling. Takeuchi and Luef [12, 20] have performed extensive studies on the influence of the base oil composition and additives. It emerged that especially low viscosities and high calcium content benefit the pre-ignition rate. The low viscosity promotes the detachment of droplets. The influence of calcium has not been clarified yet. According to Moriyoshi in [16], a chemical conversion during an exothermic reaction may be responsible. Direct proof in form of visualization of oil droplet ignition has, to our knowledge, not been published yet.

When having a closer look at ignition in the surrounding sphere of droplets as a consequence of oil and fuel mixing, several papers report on the dependency of the pre-ignition rate of the fuel-oil interaction [4, 5, 7, 13, 22]. For example, the fuel impingement on the cylinder wall varies with the injection timing. The mixture of fuel with the lubricating oil changes both the chemical as well as the physical properties of the mixture. Droplet detachment becomes more likely as a result of the reduction of the viscosity and the surface tension. Magar et al. [14] makes the assumption that the mixture of oil and fuel enriched in the top land crevice detaches due to the forces of inertia in the delay phase of the piston during the compression stroke and this subsequently could lead to a pre-ignition [8]. The rare occurrence of the pre-ignition is explained as follows. The fluid deposit in the head land must first build up again after the occurrence of a PI event.

Modelling studies on the chemical kinetics point towards a chemical change of the oil as a possible cause of the pre-ignition. These show that under the conditions of the quasi-periodical heating up and cooling down in the top land crevice out of poorly ignitable substances with advancing time, extremely reactive substances can be formed.

A review of the ignition on particles, which occur directly from oil droplets, reveals, that some authors have made the presumption that the oil droplet itself does not cause the ignition, but instead that the additives they contain are responsible [3, 6, 15, 20]. As described in [16] for the mechanism of the subsequent pre-ignitions, at high temperatures above approximately 800 °C, calcium oxide is formed, which in turn can react in an exothermic reaction with carbon dioxide to form calcium carbonate [16]. The heat generation during the oxidation is made responsible for the ignition process. Several others assume that the detached oil droplets burn diffusively and as a result carbon-rich particles, ash or solid deposits are generated. These solid particles can heat up during combustion or are also of a reactive nature and can thus cause auto-ignition in the subsequent cycles. A pre-condition for this hypothesis is, that the solid particles are not fully flushed out during the charge exchange process [10, 21]. Using optical technology, Okada shows the ignition on solid particles [18]. CFD simulations have shown that particles that flaked off from locations with deposits will largely remain in the combustion chamber. Mainly larger particles are heated up sufficiently in the subsequent regular combustion cycle, so that they can trigger further pre-ignitions [12].

Despite many high quality publications searching after the origin of initial pre-ignitions and a long list of probable causes there remain open questions and the mechanism causing an initial pre-ignition is still not completely clear.

2 Methodology

Before outlining the methodology, the research goals should be clear. The original focus of research is the investigation of the phenomenon and especially the causes of initial pre-ignition from a rather fundamental perspective. So the main research question can be formulated as follows:

Which conditions in the combustion chamber are necessary for the occurrence of an initial pre-ignition event?

As a main goal of our activities the complexity, usually characteristic of PI investigations with internal combustion engines, is to be reduced. This complexity typically comes from several processes taking place simultaneously in a gasoline engine. So which methods are to be adopted and which new ones shall be developed? Which simplifications are reasonably to be implemented?

We came to the agreement, that the interaction of the in-cylinder flow and the detachment of liquid drops from the liner shall be minimized for this baseline study. Typical fuel wall interaction possibly followed by engine oil dilution is preempted, while characteristic thermodynamic conditions for a modern boosted gasoline engine shall be maintained. Furthermore, a mainly homogeneous temperature field in the gas phase of the air fuel mixture as well as the homogeneity of the gaseous air fuel mixture itself are to be realized. As a consequence it was decided to develop a high pressure high temperature combustion chamber featuring a constant combustion volume with the aim of investigating the conditions leading to pre-ignition.

This paper shows that the recently developed high pressure high temperature combustion chamber system has a lot more features to offer (see Chaps. 2 and 3) and can be utilized for a variety of other research experiments, too.

Considering the research investigations, liquids, especially motor oil and also fuels are suspected to be a potential trigger for involuntary auto-ignition. So it is a major goal of our investigations to make a contribution to understanding, which liquids and which physical and chemical properties lead to involuntary ignition, especially to pre-ignition. Concerning the combustible mixture one can distinguish liquid and gaseous phase, homogeneous and inhomogeneous state of mixture, global or local state of mixture. Additionally, there is a variety of ignition possibilities, whereas the focus here is on hot surface ignition and pre-ignition. A multitude of different combinations of these mixture states named above plus hot surface ignition and pre-ignition are planned to be investigated in the recently developed combustion chamber. Some will be easier, some more difficult to be realized. It is important that hot surface ignition is not to be confounded with pre-ignition. The experiments presented in this paper are to be understood as startup operations, i.e. as a prestage to the pre-ignition investigations in the future.

Below the development of the test set up and first experimental results for a globally homogeneous and gaseous air fuel mixture combined with deliberate, local hot surface ignition will be presented. The comparison of two different reference fuels can be considered as a prestage to further investigations with different additives.

2.1 Development of the New Test Bed

Coming from the main question, how to facilitate the processes and parallel operations taking place in an internal combustion engine before and while pre-ignition events the decision was taken to develop a completely new test bed. Additionally, for pre-ignition investigations the new test bed needs to be extremely robust.

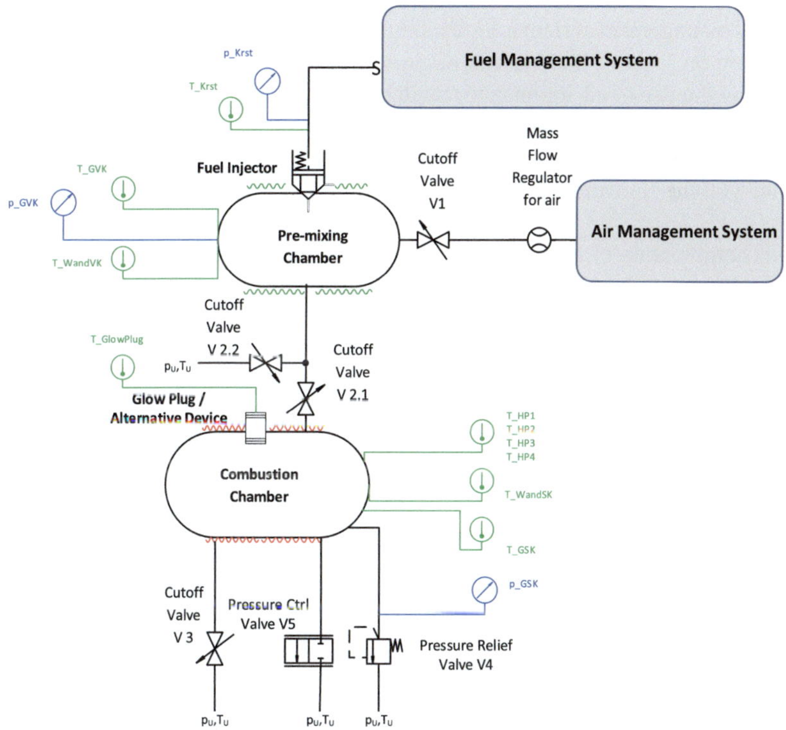

Fig. 1. Core of the combustion chamber system

A constant volume combustion cell dimensioned for a combination of high pressures and high temperatures is the centerpiece of the new test bed. The core of the combustion chamber system is shown in Fig. 1 and consists of the pre-mixing chamber for the conditioning of the air fuel mixture and the combustion chamber, in which the ignition and combustion of the mixture takes place. For managing the flow of the gases through the system and as a barrier to combustion for certain components, high pressure cutoff valves are provided. The mass of air that is allowed to pass through cutoff valve V1 into the pre-mixing chamber is controlled by the mass flow regulator. Fuel is injected into the heated pre-mixing chamber, where the mixing is supported by a magnetic stirrer. After a defined time of homogenization the mixture passes through cutoff valve V2.1 into the combustion chamber, where it is ignited either by a glow plug or an alternative device. The alternative device could be a spark plug instead of a glow plug or a droplet generator for example. With a special droplet generator the ignition in the environment of an oil droplet could be investigated in the future. Cutoff valve V2.2 gives the opportunity to release the mixture from the pre-mixing chamber into the environment and to flush it. In addition to cutoff valve V3, where the burned gases can exit, there is a pressure control valve and a pressure relief valve. When necessary, the pressure can be adjusted very precisely by the pressure control valve. The pressure relief valve is a security feature, as it opens, when the pressure in the combustion chamber exceeds 350 bar. For reasons of clarity the fuel and air management systems are simplified. The centerpiece of the fuel management system is an accumulator serving as a pressure reservoir, so that an injection pressure of 200 bar can be provided. The system can be flushed by air or nitrogen by the use of an automatized switching valve.

As the combustion chamber system should be suited for pre-ignition investigations, a maximum gas pressure of 350 bar and maximum gas temperatures of 500 °C during constant volume operation are determined in the concept phase of the development process.

This combination of high pressures and high temperatures and the additional wish for optical accesses have also been the main challenge for the design and dimensioning of the combustion chamber. Each component, sensor and each material deployed needs to be suited for these demanding requirements. Among others, one needs to use high-temperature resistant steel and provide safety features such as bursting discs at the necessary positions. In the test bed presented the relief valve opens at a pressure of 350 bar, but of course the dimensioning of the combustion chamber is based on much higher pressures in the order of magnitude of 1000 bar.

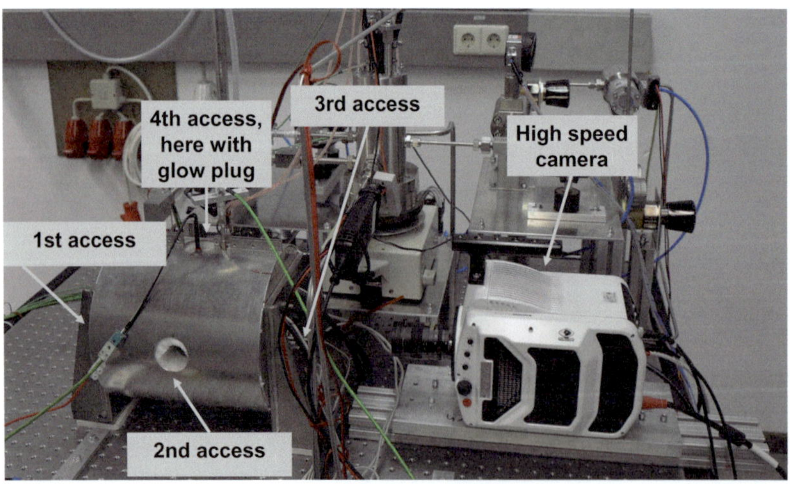

Fig. 2. Overview of the different accesses to the chamber and of the optical setup

For some components, e.g. the fuel injector or the droplet generator, it is necessary to heat and cool at the same time. On the other hand, a mainly homogeneous temperature field in the gas phase of the chambers has been a major goal of the development. Considering 500 °C maximum gas temperature, this is one of the reasons, why we opted for the pre-mixing chamber instead of positioning the fuel injector directly into the combustion chamber.

The combustion chamber has a cylindrical inner geometry with two additional cylindrical accesses (Fig. 2, 2nd and 4th access), each in a 90° angle to the main cylinder. The figure gives an overview of the three optical accesses and the access for the glow plug or some other device.

Through the positioning and orientation of the different accesses a great variability is provided. The chamber is specially designed for optical investigations at high pressures and high temperatures. However, the cylindrical accesses all have the same dimensions, so that e.g. the glow plug access and one optical access can be interchanged, if necessary. Also it is possible to mount two devices like the glow plug and a droplet generator, while using two optical accesses. As the combustion chamber is conditioned via a specially controlled wall heating, depending on the maximum temperature requirement of the device, adequate cooling might be necessary. According to the targeted experiment, one needs to design and dimension the suitable high pressure adapters from special high-temperature resistant steel.

Multiple accesses with a comparably high optical clear width of 18 mm in diameter make a variety of optical setups and investigations possible. With illumination and two optical accesses inline shadowgraphy is enabled, whereas the 90° optical access can be used for Mie scattering. Striving for droplet size detection these two methods can be of high importance. Additionally, other optical measurement techniques such as PIV (particle image velocimetry) and LIF (laser induced fluorescence) are conceivable.

Chemiluminescence measurements can be conducted without illumination in order to investigate the flame propagation. It is possible to either record the luminescence in the visible range or in the UV range with or without suitable filters and an image intensifier, if necessary.

In this paper we present the first chemiluminescence measurements in the visible range under elevated pressure conditions performed in this novel constant volume combustion chamber (see Chap. 4).

In addition to the safety requirements all these challenges, requirements, goals and features listed above are specially taken care of in the design and dimensioning of the combustion chamber system as well as in the measurement setup, which is described below.

2.2 General Measurement Setup

Sensor positions for pressure (blue) and temperature (green) measurement are depicted in Fig. 1. Control functions and driving of the different actuators (e.g. valves and mass flow regulator) as well as the recording of the pressure and temperature measurement values are realized by a modular CompactRIO System from National Instruments. The indicating system records with a frequency of 100 kHz the pressures inside the two chambers, several injection signals and the signals of the glow plug or the fast signals of the alternative devices. The optical recording system is described in Chap. 3, as this depends on the details of the experiment.

3 Experiments and Setup

While the general features of the test bed and the measurement system are described above, this chapter focuses on the experimental setup and test procedure additionally utilized for the experiments presented in this paper.

3.1 Experimental Setup

For these first experiments in the recently developed combustion chamber two different reference fuels are used. A homogeneous, gaseous mixture of air and fuel is ignited by deliberate hot surface ignition, i.e. by a glow plug with an integrated temperature sensor. Chemiluminescence images are taken.

The glow plug employed in the tests is a thermo glow plug from BorgWarner BERU Systems GmbH and is mounted in a specially designed high pressure high temperature adapter, which fits into one of the multiple accesses of the combustion chamber described above. It is a metal plug with the temperature sensor positioned inside near the tip of the glow plug. All tests are conducted with the same plug, so that the temperature measurements are reproducible and comparable.

According to pyrometry reference measurements conducted by the manufacturer, there is an offset between the temperature sensor and the temperature of the outer surface, which is around 30 K cooler than the sensor temperature. However, it is not possible to measure the exact temperature of the hot outer surface of the plug with this set up. Thus, all glow plug measurements and results presented in this paper refer to the temperature values of the integrated sensor. The temperature traces recorded by this sensor without combustion are congruent.

Two different reference fuels are utilized during these first reference experiments, which can be considered as a prestage to further experiments with additives or alternative fuels.

First pure iso-octane is used, whereas the second fuel is a blend out of n-heptane and iso-octane with a mole fraction of 0.3. Some characteristic values of pure iso-octane and pure n-heptane are shown in Table 1.

Table 1. Fuel properties

	iso-octane	n-heptane
GC-FID screening [%]	99.91	99.82
Density at 15 °C [kg/m^3]	695.9	688.0
Content of water [mg/kg]	20	<30
Boiling temperature [°C]	99	98

For the chemiluminescence imaging one optical access is used, as no illumination is necessary. A Phantom v1612 color high speed camera is combined with an optical zoom lense. Chemiluminescence images in the visible range are taken at a frequency of 10 kHz and a corresponding exposure time of 98 μs.

An event trigger is sent from the indicating system to the optical recording system, when a certain pressure threshold in the combustion chamber is reached. The image recording starts and the pre-trigger images are saved from the circular buffer. The pressure threshold is set accordingly to the operating point. Usually the pressure threshold has been set to the value of the pressure in the pre-mixing chamber shortly before overflow (see Chap. 2.1 or Fig. 3). The event trigger can be utilized as a time reference for the two measurement systems and is advantageous for evaluating the results.

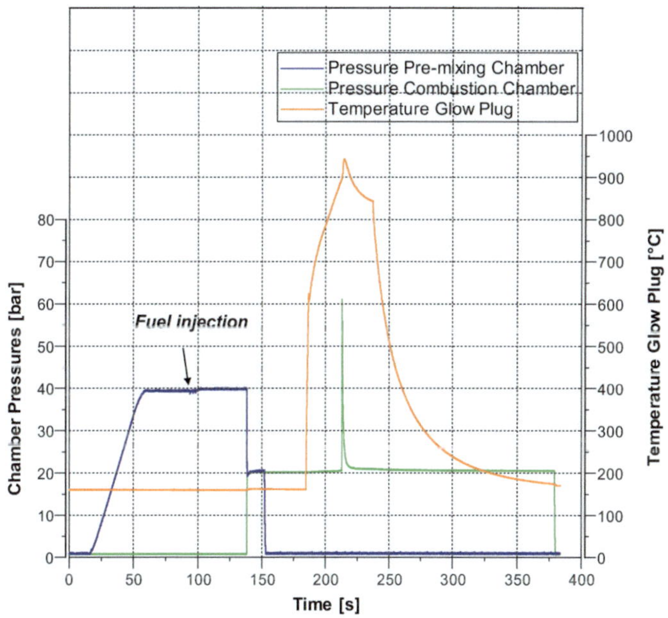

Fig. 3. Test procedure, pure iso-octane, stoichiometric mixture of air and fuel, initial pressure 20 bar

3.2 Test Procedure

The test procedure depicted below is the one which was used for the experiments presented in this paper.

Figure 3 shows the development of the pressure inside the pre-mixing chamber (blue), the pressure development inside the combustion chamber (green) and the temperature of the glow plug (orange) for an initial pressure of 20 bar. In this paper "initial pressure" means the pressure in the combustion chamber before the combustion starts. As one can see, the conditioning of the air fuel mixture starts in the pre-mixing chamber at a pressure level, which is about twice the initial pressure. Fuel is injected into a predefined mass of air by a HDEV 5.2 injector during multiple injections. The pre-mixing chamber is heated at 105 °C wall temperature. Considering the boiling temperatures of the fuels in Table 1 this should be enough. Additionally, for a better mixing there is a magnetic stirrer inside the pre-mixing chamber. After a certain mixing time cut-off valve

V2.1 is opened and the air fuel mixture flows into the combustion chamber. The combustion chamber is already heated up to a constant wall temperature of 160 °C. V 2.1 gets closed and the pressure inside the pre-mixing chamber drops, as it gets flushed. After a distinct time, during which the air fuel mixture can calm down inside the combustion chamber after overflow, the glow plug control is activated and the glow plug begins to heat up locally until the combustible mixture ignites. As a consequence one can see the peaks in glow plug temperature and in the pressure inside the combustion chamber. Afterwards the temperature declines and the pressure is released.

Of course, the times and temperatures as well as the pressures can be varied, if necessary. The temperatures are deliberately chosen not too high in order to prevent too much pre-reactions, but above the boiling temperatures of the fuels for obtaining the goal of creating a mostly homogeneous and gaseous air fuel mixture.

Alternative temperature and initial pressure levels between 10 bar and 45 bar have been tested, each for stoichiometric air fuel mixtures with pure iso-octane and the fuel mixture of n-heptane and iso-octane with a mole fraction of 0.3. For each operating point pressure indication and chemiluminescence imaging have been recorded.

4 Measurement Results and Discussion

As described above, this is the presentation of the first experimental results obtained with the recently developed combustion chamber. We selected two exemplary operating points, one at medium and one at elevated initial chamber pressures out of the test matrix given above.

First the reproducibility of the results is shown at an initial pressure of 20 bar. Afterwards the impact of two different fuels is studied in the same operating point.

At the elevated initial chamber pressure of 45 bar, the chemiluminescence measurements will be analyzed in detail for the different reference fuels. This initial chamber pressure is deliberately chosen in the dimensions of the pressure at end of compression of gasoline engines. The comparison of the two different reference fuels can be seen as a prestage to investigations with different additives.

4.1 Analysis and Outcomes for Initial Pressure of 20 Bar

This is an exemplary, rather medium level operating point recorded in the constant volume combustion chamber, but is well suited to depict some general observations that can be made based on the experiments. Starting at an initial pressure of 20 bar in the combustion chamber, the pressure traces in Fig. 4 reach their maximum after ignition at 60 bar. All pressure traces in this diagram are referenced at the same start of the first pressure rise. The experiments are conducted with a stoichiometric mixture of air and pure iso-octane. After a very steep pressure rise due to combustion, the pressure declines comparably slowly until it reaches the initial pressure level. As one can see in the figure, the pressure traces all resemble each other very well, which underlines their reproducibility.

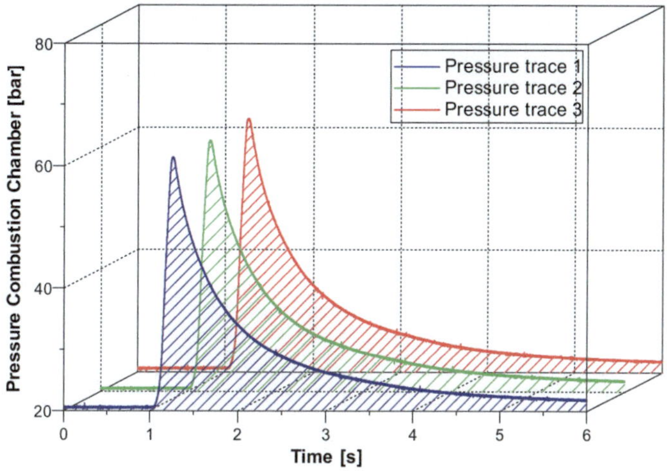

Fig. 4. Reproducibility of pressure traces, stoichiometric mixture of air and pure iso-octane, initial pressure 20 bar

Impact of different fuels

This paragraph is dedicated to the impact of different fuels. On the left hand side of Fig. 5 three pressure traces for a stoichiometric mixture of pure iso-octane and air are shown. On the right hand side three pressure traces are depicted for a mole fraction of n-heptane and iso-octane of 0.3. The injection timing is the same as for pure iso-octane. All other parameters are also held constant. Accordingly, the pressure traces of the fuel mixture on the right hand side are reproducible. Again the pressure traces are referenced at the same start of the first pressure rise.

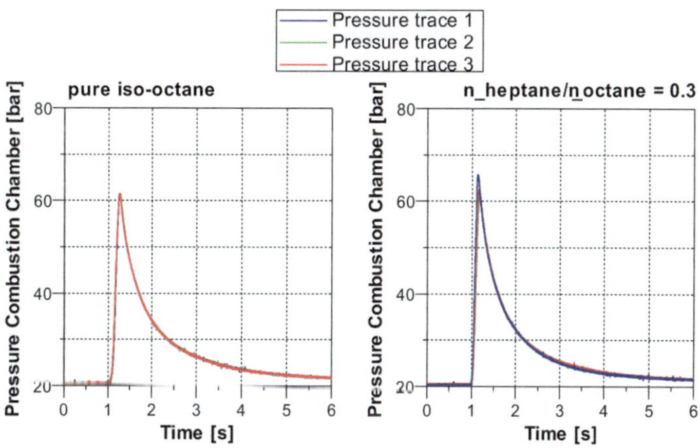

Fig. 5. Comparison of pressure traces for different fuels, stoichiometric air fuel mixtures, initial pressure 20 bar

While the process duration is comparably long for both cases, the positive pressure gradient dp/dt of the fuel mixture is higher than the one for pure iso-octane.

In Fig. 6 the temperatures, measured with the sensor that is integrated in the glow plug (please see Chap. 3.1), are plotted over time for the 20 bar operating point and the same parameters as in Fig. 5. These temperature traces are, in contrast to Fig. 5, referenced at the same start of glow plug control instead of being referenced at the same onset of pressure rise. When the fuel air mixture is ignited, one can detect an inflection point in the temperature trace.

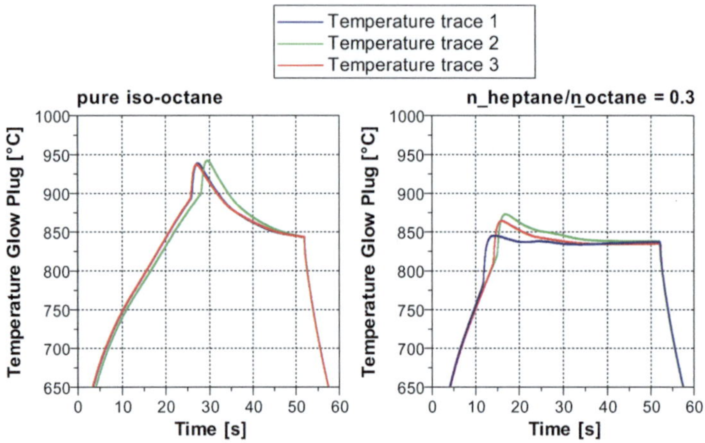

Fig. 6. Comparison of temperature traces for different fuels, stoichiometric air fuel mixtures, initial pressure 20 bar

For pure iso-octane the ignition takes place at temperatures of around 900 °C, for the heptane octane mixture at temperatures between 775 and 825 °C.

4.2 Analysis and Outcomes for Initial Pressure of 45 Bar

This operating point excels through its elevated initial pressure, which results in peak pressures between 180 and 190 bar (compare Fig. 7) inside the recently developed constant volume combustion chamber. Below also chemiluminescence results will be shown and analyzed for this operating point.

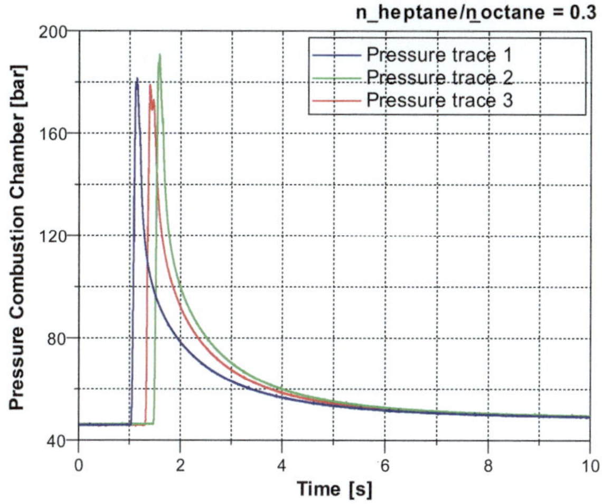

Fig. 7. Pressure traces referenced on the same start of glow plug control, iso-octane with n-heptane, stoichiometric mixture of air and fuel, initial pressure 45 bar

In Fig. 7 exemplary pressure traces for a mole fraction of n-heptane and iso-octane of 0.3 are depicted. The traces are all referenced to each other via the same start of glow plug control, so that the heating interval starts at the same relative time for the three pressure traces. However, the onset of the first pressure rise recorded in the indication data, varies slightly related to the duration of the whole process, as can be seen in the figure. The first pressure rise is defined as initial pressure plus 0.5 bar. At about 200 ms after the first pressure rise the decrease of pressure has begun. In this operating point for pure iso-octane, the ignition takes place at temperatures of 700 °C, for the heptane octane mixture the detection of those temperatures is in the order of magnitude of 625–650 °C.

Figure 8 shows high speed color images, recorded at elevated chamber pressures with a recording frequency of 10 kHz and an exposure time of 98 μs. The initial pressure before ignition is 45 bar, whereas the resulting maximum pressures vary between 180 and 190 bar. On the left hand side one can see different time steps of one characteristic combustion event taken for a stoichiometric mixture of pure iso-octane and air, on the right hand side for the fuel mixture of n-heptane and iso-octane with the mole fraction of 0.3. The experiments with the fuel mixture are conducted with the same injection timing. All images are exported from the optical software with the same resolution. Therefore the first image, when the flame can first be detected in the field of view, is comparably dark. So instead of changing the color scale of the images, we decided to highlight the flame front in these two first pictures. While the optical system is always triggered by the indicating system, when a certain pressure value is reached, the absolute start time, when the flame front is first detected in the field of view shows slight differences. For Fig. 8 the time step 0 ms was defined through evaluation of images. Compared to optical investigations in engines, the recorded processes last relatively long because of the missing turbulence. After the first detection of the flame the subsequent images are depicted in time steps of 10 ms, so that the flame propagation can be observed.

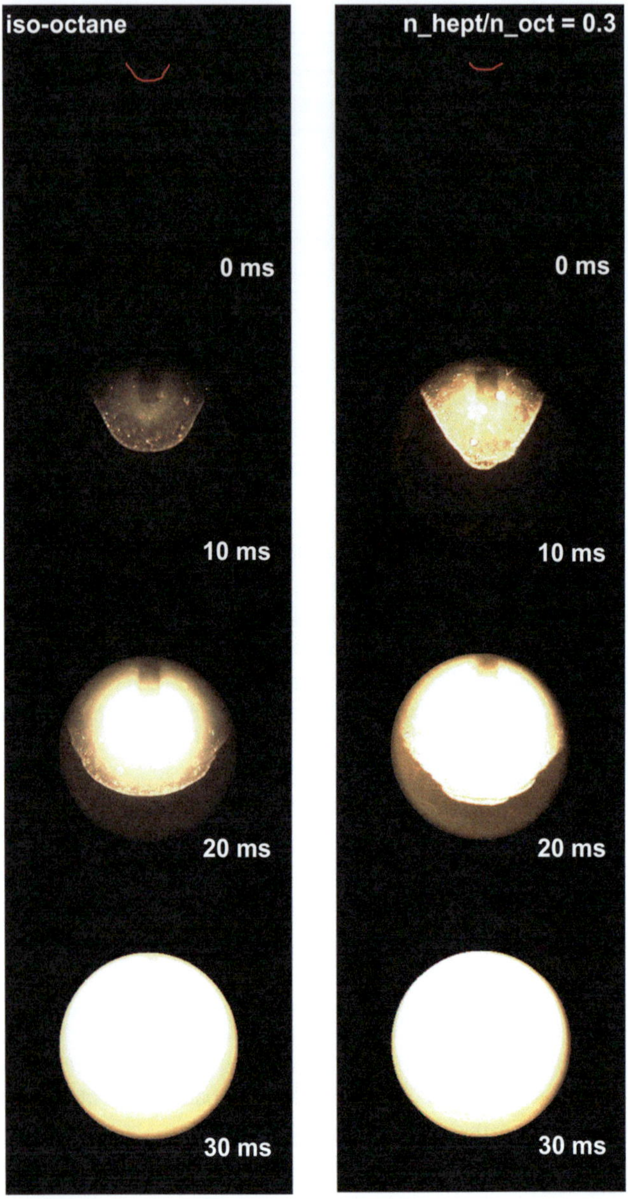

Fig. 8. Early combustion phase: Flame propagation for stoichiometric mixture of air and pure iso-octane (left) and of air and fuel mixture of iso-octane and n-heptane at the same injection timing (right), initial pressure 45 bar

Optical analysis - Impact of different fuels

Looking at the images, it is obvious that in both cases one can easily detect a flame front, so that a global ignition of the gaseous phase of fuel and air can be ruled out. On the first impression the flame propagation looks similar both for iso-octane and the fuel mixture. Taking a closer look at the records of 10 ms and 20 ms, the flame front of the fuel mixture looks slightly more folded and rather like a “v”-shape, which seems characteristic for this operating point. For pure iso-octane the flame front could rather be described as a “u”-shape and is comparably smooth. In the last picture at 30 ms for both fuels the field of view is completely illuminated.

Afterwards several intensity maxima can be detected, for which we were first not sure, if those are blooming effects of the camera. However, in the pressure trace two maxima can be found with a time difference to each other corresponding to the time difference of two of the intensity maxima. So we will have a closer look in the future and focus on the posterior combustion phases, while we focused here on the early phases of the combustion. Comparing the pressure traces recorded at 100 kHz and the optical data recorded at 10 kHz, the flame’s first detection in the visible range is significantly earlier than the first pressure rise. The first pressure rise here is defined as initial pressure plus 0.5 bar. When zooming into the pressure data, after about 200 ms the most important part seems to be over, while after a much longer time the optical data still shows diffuse illumination, before finally fading completely away.

Based on the optical data for this elevated operating point the flame propagation velocities have been calculated. It is not meant to be an exact determination of laminar burning velocities with high temporal resolution, as the combustion chamber has been developed for thoroughly investigating the conditions and origins of pre-ignition instead of exact evaluation of laminar burning velocities. For the flame speed calculations the differences of the flame fronts in the top down direction at certain time steps are analyzed and related to the according time steps. Thus, it is an average velocity evaluated for high pressure high temperature conditions with the assumption of a linear motion of the flame front into this direction, which illustrates the propagation of the flame. This is an add-on feature of the recently developed combustion chamber.

For pure iso-octane the propagation velocities of the flame front stand out due to the exact same values for the three repeated measurements in this 45 bar operating point and amount to ca. 1 m/s in the first section and around 0.6–0.7 m/s in the second section. When the flame comes closer to the chamber walls, the velocity decreases to ca. 0.3–0.4 m/s. For this latter case, the flame front is sometimes already out of the observation area.

The order of magnitude for the measurements with the octane heptane mixture is the same, whereas in the first section the velocities between the single experiments vary between 1.1 and 1.3 m/s. In the second section the three tests show a flame speed of 0.6 m/s and do not differentiate much from each other and from the measurements conducted with pure iso-octane.

It has not been completely clear, how this relatively similar tendencies in flame propagation speed correlate with the steeper rise in pressure for the fuel mixture with n-heptane content. A possible explanation is, that the chemiluminescence imaging is focused on the early phase of combustion, while the pressure rise is a rather global effect of the combustion. So there does not definitely need to be a correlation between those two observed effects. As implied above, additional chemiluminescence imaging is conceivable with either further closing the aperture and thus focusing on the later phases of combustion or even with a second camera in a 90° angle for 3D chemiluminescence imaging or in order to detect the time step, when the flame leaves the sharpened observation volume. Furthermore one can try to increase the depth of focus by using a special lens or lens system.

5 Summary and Outlook

First, the targets and development of a new, high pressure high temperature combustion chamber system for pre-ignition research are outlined. One major issue has been the simplification of the interactions taking place in a gasoline engine in order to reach the goal of submitting rather fundamental research instead of statistics evaluation of internal engine PI tests. An overview of the features of the combustion cell, especially the multiple interchangeable accesses for optical windows and alternative components is given. During the experiments presented in this paper homogeneous air fuel mixtures are ignited by deliberate hot surface ignition, i.e. by a temperature glow plug and high speed chemiluminescence images in the visible range are taken.

These first experiments in the specially designed combustion chamber have been successful at elevated initial pressures of 45 bar and thus comparably high peak pressures of the order of 190 bar. During pre-ignition events significantly higher pressures are expected. Thus, the combustion chamber is designed and dimensioned for even higher pressures and temperatures, namely 500 °C gas temperature before combustion combined with 350 bar pressure for constant volume operation. If constant volume operation during combustion can be renounced, the initial pressures can be increased at levels way beyond typical pressures at the end of compression for gasoline engines.

Originally, the combustion cell is designed and dimensioned for investigating preconditions of PI instead of analyzing phenomena and effects during combustion. However, this has been possible, which is a further asset of the combustion chamber system in addition to the ruggedness of the combustion cell.

In this paper the first measurement results with the recently developed combustion cell are presented. Two exemplary operating points are shown, one at a medium, initial chamber pressure of 20 bar and one an elevated initial pressure of 45 bar.

Both the reproducibility of the pressure traces and the impact of two different fuels are shown in the same medium level operating point. Compared to a stoichiometric mixture of pure iso-octane and air the positive pressure gradient dp/dt is higher for the fuel mixture containing n-heptane. For this heptane mixture the hot surface ignition occurs earlier and at lower glow plug temperatures.

For the experiments resulting in peak pressures of around 190 bar, chemiluminescence images are analyzed in detail and compared for two different reference fuels.

For both cases the propagating flame front is easily detectable, so that a global ignition of the gaseous phase of fuel and air can be ruled out. The appearance of the flame front in the visible range shows slight differences for the two different fuels. Additionally, flame propagation velocities are deduced from the chemiluminescence results.

This comparison of the two different reference fuels can be seen as a prestage to investigations with different additives. These first results show the great potential of the recently developed combustion chamber and should give a pretaste for various experiments to come in the future.

Considering experiments with homogeneous air fuel mixtures and local hot surface ignition further investigations with a special ceramic temperature glow plug will be conducted. As mentioned above, besides further optical investigations chemiluminescence recording in the visible range from two different directions for 3D visualization is conceivable as well as chemiluminescence imaging within the UV spectrum.

If necessary, different test procedures are possible. Pressure and temperature levels are variable as well as the air fuel ratio and the timing of the consecutive process steps. Several fuels can be tested.

Concerning the combustible mixture one can distinguish liquid and gaseous phase, homogeneous and inhomogeneous state of mixture and global or local state of mixture. In addition there is a variety of ignition possibilities, whereas the focus here is on hot surface ignition and pre-ignition. A multitude of different combinations of these mixture states listed above plus hot surface ignition and pre-ignition are planned to be investigated with the recently developed high pressure high temperature combustion chamber in the future.

Acknowledgements. The authors express their special thanks to Daimler AG for funding this research project. The authors would also like to thank the students Tobias Bezner, Mario Frommherz, Thomas Gal, Patrick Härdtner, Martin Prinzen and Fabian Weitz for their enthusiasm, effort and research spirit.

References

1. Dingle, S.F., Cairns, A., Zhao, H., Williams, J., Williams, O., Ali, R.: Lubricant induced pre-ignition in an optical SI engine. SAE Technical Paper 2014-01-1222 (2014). https://doi.org/10.4271/2014-01-1222
2. Döhler, A., et al.: A contribution to better understanding the pre-ignition phenomenon in highly charged internal combustion engines with direct fuel injection. In: 4th International Conference on Knocking in Gasoline Engines (Berlin), pp. 41–61 (2013)
3. Fujimoto, K., Yamashita, M., Hirano, S., Kato, K., Watanabe, I., Ito, K.: Engine oil development for preventing pre-ignition in turbocharged gasoline engine. SAE Int. J. Fuels Lubr. **7**(3), 869–874 (2014). https://doi.org/10.4271/2014-01-2785
4. Gohl, M., Brandt, S., Wittler, M., Budde, M., et al.: Influence of the mixture formation on the lubrication oil emission of combustion engines. SAE Int. J. Fuels Lubr. **3**(1), 733–744 (2010). https://doi.org/10.4271/2010-01-1275
5. Han, L., Zhu, T., Qiao, H., Zhang, D., et al.: Investigation of low-speed pre-ignition in boosted spark ignition engine. SAE Technical Paper 2015-01-0751 (2015). https://doi.org/10.4271/2015-01-0751
6. Hirano, S., Yamashita, M., Fujimoto, K., Kato, K.: Investigation of engine oil effect on abnormal combustion in turbocharged direct injection - spark ignition engines (Part 2). SAE Technical Paper 2013-01-2569 (2013). https://doi.org/10.4271/2013-01-2569
7. Inoue, T., et al.: Abnormal combustion in a highly boosted SI engine - the occurrence of super knock. SAE 2012-01-1141 (2012)

8. Kassai, M., Hashimoto, H., Shiraishi, T., Teraji, A., et al.: Mechanism analysis on LSPI occurrence in boosted S. I. ENgines. SAE Int. J. Engines **9**(1) (2016). https://doi.org/10.4271/2015-01-1867
9. Palaveev, S., Magar, M., Kubach, H., Schiessl, R., Spicher, U., Maas, U.: Premature flame initiation in a turbocharge DISI engine - numerical and experimental investigations. SAE Technical Paper 2013-01-0252 (2013). https://doi.org/10.4271/2013-01-0252
10. Kuboyama, T., Moriyoshi, Y., Morikawa, K.: Visualization and analysis of LSPI mechanism caused by oil droplet, particle and deposit in highly boosted SI combustion in low speed range. SAE Int. J. Engines **8**(2), 529–537 (2015). https://doi.org/10.4271/2015-01-0761
11. Lauer, T., et al.: Model approach for pre-ignition mechanisms. MTZ Worldwide **75**(1), 44–49 (2014)
12. Lauer, T., Heiß, M.: Analysis of particle separation with respect to pre-ignition in an SI-Engine. In: 6th European Conference on Computational Fluid Dynamics (ECFD VI) (2014)
13. Luo, X., Teng, H., Hu, T., Miao, R., Cao, L.: An experimental investigation on low speed pre-ignition in a highly boosted gasoline direct injection engine. SAE Int. J. Engines **8**(2), 520–528 (2015). https://doi.org/10.4271/2015-01-0758
14. Magar, M., Spicher, U., Palaveev, S., Gohl, M., et al.: Experimental studies on the occurrence of low-speed pre-ignition in turbocharged GDI engines. SAE Int. J. Engines **8**(2) (2015). https://doi.org/10.4271/2015-01-0753
15. Miyasaka, T., Miura, K., Hayakawa, N., Ishino, T., Iijima, A., Shoji, H., et al.: A study on the effect of a calcium-based engine oil additive on abnormal SI engine combustion. SAE Int. J. Engines **8**(1), 206–213 (2015). https://doi.org/10.4271/2014-32-0092
16. Moriyoshi, Y., Yamada, T., Tsunoda, D., Xie, M., et al.: Numerical simulation to understand the cause and sequence of LSPI Phenomena and suggestion of CaO mechanism in highly boosted SI combustion in low speed range. SAE Technical Paper 2015-01-0755 (2015). https://doi.org/10.4271/2015-01-0755
17. Ohtomo, M., Miyagawa, H., Koike, M., Yokoo, N., et al.: Pre-ignition of gasoline-air mixture triggered by a lubricant oil droplet. SAE Int. J. Fuels Lubr. **7**(3), 673–682 (2014). https://doi.org/10.4271/2014-01-2627
18. Okada, Y., Miyashita, S., Izumi, Y., Hayakawa, Y.: Study of low-speed pre-ignition in boosted spark ignition engine. SAE Int. J. Engines **7**(2), 584–594 (2014). https://doi.org/10.4271/2014-01-1218
19. Palaveev, S., Magar, M., Disch, C., Schießl, R., Kubach, H., et al: Simulations and experimental investigations of intermittent pre-ignition series in a turbocharged DISI-engine. 4. IAV Tagung "Ottomotorisches Klopfen – Irreguläre Verbrennung", 9–10 Dez. 2013, Berlin (2013)
20. Takeuchi, K., Fujimoto, K., Hirano, S., Yamashita, M.: Investigation of engine oil effect on abnormal combustion in turbocharged direct injection - spark ignition engines. SAE Int. J. Fuels Lubr. **5**(3) (2012). https://doi.org/10.4271/2012-01-1615
21. Wang, Z., Qi, Y., Liu, H., Long, Y., Wang, J.: Experimental study on pre-ignition and super-knock in gasoline engine combustion with carbon particle at elevated temperatures and pressures. SAE Technical Paper 2015-01-0752 (2015). https://doi.org/10.4271/2015-01-0752
22. Zahdeh, A., Rothenberger, P., Nguyen, W., Anbarasu, M., et al.: Fundamental approach to investigate pre-ignition in boosted SI engines. SAE Int. J. Engines **4**(1), 246–273 (2011). https://doi.org/10.4271/2011-01-0340

Optical Diagnostic Tools for Detection and Evaluation of Glow Ignitions

Arndt Döhler(✉) and Peter Schaffner

Advanced Engineering Gasoline Engine Technologies, Opel Automobile GmbH, Rüsselsheim am Main, Germany
Arndt.doehler@opel.com

Abstract. Downsizing as the development path chosen in recent years allows for a significant efficiency improvement of current gasoline engine concepts. However, with power densities beyond 20 bar BMEP the likelyhood of combustion irregularities is increasing.

State-of-the-art is the detection of these abnormal combustion phenomena by indication measurement technology. However, this measuring method reaches its limits when the point of origin of abnormal combustion events must be detected. The use of optical measuring systems makes it possible to generate the necessary location information.

In this study the use of optical diagnostic tools for the detection and evaluation of auto ignitions in charged direct-injection gasoline engines is evaluated. The application of "low-speed" and "high-speed" camera systems by means of combustion chamber endoscopy as well as fiber-optic measuring methods with optical access via the spark plug is presented. In addition, the use of infrared measurement technologies is discussed to enable a thermography of the combustion chamber surfaces.

It is explained how these analysis tools can be used for the detection of auto ignitions. Furthermore, the application limits of the measurement methods are discussed as well as the specific strengths of the individual methods, e.g. Low-Speed and High-Speed Video Endoscopy. Finally, an assessment is provided to explain to which extent the presented optical tools are suitable to provide the required location information.

1 Preface

Downsizing represents a technology to increase efficiency of SI engines significantly. Unfortunately, these engines are prone to abnormal combustion phenomena driven by the achieved high power densities, e.g., at high speed/high load conditions high combustion chamber surface temperatures can lead to glow ignitions. It is not known to the authors that a glow ignition like phenomenon can occur at low speed/high load too. With the thermodynamic conditions as well as the surface temperatures close to firing TDC being far below the limits for typical glow ignitions there is no obvious explanation for this phenomenon available.

M. Günther and M. Sens (eds.), *Knocking in Gasoline Engines*,
https://doi.org/10.1007/978-3-319-69760-4_4

Since any kind of auto ignitions can cause severe engine failures understanding of the phenomena is crucial as a prerequisite for the development of appropriate mitigation measures.

2 Description of the Phenomenon

During the development of the 1.6 L Midsize Gasoline Engine with direct injection introduced to market for the first time in 2012, engine damages occurred under specific test conditions typical for Non-European markets.

Precondition for the occurrence of the damages was a certain combination of operational and environmental parameters: very high charge air temperatures, low engine and vehicle speeds with simultaneous driver demand for high engine loads and use of fuel with lowest octane numbers (<RON91). As a result, damages as typically known from glow ignitions occurred at the spark plug or the exhaust valves. Even total engine write-offs arose due to ring-land cracks.

It was found from the readouts of the fault memory and from the damage reports of the development engineers that the phenomenon occurred in gradations. Strongly knocking combustion cycles were detected. In addition, stationary auto ignitions with an early combustion phasing and knocking combustion occurred. Furthermore, stationary auto ignition events with a very early combustion phasing took place without knocking; see Fig. 1. A reliable detection of these auto ignitions by the knock control system was not guaranteed since the events occurred outside the knock detection window or the maximum permissible knock adjustment had already been used up.

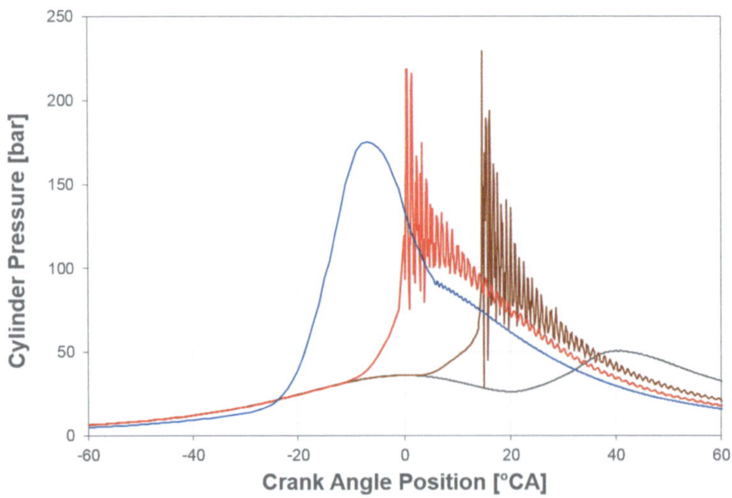

Fig. 1. Pressure traces containing auto ignitions of different intensity/combustion phasing

3 Development of Test Procedure

Reproducible tests require a test procedure which enables the development engineer to transfer the phenomenon occurring in the vehicle to the engine test stand. As described in Sect. 2, the boundary conditions have been a heavily heated engine, dynamic engine operation, high load at low speed and poor fuel quality. A qualitative description of the procedure is shown in Fig. 2. That is a ramp with decreasing speed and increasing engine load up to WOT at the same time. At this test, starting at about 1800 rpm, auto ignitions occur sporadic with increasing frequency at decreasing speed. The frequency and intensity of these auto ignitions can be varied at the test-bed by fuel quality and charge air temperature. No auto ignition occurs with fuel according to EN228.

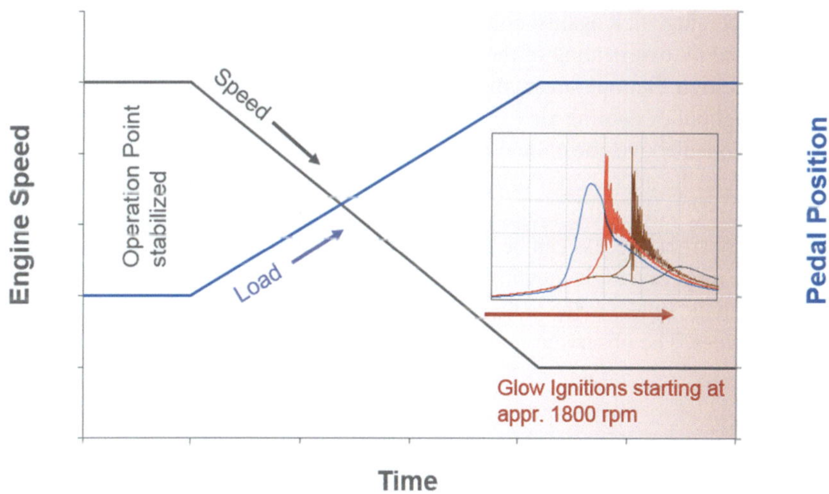

Fig. 2. Test procedure for the reproduction of auto ignitions on the engine test stand

The load/speed profile in combination with the above-mentioned boundary conditions high charge air temperature and poor fuel quality results in very late combustion phasing and enriched air-fuel mixture. At the lower end of the ramp shown in Fig. 2 an extremely retarded spark advance of approximately 20 °CA after TDC has been reached. Therefore, a long reaction time is available for self-ignition of the mixture.

4 Test Engine

Opel's 1.6 L SIDI (spark-ignited direct injection) MGE (midsize gasoline engine) used for the investigations applies a homogeneous combustion process and exhaust gas turbocharging. The engine has a capacity of 1598 ccm with a bore of 79.0 mm and a stroke of 81.5 mm. In the upper range of the scatter band for this engine concept is the compression ratio of 10.5: 1. The centrally mounted multi-hole injector is inclined by 5° and the spark plug by 21° relative to the cylinder axis. A strong in-cylinder tumble motion is generated by the inlet port. The piston has a small bowl for improved cold start performance. A magnetic injector with a 6-hole design and magnetic coil injects fuel with a maximum pressure of up to 200 bar. The pressure control of the return-less fuel system is performed at the fuel pump. Further details can be found in [1].

4.1 Optical Access

The test carrier was equipped with two optical accesses. Cylinder 4 on the output side of the engine has been chosen for installation of these accesses due to design limitations. Based on the experience within preceding projects the design of the combustion chamber windows allows for investigations of combustion peak pressures above 300 bar. Furthermore, that design shields the optical access for the endoscope as well as the access for the light dispersing stick against combustion gases to avoid glow ignitions or pre-ignitions triggered by overheating of the optical equipment, for details refer to [2]. Due to the expected high thermal stress, the cooling system has been refined to cope with a higher throughput of cooling air.

As mentioned above, the optical system consists of two optical accesses. One access is used to house the endoscope during measurement. The second bore contains the light dispersing stick. All of the optical components are still interchangeable, i.e. every combustion chamber access can be used for either lighting or recording.

Since the most favorable viewing direction for the detection of the phenomenon was the sight from the intake to the exhaust valves all following results contain only images from that access. The observation angle of the camera comprises of (see Fig. 3):

- Piston bowl,
- Spark plug,
- Injector and
- Exhaust valves.

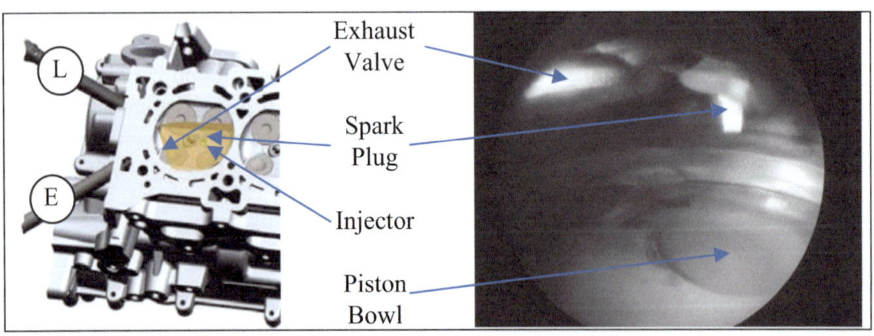

Fig. 3. Cylinder head, viewing direction on cylinder No. 4 (L = Light, E = Endoscope)

5 Motivation for the Application of Optical Measurement Tools

The occurrence, frequency and intensity of auto-ignitions can be clearly detected using standard measurement methods such as combustion chamber pressure indication. However, this method shows limitations in the analysis of the issue described above. It does not provide any information about the root cause and the location where self-ignition starts in the combustion chamber. A direct determination of sources typical for abnormal combustions, e.g. hot components in the combustion chamber and their respective temperatures, would be extremely helpful for a profound analysis. At this point, optical measurement tools with access to the combustion chamber are the means of choice since they enable a detailed analysis during steady state and transient engine operation.

Since the described phenomenon occurs on the multi-cylinder engine, it is necessary that optical measurement tools are applicable to the multi-cylinder engine too. The optical measurement tools should as little as possible affect the boundary conditions in the combustion chamber and the combustion process. Therefore, any modification of the combustion chamber at the test carrier due to the access bores was limited to a minor weakening of the water jacket and oil circuit. In any case, the influence of the measurement technique needs to be taken into account when interpreting the results.

For the analysis of the glow-ignition phenomenon image based measurement tools were used in the first step. Images and videos of the combustion process have been recorded by using endoscopes via optical accesses to the combustion chamber. In particular, this method provides information of the location in the combustion chamber where self-ignitions have their origin.

Another appropriate measurement tool detects the light intensity of the flame via optical lenses and fibers incorporated in the spark plug. It can be used on all cylinders which represent the greatest advantage of this technology. This tool provides statistical information about frequency and distribution of events supplementary to the determination of the location of origin of the auto ignition. Nevertheless, the use of the optical spark plug influences boundary conditions in the combustion chamber and therefore the combustion process itself by heat value and spark protrusion.

Temperatures of critical components are of particular interest once the origin of the auto ignition is known. They can be determined by measuring the radiation of the components in the infrared wavelength range. The access to the combustion chamber is realized via bores into the cylinder head too. An endoscope with transmittance for infrared radiation is mandatory.

6 Application of Image Based Tools for the Detection of Glow Ignitions

Image based combustion diagnostics allows for recording of images of the combustion chamber and combustion process at fired engine operation via an optical access in the cylinder head. A distinction must be made between so-called low-speed systems, e.g. AVL Visioscope®, and high-speed cameras. It is possible to record only one image per cycle with the low-speed system. Therefore, videos or image sequences are composed of recordings of several combustion cycles. The system is mainly suitable for the investigation of periodic processes. Greatest advantage of the low-speed endoscopy, which makes an application nevertheless meaningful, is the simple application on the engine. Easy handling due to a small camera, bright pictures and low data volume are additional benefits.

A high-speed camera, on the other hand, is capable to take continuous recordings due to its high frame rate. Its temporal resolution extends into the crank angle range. The system uses the same optical access in the cylinder head and the same endoscope as the low-speed system.

6.1 Low Speed Endoscopy

Requirements of the System and Test Conditions

Since the low-speed system allows to record only one picture per cycle, the recordings must be restricted to a constant crank angle position. Several combustion cycles are recorded at a fixed time in the cycle. The recording time is chosen in such a way that the early flame front, a few degrees crank angle after the inflammation, is recorded. In case the onset of the auto ignitions varies during the test, e.g. by changing of engine speed, several measurements at different crank angle positions are necessary. The camera system and the indication system must be synchronized in order to establish a correlation between the cylinder pressure traces and the respective images.

The early flame front is a weakly glowing event. Therefore, settings and parameters of the measurement system must be optimized with regard to light output. Parameters are the type of the camera (black & white or color), elimination of external illumination of the combustion chamber and of course exposure time. The exposure time has to be a compromise between brightness of the images and temporal resolution.

Detected Origins of Glow Ignitions with Low Speed Endoscopy

Images taken with low speed endoscopy are showing the origin of auto ignitions in the area of the exhaust valves. An example at 1556 rpm is presented in Fig. 4. The recording time of the image is 6.0 to 8.5 °CA aTDC. Auto ignition starts 9 °CA before the ignition set point at 15 °CA aTDC with the flame front staring underneath the left exhaust valve. At the time of recording, no ignition can be detected near the spark plug.

Statistical relevance is ensured by a sufficiently high number of recorded cycles containing auto ignitions. In all cases auto ignitions have their origin in the vicinity of the exhaust valves. Auto ignitions starting directly at the spark plug have not been observed. In most cases the source of the flame front is the left exhaust valve, i.e. exhaust valve #8. More rarely, two flame fronts are formed simultaneously at both exhaust valves.

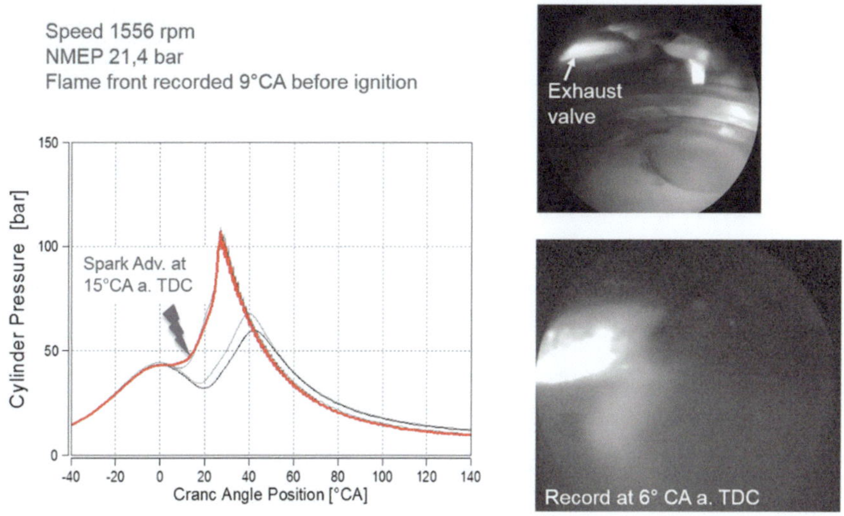

Fig. 4. Auto ignitions at the exhaust valve

6.2 High Speed Endoscopy

Requirements of the System and Test Conditions

In contrast to the frame-by-frame images taken with the low-speed system, the high speed camera is capable to record high-resolution videos of consecutive combustion cycles. Especially the brightness of the images is a compromise between exposure time and temporal resolution. In order to achieve the highest possible temporal resolution, an electronic image intensifier can be connected to the high-speed camera. In the presented study, however, this approach was not possible due to strong vibrations driven by the high engine speeds and highly dynamic engine operation. The selected recording rate was 3000 frames per second. At an engine speed of 2000 rpm this corresponds to an exposure time of approximately 333 µs per image, respectively 4 °CA.

The measurements with the high-speed camera are time-based. There is no correlation to the crank angle position. Synchronization with the combustion chamber pressure is achieved via a synchronized start of both measuring systems. The identification of cycles with auto ignitions is based on the cylinder pressure traces. Further details about the application of the camera system to the engine are described in [2].

Glow Ignitions at the Exhaust Valve

The records of the high-speed camera confirm the results obtained with the low-speed system in more detail. The source of auto ignition is mainly located in the area below the exhaust valves. Figure 5 shows an example of an auto ignition starting underneath the left exhaust valve. The flame propagation from the left side of the combustion chamber towards the center of the combustion chamber is clearly visible.

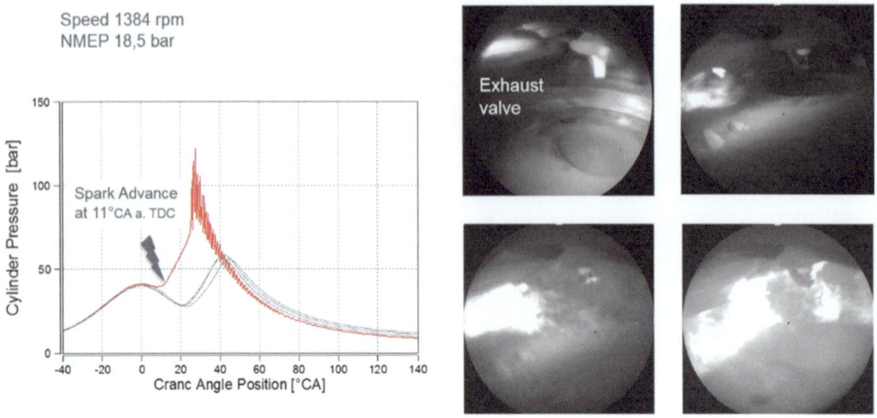

Fig. 5. High-Speed imaging of auto ignition at the exhaust valve

7 Detection of Glow-Ignitions with Fiber Optic Based Tools

Fiber optic based measurement tools have been originally designed for the evaluation of early flame kernel propagation as well as the determination of the origin of knocking combustions cycles. The system detects the self-lighting of the flame. Lenses integrated into a spark plug sensor detect light signals transmitted to photodiodes via optical fibers. The measured signal is proportional to the luminous intensity of the flame and recorded as a function of crank angle position. The location information of the detected phenomenon is derived from the known geometric alignment of the lenses.

No modifications of the cylinder head for optical accesses are necessary. Measurements are possible on all cylinders and a high number of cycles can be recorded. A minor disadvantage of the measuring system is the influence on the combustion process by replacing the original spark plug.

7.1 Sensor Variants and Evaluation Methods

Figure 6 shows the spark plug sensor used for the investigations. It is fitted with 35 optical channels, 28 channels aligned to the circumference of the combustion chamber (knock signals), seven in parallel to the spark plug with a direct view to the piston crown (flame signals). Thus is it possible to detect the origin of auto ignitions on the circumference of the combustion chamber, the piston crown as well as on the spark plug.

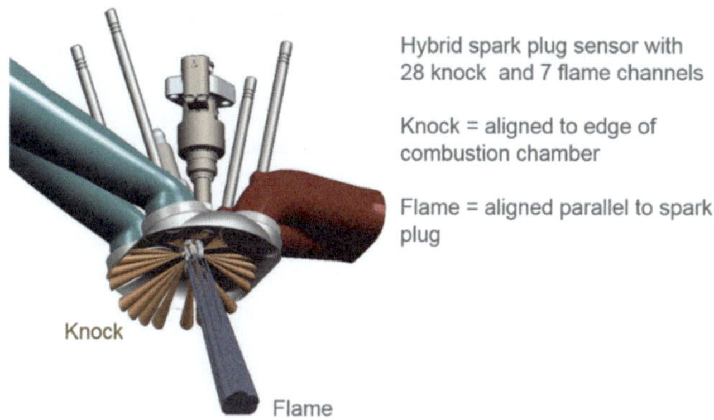

Fig. 6. Fiber optic based spark plug sensor mounted in the combustion chamber

All signal traces are high-pass filtered and plotted in a waterfall diagram as a function of the crankshaft position (see top right of Fig. 7). Here the top eight lines contain the flame signals recorded at the piston crown and the following 28 light lines show the signals coming from the combustion chamber circumference. The brightness gradients shown are used to determine the origin of the auto ignitions, i.e., the auto ignition starts at the point in the combustion chamber where the light signal trace deviates from the background noise for the first time. The known orientation of the corresponding lens provides information on the origin of the glow-igniting event.

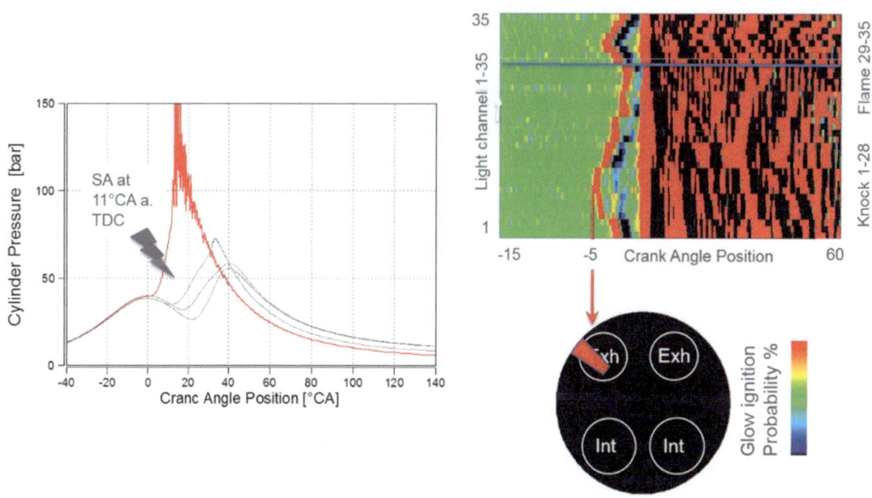

Fig. 7. Evaluation of light signals

In Fig. 7 the auto ignition is recorded for the first time 15 °CA before the ignition set point on channel 11 which correspond to a position underneath the left exhaust valve. Through the evaluation of a sufficiently high number of cycles, statistical information about the frequency and occurrence probability of auto ignition events at the corresponding positions in the combustion chamber are possible.

7.2 Detected Origin of Glow-Ignitions by Means of Fiber Bases Tools

The results obtained with image based methods could be confirmed with the fiber optic based measurement tool. On the fourth cylinder, the highest occurrence probability and frequency of auto ignitions is in the area around or underneath the left exhaust valve. On the other cylinders, auto ignitions occur as well only in the area around the exhaust valves (see Fig. 8).

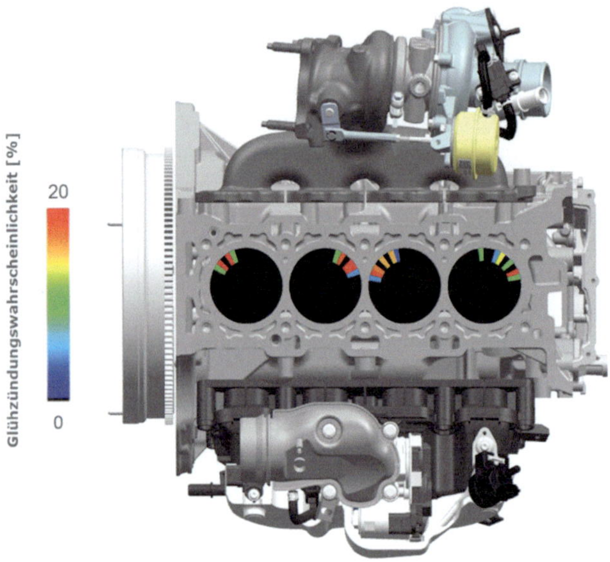

Fig. 8. Glow ignition probability at all cylinder

Auto Ignitions starting at the spark plug were not detected. This is consistent with the previous results and it seems reasonable due to the low heat value of the measuring spark plug.

8 Measurement of Critical Component Temperatures in the Combustion Chamber

All the measuring methods described above indicate the area around the exhaust valves as the point of origin of the auto ignitions. This raises inevitably the following questions regarding the temperature of the exhaust valves:

- Are exhaust valves hotter under the given operating conditions compared to WOT steady state operation?

- What is the influence of retarded ignition timing on the exhaust valve temperature due to poor fuel quality?
- What is the effect of the increased charge air temperature?
- How are the exhaust valve temperatures compared to the spark plug temperature at steady state operation and in the transient test ramp?

Measurement of component temperatures in the combustion chamber is possible by detecting the intensity of radiation of the components in the infrared spectrum between 700 and 2500 nm. The design of the optical accesses to the combustion chamber is identical to those used for the low-speed and high-speed cameras. However, the used endoscope must be permeable to infrared radiation. A specific camera is required, which can detect the heat radiation in the infrared wavelength range. The data recording system is the same as used for the low speed camera. For this reason only one record respectively temperature measurement per engine cycle is possible with this application.

Figure 9 shows exemplarily a record with open exhaust valves taken at 240 °CA aTDC. Exhaust valve temperatures can be measured at open and closed valves. In this context the temperature of the closed valves just before the intended ignition set point is relevant. The temperature level of the closed exhaust valves is close to the lower measurement limit of the infrared system. Since the lower temperature limit is mostly defined by exposure time, the chosen exposure time must be a compromise between the longest possible exposure duration and the necessary temporal resolution. Therefore, the selected exposure time is two milliseconds in the present case. The temperature measurements have been performed at 40 °CA before the ignition TDC in order to ensure a sufficient distance to the start of combustion, in particular in case of auto ignitions.

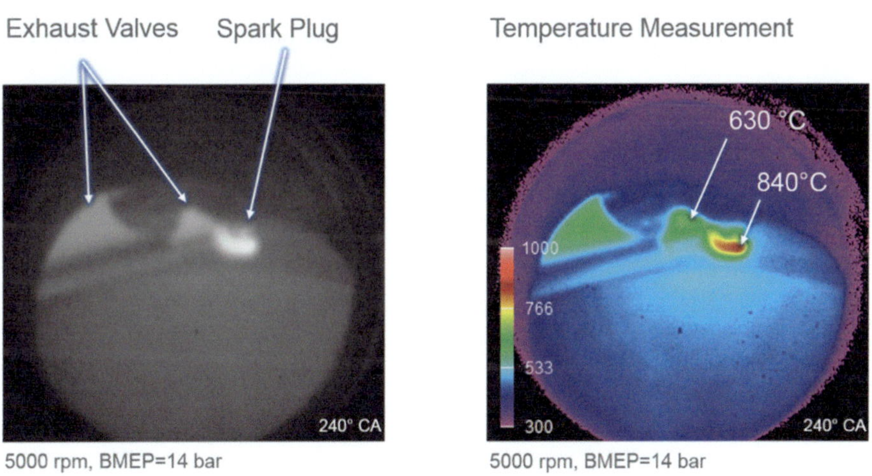

Fig. 9. Infrared record of combustion chamber at open exhaust valves

A parameter leading to an increased occurrence probability of glow ignitions is the ignition retard due to low octane fuels, high charge air temperature and transient operating conditions. Compared to steady-state operation at WOT with, e.g., RON95 pump fuel the exhaust valve temperature increases by 50 °C in case of maximum retarded ignition timing. This ignition retard increases the exhaust gas temperature by up to 240 °C (see Fig. 10, left).

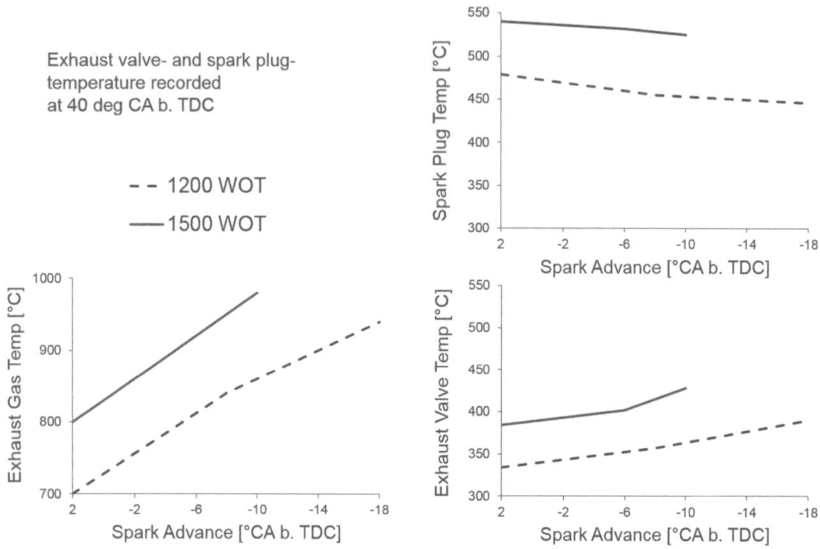

Fig. 10. Exhaust valve and spark plug temperature at WOT steady state operation

The temperature of the spark plug mass electrode remains rather stable. At most the temperature is slightly decreasing when retarding the ignition timing and combustion

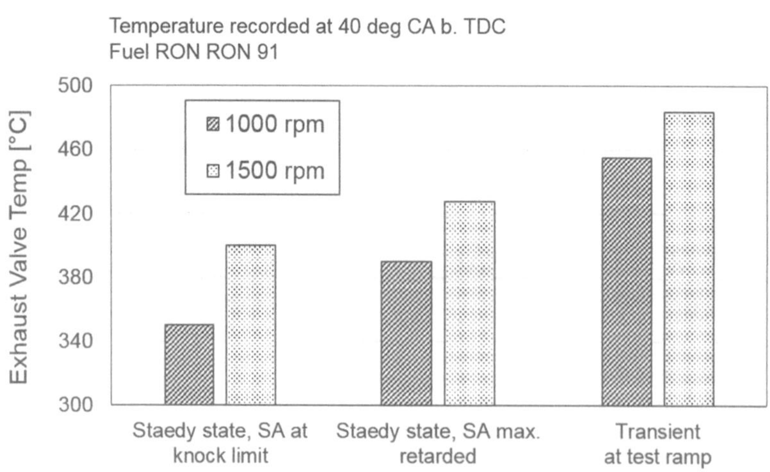

Fig. 11. Exhaust valve temperature at test ramp

phasing. However, the spark plug is always hotter than the exhaust valves with a difference of about 150 °C at regular ignition timing. At maximum retarded ignition set point that difference is reduced to 60 °C (see Fig. 10, right).

During transient operation of the test ramp the outlet valves are again 50° to 60 °C hotter compared to steady state operation at maximum retarded combustion phasing. That means the exhaust valve temperature during the test ramp is approximately 100° to 120 °C hotter compared to steady state WOT operation with optimized parameter settings Fig. 11. It means that the described combination of parameters leads to temperature conditions in the combustion chamber which are not reached during steady-state WOT operation. A further rise of exhaust valve temperature during the test ramp is driven by the temperature storage capacity of the exhaust valves.

At steady state operation, the temperature at the spark plug mass electrode is always significantly higher than the temperature of the closed exhaust valves. During retarded combustion phasing, this is the case too. During the test ramp conditions can be reversed and the exhaust valves are hotter than the spark plug (Fig. 12). The inertia of the spark plug electrode against temperature changes is significantly lower than that of the much larger exhaust valves. Therefore, the spark plug temperature changes relatively quickly during transient operation.

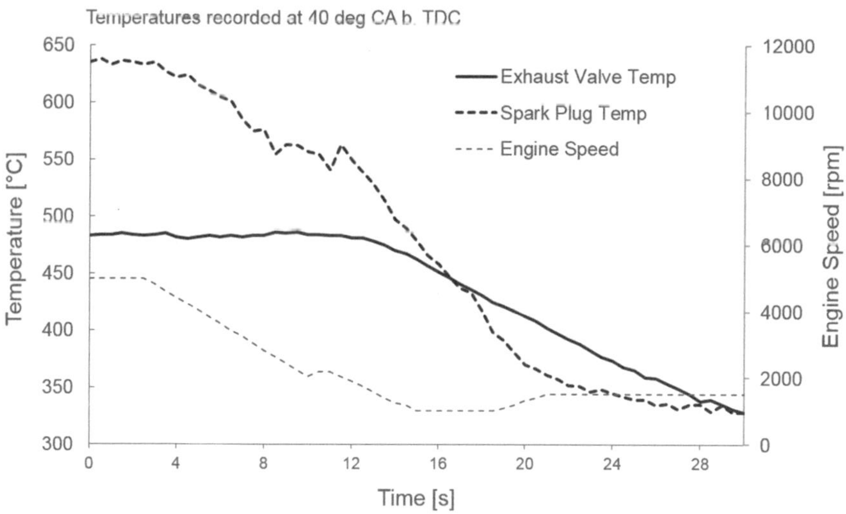

Fig. 12. Exhaust valve- and spark plug temperature at test ramp

The comparatively large and hot exhaust valves cause a much higher heat input into the combustion chamber than the small spark plug. Therefore, it is understandable that the auto ignitions originate from the circumference of the combustion chamber underneath the exhaust valves and not from the vicinity of the spark plugs.

9 Simulation

Opel uses an in-house simulation model to assess the tendency to knock for a specific combustion chambers design during the development process. With further progress of the development, this model is repeatedly validated and calibrated taking into account the results of other simulation tools (e.g. GT-Power) and results from engine measurements (e.g. engines equipped with temperature measuring systems).

In addition, the knock model has been validated with the location and temperature information derived from optical measurements. The goal was to evaluate the critical area of the combustion chamber around the exhaust valves in terms of knock probability due to residual gas content as well as mixture and temperature distribution in the relevant load and speed ranges.

Figure 13 shows the probability of knocking combustion events at a given location within the combustion chamber. It can be derived from Fig. 13 shown below, that the probability for knocking cycles and auto ignitions is significantly increased in the area underneath the exhaust valves. This simulation result correlates well with the experiment; see Figs. 8 and 13.

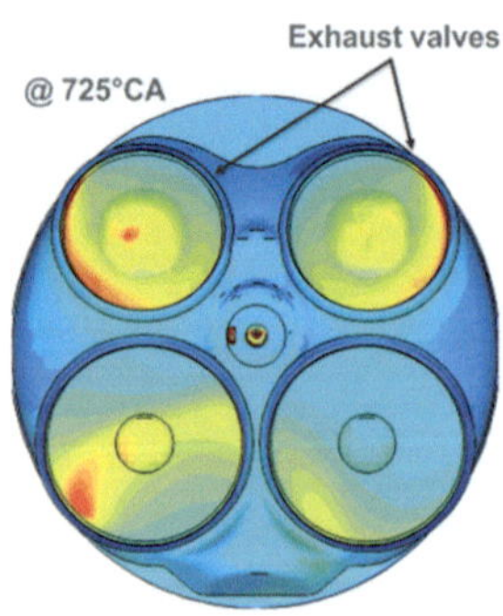

Fig. 13. Knock probability at cylinder No. 1

10 Discussion of Results

The phenomenon of auto ignitions in the low speed range described in detail in Sect. 1 can be a problem especially in "hot" markets. Charged engines with direct fuel injection and thus relatively high compression ratio - in the present case with a compression ratio of 10.5: 1 - have been introduced in these markets in recent years. Therefore, no validated development procedure tailored to the specific issue of low-speed auto ignitions was available. The test procedure described in Sect. 2 reproduces the customer behavior under standardized, repeatable test conditions. This reproducibility is a fundamental prerequisite for the comparison of the used measurement methods and the results achieved with them. In the future, the developed test procedure enables an early validation of the product under development as part of the standard development process.

As described in Sects. 6 to 8, all measurement methods deliver similar results. The optical results in the visible range of the spectrum enable in any case the allocation of the point of origin of the auto ignitions to an area below the outlet valves. This result is supported by the surface temperatures of the exhaust valves determined by means of infrared measurement technology. With temperatures of about 400 °C the exhaust valves represent by far the hottest parts in the vicinity of the combustion chamber roof. These results lead to the conclusion that the exhaust valves are the most likely source of ignition.

However, these temperatures are still significantly lower than the component temperatures that typically lead to glow ignitions at high engine speeds. The question arises whether the described phenomena is a glow-ignition in the most traditional sense or not. An additional temperature increase through the "hot" outlet valves leads to a thermodynamically critical state of charge. The "hot" exhaust valves provide enough locally limited activation energy to ignite the cylinder charge. This inflammation process is promoted by the high retention time of the charge in a thermodynamically critical state.

11 Summary

It can be derived from the above that numerous measurement methods are available to conduct investigations of abnormal combustion phenomena. Advantages and disadvantages of these tools are shown in Fig. 14.

Measurement System	Time Information	Location Information	Temperature Information	Application and Handling Effort	Measurement and Evaluation Effort
Low-Speed Camera (visible spectrum)	-	+	-	O	O
Low Speed Camera (Near Infrared)	-	+	+	O / -	O
High-Speed Camera	+	+	-	-	-
Optical Spark Plug	+	O	+ / O	+	+

Fig. 14. Comparison of available measuring tools

The use of an infrared measurement system is essential if the determination of temperature information is in focus. Optical spark plugs and low-speed infrared cameras are available for this specific task. Beneficial is the simple handling of the optical spark plug on a multi-cylinder engine. However, temperature information can be obtained only in a very small viewing area when using this measuring system. In case a temperature distribution over larger areas of the combustion chamber surface needs to be determined, an infrared camera is the tool of choice.

The great strength of high-speed cameras lies in the measurement of highly dynamic processes. However, these high-speed investigations are limited to the visible spectral range due to the necessary short exposure times. At the same time, the localization of phenomena is easily possible due to the large field of view. In contrast, the considerable handling and evaluation effort needs to be considered. Minimal application effort in conjunction with acceptable local resolution is the strength of the optical spark plug, which can be applied to any cylinder without modifications.

Acknowledgements. The authors would like to thank the peers at Opel Automobile GmbH supporting the conducted investigations with design, simulation and execution of testing as well as valuable discussions: Gerwin Tittes, Stefan Gumprich, Bernd Veith and Achim Königstein.

References

1. Frensch, M., et al.: The new midsize gasoline engine with direct injection and turbocharging from GM. In: 21st Aachen Colloquium Automobile and Engine Technology, Aachen, pp. 161–187 (2012)
2. Döhler, A., Pritze, S.: A contribution to better understanding the pre-ignition phenomenon in highly charged internal combustion engines with direct fuel injection. In: 4th Conference on Knocking in Gasoline Engines, Berlin, pp. 41–61 (2013)

Study of HSPI/LSPI from Spark Plugs on Turbocharged Gasoline Engines

Tomohiro Iwatsuka[1(✉)], Masaru Kano[1], Kiyoteru Mori[1], Waldemar Werner[2], and Stefan Schulte[2]

[1] NGK Spark Plug Co., Ltd., Nagoya, Japan
t-iwatsuka@mg.ngkntk.co.jp
[2] NGK Spark Plug Europe GmbH, Ratingen, Germany
S_Schulte@ngkntk.de

Abstract. Pre-ignition caused by the spark plug was investigated with in-cylinder visible engine. As a result, two types of pre-ignition were observed; one was heat surface ignition at a high revolution speed and the other was non-heat surface ignition at a low revolution speed. As for the latter phenomenon, a presumption of the ignition location, investigation of the associated engine parameters, and comparison of the occurrence conditions for LSPI caused by red-hot particles were carried out to meticulously classify pre-ignition.

1 Introduction

Nowadays, improvement in fuel efficiency and reinforced emission regulations should be one of the most urgent issues to be addressed. Therefore, highly boosted, downsized engines [1–2] with high compression ratios have been developed and put into the market. During the development phase, there was an abnormal combustion phenomenon called Low Speed Pre-Ignition (LSPI) which occurs at low revolution, high load engine conditions. This is a known problem which should be avoided [3].

It has been reported that LSPI is caused by red-hot particles which are made from deposits or oil droplets [4–9]. It has also been reported that the spark plug design doesn't affect the occurrence of LSPI caused by the red- hot particles. However, it has also been confirmed that the frequency of LSPI changes in accordance with spark plug designs during the actual engine development phase. Therefore, there might be a possibility that the spark plug is one of the causes of LSPI, in addition to the red-hot particles.

On the other hand, pre-ignition caused by the spark plugs is recognized as surface ignition from hot spots on the spark plug which are overheated by high temperature combustion gas during high revolution and high load engine conditions, (hereinafter HSPI: High Speed Pre-Ignition) [11]. However, the spark plug temperature range in LSPI occurrence region is lower than that of HSPI, therefore the cause of the pre-ignition may be different.

Described above, LSPI caused by spark plug can be considered. Therefore, there is a need to understand the mechanism. If the root cause could be determined and LSPI

M. Günther and M. Sens (eds.), *Knocking in Gasoline Engines*,
https://doi.org/10.1007/978-3-319-69760-4_5

caused by spark plug was effectively avoided or reduced, then the improvement in efficiency of downsized engines can be anticipated. In this study, the ignition points of the spark plug where HSPI and LSPI occur in the engine are investigated and compared visually.

2 Experimental Equipment and Test Procedure

For the experiments, a commercial 4-cylinder turbo- charged gasoline engine was used. The specification of engine is shown as Table 1. An endoscope was installed to observe the combustion event at high load condition at the 4th cylinder. During the investigation, all the other cylinders were normally operated.

Table 1. Engine specification

Displacement	1618 cc
Design and configuration	Straight-four
Combustion process	Turbocharged gasoline engine with direct injection
Compression	9.5:1
Bore × Stroke	79.7 × 81.1

Figure 1 shows the experimental configuration. In order to observe pre-ignition caused by a spark plug, the field of view of the endoscope (KARL STORZ: 93670AX2) was set per Fig. 2, and recorded by a high speed camera (Photron: FASTCAM SA-Z). At the same time, the in-cylinder pressure was measured by pressure sensor (KISTLER: 6056AU20) and the combustion analysis device (KISTLER: KiBox). The filming condition is shown per Table 2.

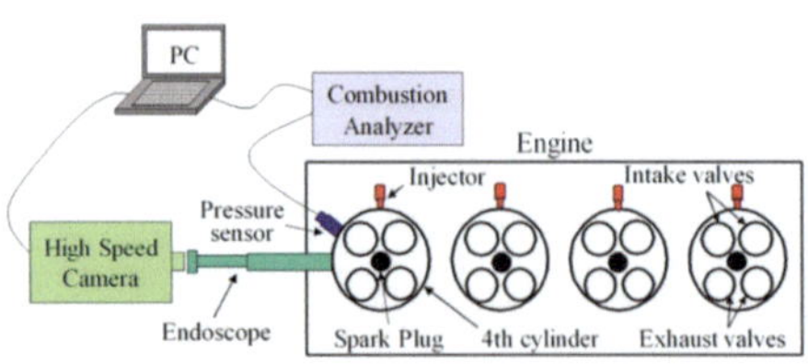

Fig. 1. Experimental configuration

The spark plug was installed so that the ground electrode was directed toward the Intake side shown as Figs. 2 and 3. Occurrences of pre-ignition were judged by comparing the ignition timing, and the timing of the ionic current flow from the combustion gas detected between spark plug electrodes (hereinafter spark gap). For the detection, a positive polarity ignition coil and a positive detection voltage on the spark gap were applied.

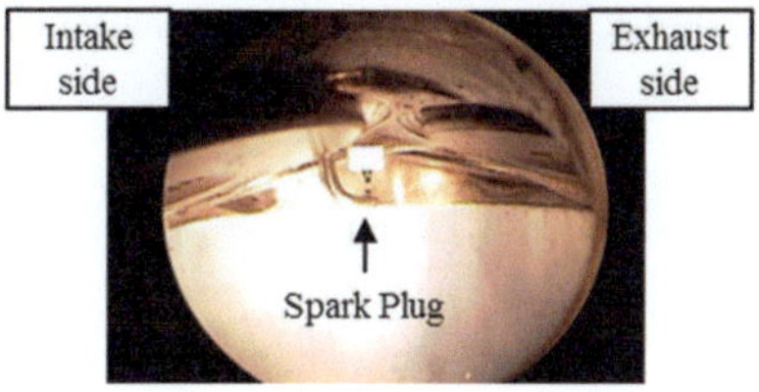

Fig. 2. View from installed endoscope

Table 2. Filming condition

Filming condition	
Frame rate	20000 fps
Shutter speed	1/20409 s
Resolution	384 × 256

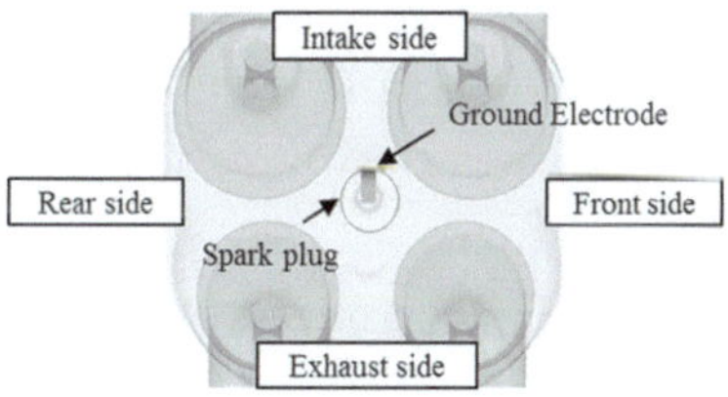

Fig. 3. Bottom view of combustion chamber

3 Results

3.1 Visualization of Pre-ignition Caused by Spark Plugs

The conditions of visual test are shown on Table 3. To observe the HSPI and LSPI event, high revolution condition (6000 rpm for HSPI) and the low revolution condition (2000 rpm for LSPI) were used. Obtained image was treated to enhance the observation of flame propagation.

Figure 4 shows the visually recorded results of HSPI caused by the spark plug at high revolution condition. From the recorded images, it was confirmed that the insulator and electrodes were overheated which sparked the ignition from the insulator tip before ignition timing. Therefore, the HSPI was confirmed as a result of surface ignition originated from the overheated insulator tip.

Figure 5 shows the visually recorded result of LSPI caused by the spark plug at low revolution condition. From the recorded images, it was confirmed that the flame propagated from the clearance between metal shell and insulator of the spark plug (hereinafter Gas-volume) before ignition timing. This was different compared with the surface ignition from the overheated insulator tip of the spark plug in the high

Table 3. Test condition for visualization

Condition	HSPI	LSPI
Revolution (rpm)	6000	2000
Intake air pressure (kPa abs.)	110	194
Ignition timing (°BTDC)	36	4
A/F	13.5	14.0
Start of injection (°BTDC)	240	330
Water temperature (°C)	90	
Oil temperature (°C)	100	
Fuel	RON90	

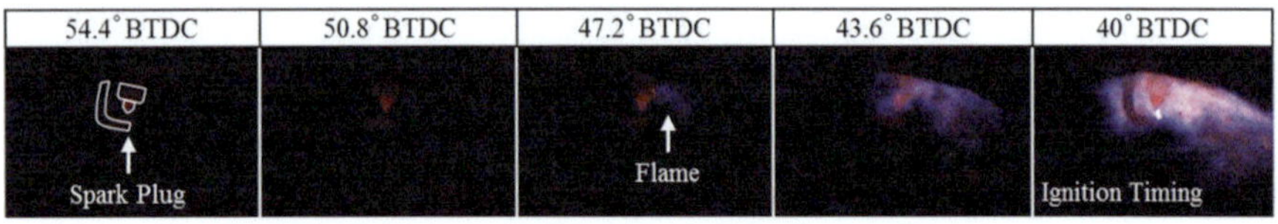

Fig. 4. Sequential image of HSPI from spark plug

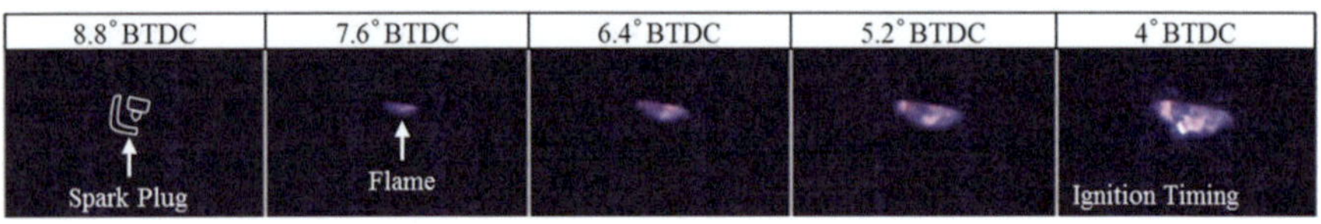

Fig. 5. Sequential image of LSPI from spark plug

revolution condition. Therefore, the LSPI caused by the spark plug was not originated from a surface ignition, but an auto-ignition from inside the Gas-volume.

Figure 6 shows the visually recorded result of LSPI caused by the red-hot particles. Same results were obtained in the previous studies [5–9]. It was confirmed that LSPI caused by red-hot particles occurred in the same condition as LSPI caused by the spark plug.

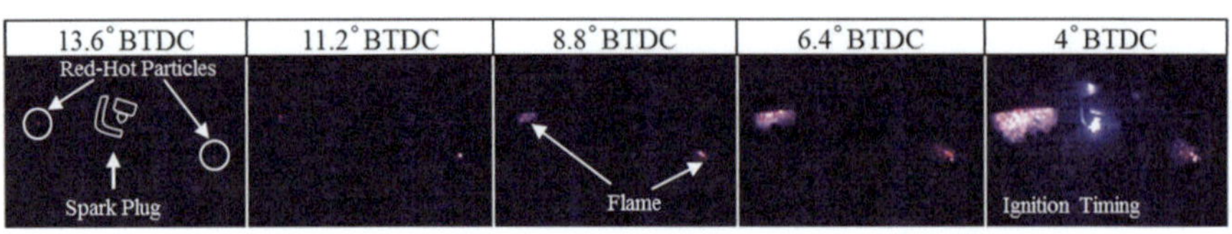

Fig. 6. Sequential image of LSPI from red-hot particles

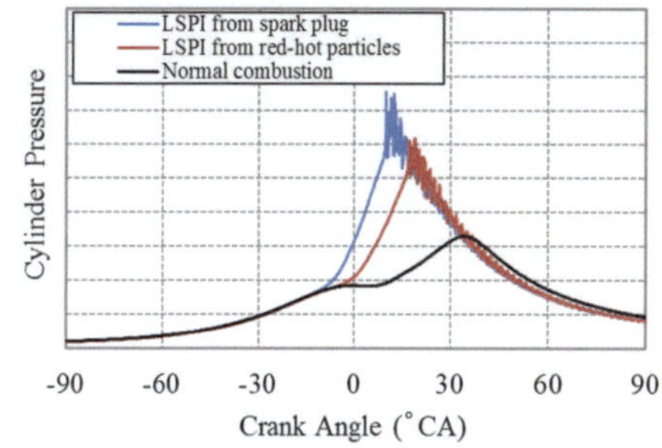

Fig. 7. Cylinder pressure vs. Crank angle for different LSPI

As shown in Fig. 7, it was difficult to distinguish different LSPI causes by monitoring only the cylinder pressure. However, it can be visually by endoscope.

From the observation results described above, it was confirmed that there were two types of pre-ignition caused by a spark plug. One was surface ignition from the insulator tip. And the other was assumed to be auto-ignition from the Gas-volume.

The former can be easily understood. Reducing the temperature of an insulator tip are an effective way to prevent surface ignition. There were many prevention methods were discussed, and using colder type spark plugs was most commonly used.

On the other hand, it was confirmed the latter wasn't caused by surface ignition. It is possible the ignition could be caused by high compression in the narrow spaces or the red-hot particles flew into the Gas-volume. However, it is too early to determine the countermeasure because the root cause was yet to be determined. Therefore, additional tests were performed to clarify the location and the root cause of a pre-ignition from the Gas-volume.

At first, the engine conditions influences were investigated by comparing the occurrence tendencies of LSPI caused by red-hot particles and the spark plug. Then, gas flow and temperature inside combustion chamber was calculated by CFD to estimate the occurrence location. Furthermore, it was investigated to verify the actual point of auto-ignition where it had been preliminary estimated by CFD analysis. And then, the influence of spark plug designs was examined.

3.2 The Evaluation of Engine Conditions

The influences of engine conditions leading to LSPI caused by the spark plug were investigated. In order to avoid LSPI caused by red-hot particles, a specific engine oil (SN 0W-20) was used. This oil is able to avoid LSPI occurrences caused by red-hot particles and is produced by Tonen General Sekiyu. The experimental conditions were set per Table 4, and the parameters were changed respectively from the standard engine condition. In this section, the focus was on the results of varied boost pressures and coolant temperatures as this showed interesting trend.

The LSPI tendency vs boost pressure is shown per Fig. 8(a). In accordance with the increase of boost pressure, it was confirmed that the frequency of LSPI caused by the spark plug increased. It was observed that at the condition of 210 kPa absolute, frequent

Table 4. Test condition for sensitivity of engine parameter

Condition	LSPI
Revolution (rpm)	2000
Boost pressure (kPa abs.)	200
Ignition timing (˚BTDC)	TDC
A/F	13.5
Start of injection (˚BTDC)	330
Water temperature (˚C)	90
Oil temperature (˚C)	100
Fuel	RON90

and continuous LSPI events occurred. This phenomenon was different from the LSPI caused by red-hot particles which was supposed to stop when deposits disappear [10]. The LSPI event caused by spark plug Gas-volume was continuous and will not disappear which was different than the same phenomenon caused by red-hot particles.

The LSPI tendency vs coolant temperature is shown per Fig. 8(b). In accordance with the increase of coolant temperature, it was confirmed that the frequency of LSPI caused by spark plug increased as well. This result showed opposite tendency from the conventional tendency of LSPI caused by red-hot particles [3]. The ignition source of LSPI caused by the spark plug is something stimulated by a temperature rise of the spark plug itself or the cylinder head, and so on.

From these results, it is considered that LSPI caused by the spark plug is a different phenomenon from LSPI caused by red-hot particles because this LSPI shows different behaviors (than from the LSPI caused by red-hot particles).

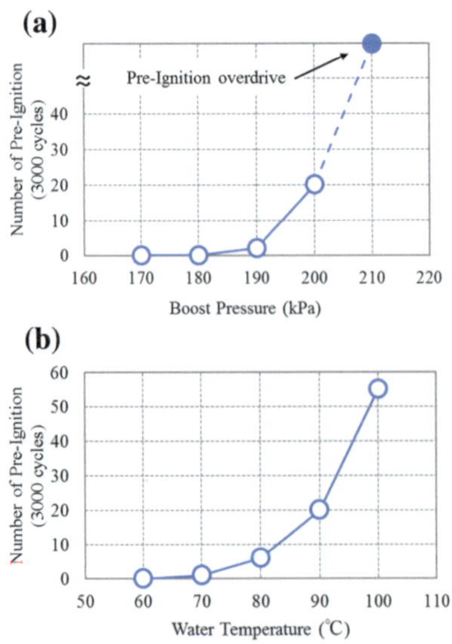

Fig. 8. (a) Influence of boost pressure. (b) Influence of water temperature

3.3 CFD Calculation of In-Cylinder Temperature and Gas Flow

The tested engine model was constructed and calculations of the in-cylinder temperature and gas flow were carried out in order to estimate the root cause of LSPI caused by the spark plug. The engine model is shown per Fig. 9. For the analysis, CONVERGE Version 2.2 (CONVERGENT SCIENCE) was used.

Fig. 9. Outlook of analysis engine model

The calculated condition was based on LPSI conditions described in Table 3. The applied initial temperature distribution to the spark plugs was calculated from exhaust top dead center to compression top dead center. The set direction of spark plugs is shown per Fig. 3. In the CFD, a homogeneous mixture of air and ethane was used instead of modeling the direct fuel injection of these mixtures into the combustion chamber. During the calculation, the spark plugs were not ignited.

Figure 10 shows the temperature distribution of the cross sectioned combustion chamber at 10 °BTDC. It was confirmed that the temperature of the Gas-volume increased. And, 3D flux line in the Gas-volume and the area above 850 K at 10 °BTDC are shown per Fig. 11. It was confirmed that gas retention was seen in the Gas-volume, and the temperature increased in that area. That was centrically located on the front side of insulator nose, intake side of insulator, and root intake side of insulator.

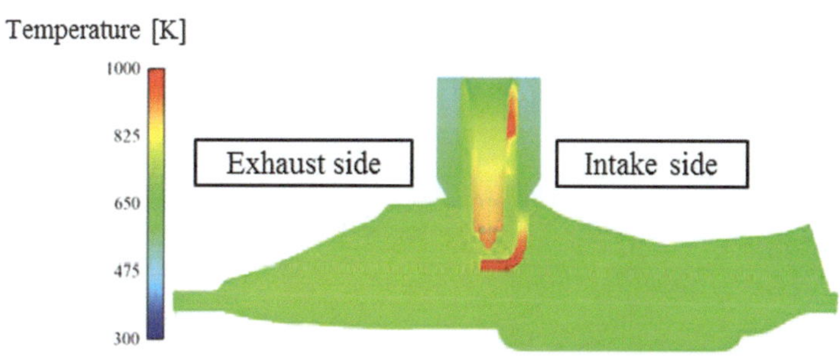

Fig. 10. Calculated distribution of temperature of combustion chamber

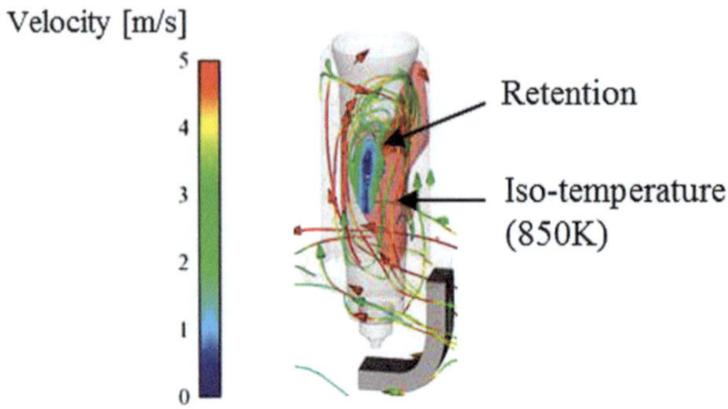

Fig. 11. Calculated iso-temperature distribution (850 K) and 3D flow analysis inside spark plug

From these CFD results, it was clear that LSPI could occur due to the high temperature point.

3.4 Verifying the Points of Auto Ignition Using a Test Engine

From the CFD analysis, it was confirmed that the temperature at the flow retention point in Gas-volume increased. For the purpose of verifying whether the high temperature area of the Gas-volume was the very point of auto-ignition in cylinder or not, special spark plugs were prepared to be able to detect auto-ignition inside the Gas-volume. The surface of the spark plug insulators were partially coated with conductive metal paste, and the points where metal pastes contact the metal shell were covered with insulation material. This made it possible to arrange the independent electrodes in the Gas-volume. The auto-ignition points were investigated by detecting the timing order of ionic current flow from flame which propagated between the independent electrodes and metal shell. In addition, the actual auto-ignition location was confirmed to be in the location similar found in the CFD analysis.

Table 5. Spark plugs for investigation of Auto-ignition point

Condition	Height	Orientation①	Orientation②
Paste position	-	Intake, Exhaust	Front, Rear
Plug image			

The spark plugs used for the investigation are shown on Table 5. In this experiment, two types of spark plugs were prepared: the first type was for height investigation, the second type was for orientation investigation. As for the blue and red lines shown in the table, the samples pasted electrodes for the height investigation were arranged in the middle portion of insulator and in the root portion of insulator. On the other hand, the samples pasted electrodes for the orientation investigation were arranged ① 0° and 180° (Intake side and Exhaust side for engine) ② ±90° against ground electrode (Front side and Rear side for engine). The ionic current detection

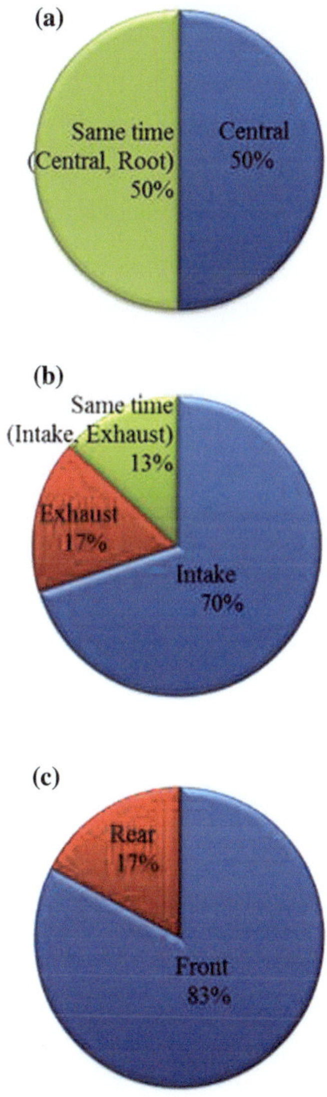

Fig. 12. (a) Result of height. (b) Result of Orientation①. (c) Result of Orientation②

circuits were connected with these two paste electrodes independently. A center electrode was also connected to a detection circuit in order to investigate the detection order. The results are shown on Fig. 12.

The results of the height investigation are shown on Fig. 12(a). It was confirmed that the auto-ignition points tended to be from the central portion of Gas-volume.

The 1st round of results for the orientation investigation are shown on Fig. 12(b). It was confirmed that the auto-ignition points tended to be from the Intake side.

The 2nd round of results for the orientation investigation are shown on Fig. 12(c). It was confirmed that the auto-ignition points tended to be from the Front side.

During the series of investigations, the fast ionic current detection had never been seen in the spark gap.

From these results described above, we suppose that the auto-ignition point in the Gas-volume should be described as shown by Fig. 13.

The comparison results of the auto-ignition point estimated from the engine test and from CFD is shown on Fig. 14, both results were roughly corresponded. Therefore, the estimation of the occurrence point by CFD was reasonable. And we are able to mention that the root cause of LSPI caused by spark plugs is the high temperature gas retention inside the Gas-volume portion.

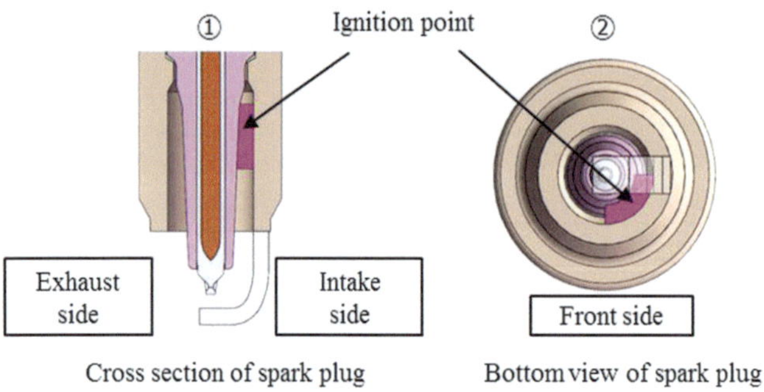

Fig. 13. Auto-ignition point

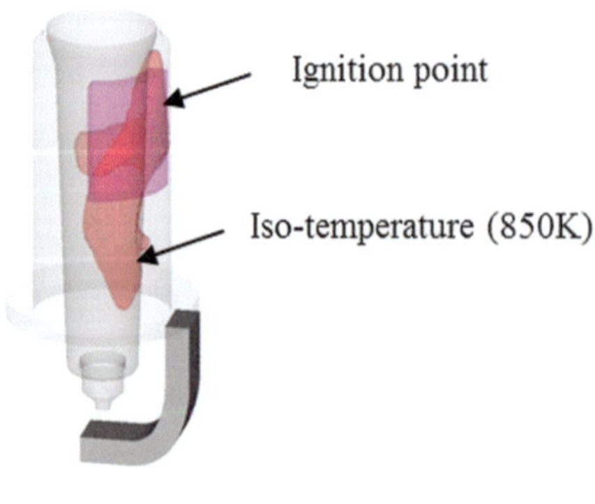

Fig. 14. Compared Auto-ignition points by engine test to iso-temperature distribution (850 K) by CFD

3.5 The Evaluation of Spark Plug Insulator Design

As the verification result of auto-ignition point, influence of insulator temperature of spark plug is evaluated to consider the way for inhibiting LSPI caused by spark plug. And, influence of gas retention is also evaluated with varying depth and width of Gas-volume. Each evaluation is done with the condition showed in Table.

3.5.1 The Influence of Insulator Temperature and Depth of Gas-Volume

Influence for frequency of LSPI according to insulator temperature and depth of Gas-volume were evaluated. Table 6 shows the evaluated samples.

Table 6. Spark plugs for sensitivity of insulator temperature and depth of Gas-volume

	Plug A	Plug B	Plug C
NGK Heat Rating	6	-	5
Copper Core inside center electrode	Applied	Not applied	Applied
Insulator shape	a	a	b
Gas-volume depth	Shallow	Shallow	Deep
Appearance	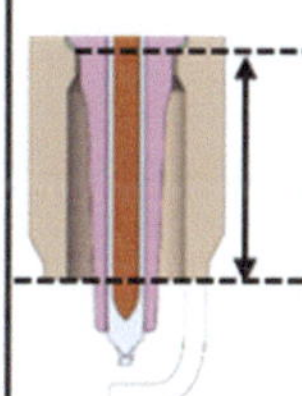	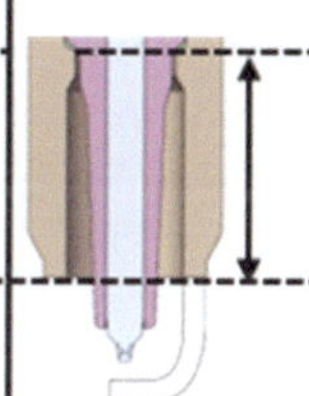	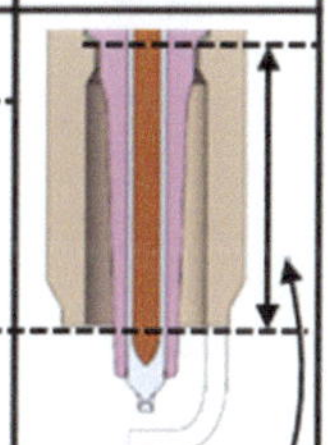

Gas-volume depth

The influence of insulator temperature was evaluated by using Plug A and Plug B. Cooling performance of Plug A was superior to Plug B because of its copper core in center electrode.

The influence of depth of Gas-volume was evaluated by using Plug A and Plug C. Plug A is colder heat range type, and Plug C is hotter heat range type. Temperature of Plug C rises easier than Plug A because it has wider insulator surface. And, it was estimated that the mixture was retained easily due to the deep structure of Gas-volume.

Figure 15 shows the result of engine test, and Fig. 16 shows the CFD result. LSPI was occurred more frequently in Plug B compared to Plug A, and the high temperature region inside Gas-volume was also wider. Then, LSPI was occurred more frequently in

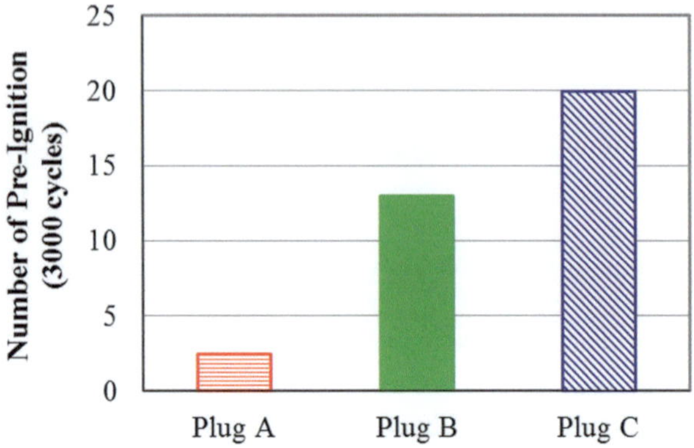

Fig. 15. Spark Plug sensitivity of spark plug LSPI by engine test

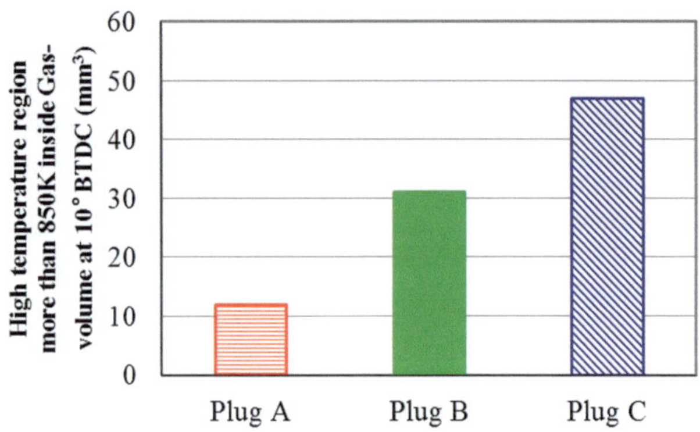

Fig. 16. Spark Plug sensitivity of spark plug LSPI by CFD

Plug C compared to Plug A, and the high temperature region inside Gas-volume was also wider.

As the result above, both insulator temperature and depth of Gas-volume affected to occurrence frequency of LSPI.

3.5.2 The Influence of Width of Gas-Volume

Influence for frequency of LSPI according to width of Gas-volume is evaluated. Table 7 shows the evaluated samples.

Insulator of Plug X has one step taper shape. On the other side, Plug Y and Z have two step taper shape. It was estimated that the gas retention was inhibited by the better scavenging performance due to its wider Gas-volume. Additionally, Plug Z was aimed at inhibiting gas retention by changing insulator diameters for much wider Gas-volume.

Figure 17 shows the result of engine test, and Fig. 18 shows the CFD result. LSPI was occurred more frequent in Plug X compared to Plug Y, and the high temperature region inside Gas-volume was also wider. LSPI was occurred more frequently in

Table 7. Spark plugs for sensitivity of width of Gas-volume

	Plug X	Plug Y	Plug Z
Insulator shape	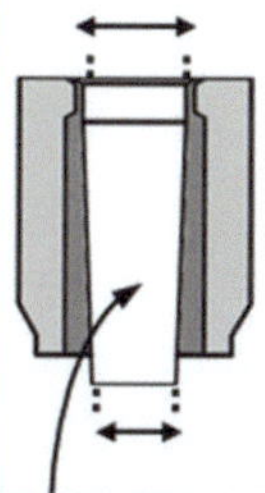 1 Step taper	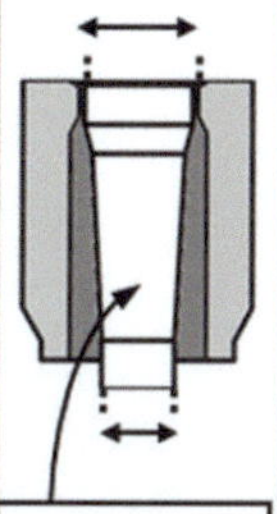 2 Step taper	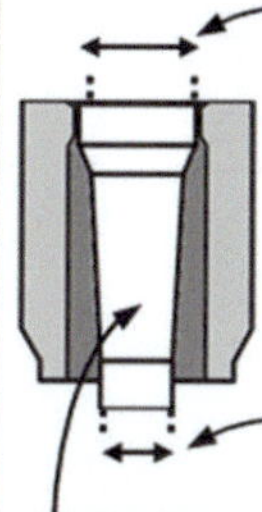 2 Step taper
Insulator root dia.	Thin	Middle	Wide
Insulator tip dia.	Wide	Middle	Thin
Gas-volume	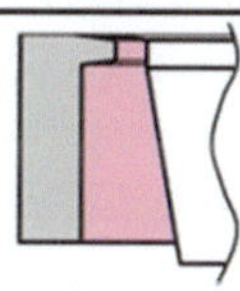	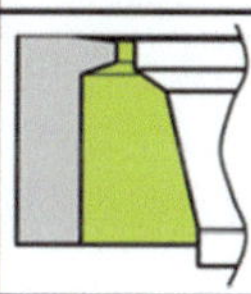	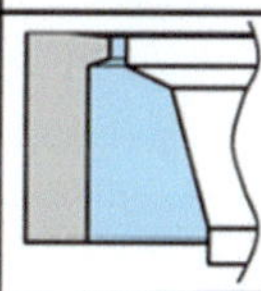

Insulator root dia.

Insulator tip dia.

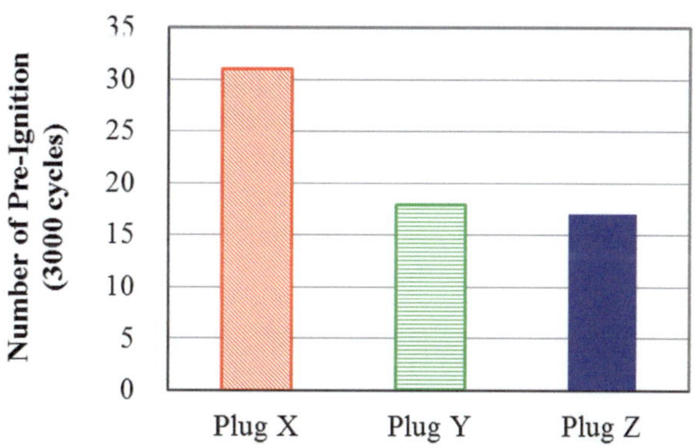

Fig. 17. Spark Plug sensitivity of spark plug LSPI by engine test

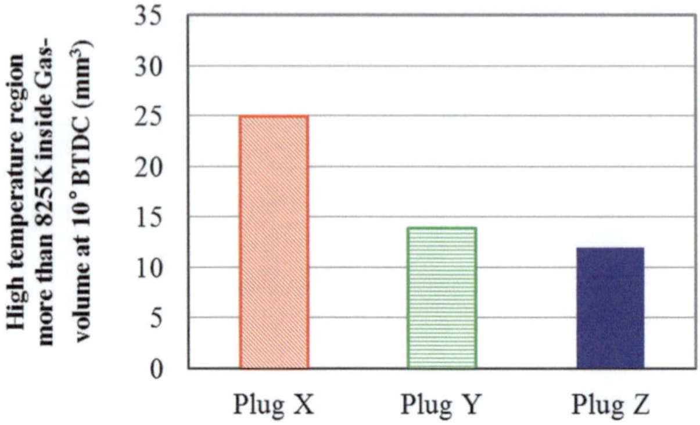

Fig. 18. Spark Plug sensitivity of spark plug LSPI by CFD

Plug Y compared to Plug Z, and the high temperature region inside Gas-volume was also wider.

As the result above, width of Gas-volume affect to occurrence frequency of LSPI.

4 Conclusions

The evaluate HSPI and LSPI was successful, and from this investigation of the root cause, the conclusion can be summarized in below.

1. Two types of Pre-ignition caused by spark plug exist.
   (a) Surface ignition from insulator tip surface.
   (b) Auto-ignition from the inside of the Gas-volume.
2. LSPI caused by the spark plug is a different phenomenon from LSPI caused by red-hot particles because the tendencies were different.
3. As the result of the CFD and auto-ignition point survey, we can conclude that LSPI caused by spark plug is an auto-ignition of retention gas inside the Gas-volume.
4. Occurrence frequency of LSPI caused by spark plug is influenced to spark plug temperature and its Gas-volume shape.

The occurrence of LSPI caused by spark plug could be reduced by changing its heat range into colder as is the case with HSPI. Moreover, the optimization of Gas-volume shape could contribute to the improvement of the LSPI to improve the efficiency of the downsized engine applications.

References

1. Tomoaki, S., et al.: The New Toyota 2.0L I4 downsized turbocharged gasoline direct injection engine. In: 2014 JSAE Annual Congress (Autumn), No. 94–14, pp. 1–4 (2014)
2. Mitsutaka, J., et al.: New 2.0L 4-cylinder gasoline direct injection turbocharged downsizing engine. In: 2015 JSAE Annual Congress (Autumn), S002, pp. 6–10 (2015)

3. Koji, M., et al.: Study of Highly Boosted Gasoline Engine to Enhance a Limit of Downsizing Concept (First Report: Understanding, Analysis and Avoidance of LSPI Phenomena and Achievement of Very High BMEP). The 25th Internal Combustion Engine Symposium, Paper No. 1 (2014)
4. Yasuo, M., et al.: A Study of a Highly Boosted Gasoline Engine in Low Speed Region Aiming at Extension of the Limit of Downsizing Concept (Second Report: Numerical Analysis of LSPI Mechanism and Auto ignition Caused by CaO Formation). The 25th Internal Combustion Engine Symposium, Paper No. 2 (2014)
5. Toshiyasu, S., et al.: A Study of Highly Boosted Gasoline Engine Aiming at Extension of the Limit of Downsizing Concept. Third Report Experimental Analysis and Visualization of LSPI, Paper No. 3 (2014)
6. Amann, M., et al.: Engine operating condition and gasoline fuel composition effects on low-speed pre-ignition in high-performance spark ignited gasoline engines. SAE Int. J. Engines **4**(1), 274–285 (2011). (SAE paper 2011-01-0342)
7. Optical Analysis of Low-Speed Pre-Ignition on Highly Boosted SI Engine. In: 2014 JSAE Annual Congress (Spring), No. 54–14, pp. 1–6 (2014)
8. Yoshihiro, O., et al.: Study of LSPI occurring mechanism form deposit. In: 2014 JSAE Annual Congress (Autumn), Paper No. 3-20145633, pp. 11–16 (2014)
9. Masaharu, K., et al.: Mechanism Analysis on Sudden Occurrence of LSPI in Boosted DISI Engines S276, pp. 1439–1444 (2015)
10. Inoue,T., et al.: Abnormal Combustion in a Highly Boosted SI Engine - The Occurrence of Super Knock, SAE2012-01-1141 (2012)
11. Kanemitsu, N., Yoshio, N.: Knowledge and Characteristics of the Spark Plug. Sankaido, Tokyo (1984). 241p

Simulation

Fundamental Mechanism Analysis on the Underlying Processes of LSPI Using Experimental and Modeling Approaches

Masaharu Kassai(✉), Taisuke Shiraishi, and Toru Noda

Powertrain Engineering Division, Powertrain Advanced Engineering Department, Nissan Motor Co., Ltd., 560-2, Okatsukoku, Atsugi-shi, Kanagawa 243-0192, Japan
m-kassai@mail.nissan.co.jp

1 Introduction

In view of growing global mobility requirements, energy security and climate change remain prominent issues. Consequently, there are strong drivers for further improved efficiency of vehicle powertrains, especially for internal combustion engines, which represent today's mainstream technology.

The three main approaches to achieving thermal efficiency improvement of SI engines are: 1. High compression ratio, 2. High dilution, 3. Loss reduction. As the dominant technology via a loss reduction approach, downsizing technology has been the focus of recent attention and widely applied by many automotive companies.

For such downsized engines, operating with higher boost pressures at lower engine speeds is an essential technical challenge to achieving both fuel economy and driving performance like torque response. However, under these conditions the likelihood of an abnormal combustion phenomenon (LSPI, low speed pre-ignition) is increased, and combating this issue is critical to developing such engines with improved performance.

Figure 1 shows a typical pressure profile of LSPI (suddenly occurred 1st cycle). LSPI can be manifested as heat release before spark ignition timing and subsequent extreme pressure oscillations. It is accepted that LSPI does not arise directly from fuel autoignition, but from hot spots containing material with particularly short autoignition delay times.

Previously, various parameters including engine conditions, lubricant composition and fuel properties have been reported to affect LSPI occurrence frequency. These observations have led to various hypotheses for pre-ignition causes and mechanistic pathways, which have been validated directly or indirectly using various experimental approaches including ignitability tests and endoscopic visualization in engines.

Figure 2 summarizes hypothesized mechanistic pathways for LSPI as reported previously. Much evidence indicates that the primary steps in LSPI are influenced by the extent of fuel adhesion on the liner wall and the properties of the lubricant oil that it interacts with on the liner wall [1–3]. Although many pathways to LSPI have been reported [4–12], common to all is the involvement of lubricant oil derived contents as pre-ignition sources.

M. Günther and M. Sens (eds.), *Knocking in Gasoline Engines*,
https://doi.org/10.1007/978-3-319-69760-4_6

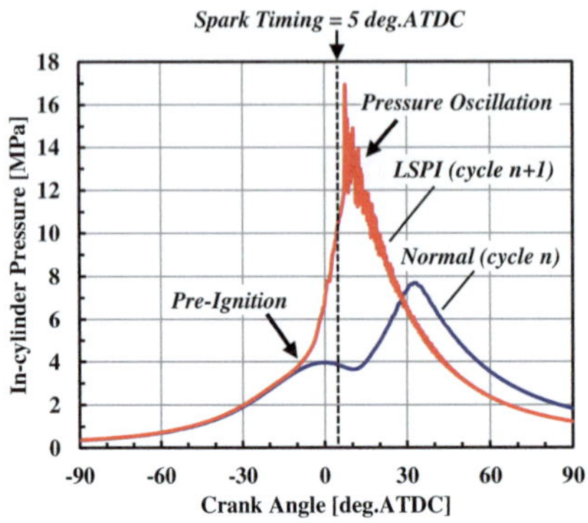

Fig. 1. Pressure profile of LSPI (1st cycle) and normal cycle

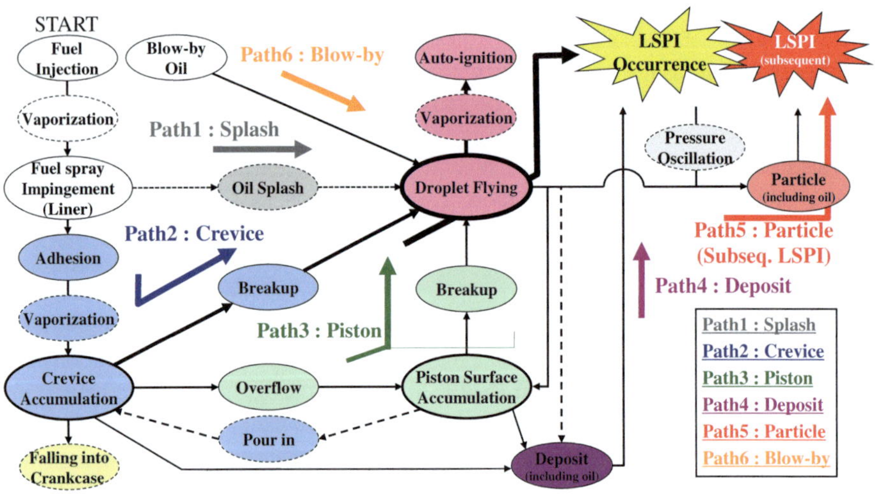

Fig. 2. Summary of proposed mechanistic pathways for LSPI occurrence

In this work, we follow the well-known mechanism hypothesis which assume lubricant oil containing droplets from piston crevice as a candidate of pre-ignition source. The mechanism hypothesis examined is shown in Fig. 3: Droplet of lubricant oil and fuel mixture, caused by adhesion of fuel spray on liner wall, will fly and pre-ignite before spark ignition.

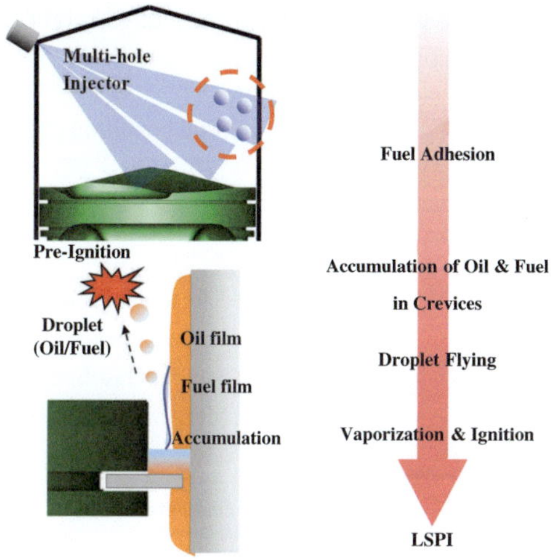

Fig. 3. Mechanism hypothesis of droplet (from piston crevice) triggered LSPI

2 Endoscopic Visualization of Droplet Triggered LSPI

2.1 Experimental Methodology: Single Cylinder Engine with Endoscope

To confirm the droplet triggered LSPI phenomenon, endoscopic visualization of LSPI was conducted on a single cylinder engine. Table 1 shows the specification of the single cylinder engine for this work. Experimental condition is shown in Table 2. Single cylinder engine based on downsized SI engine with gasoline direct injection (prototype) was used for this study. Using an external supercharger and intake air heater, boost pressure and intake air temperature was kept constant. Steady operation at low engine speed and highly boosted conditions was performed.

Table 1. Specification of single cylinder engine

Engine specification	–	Single cyl. engine (prototype)
Displacement	cc	400
Bore/Stroke	mm	79.7/81.1
Fuel preparation	–	GDI (Lateral)
Fuel pressure	MPa	20
Injection type	–	Multi-hole (6 holes)

Figure 4 shows the abstract of the visualization system with endoscope. An endoscope with high pressure resistance ("Bore scope" by NAC Image Technology, φ8.5 mm, pressure resistance up to 30 MPa) was mounted to the side part of combustion chamber. In-cylinder images were visualized and recorded with using relay lens and high speed camera (NAC Image Technology, MEMRECAM GX-8/K4). Flame speed was set to 10000 fps for all experiment. As well, this visualization system can observe almost all area of in-cylinder with changing front lens (straight and lateral view) and mounting angle of the endoscope.

Table 2. Experimental setting and condition (SCE)

Engine speed	rpm	1600
BMEP	MPa	1.8–2.2
Intake air temp.	K	333.15
Equivalence ratio	–	1.0 (Stoichiometric)
Ignition timing	deg. ATDC	Trace Knock
Fuel	–	Market Gasoline (RON100)
Oil	–	Market Oil (5 W-30)

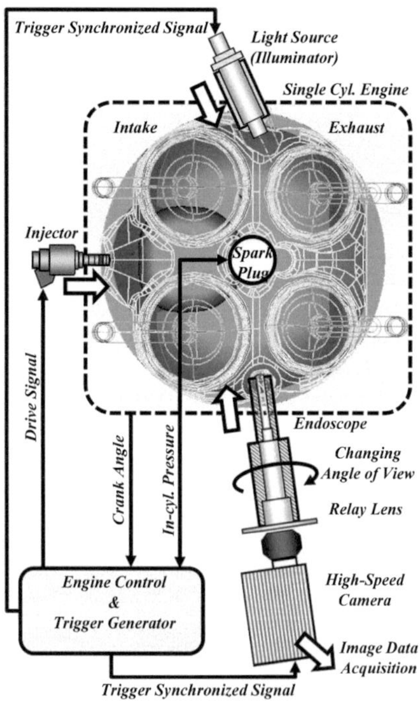

Fig. 4. Endoscopic visualization system apparatus

2.2 Visualized Droplet-Triggered LSPI (1st Cycle)

Figure 5 shows the 1st abnormal cycle of LSPI, visualized by the endoscope on the single cylinder engine. Start of injection was set to 280° BTDC.

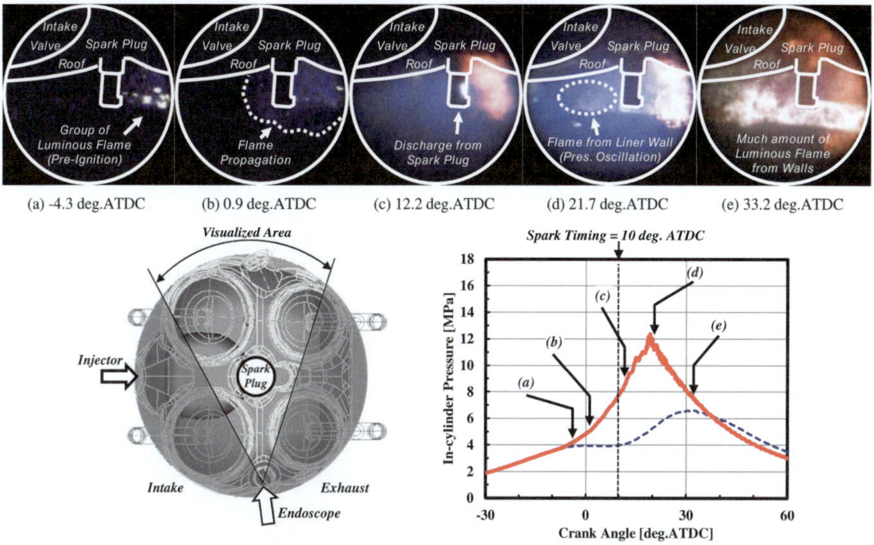

Fig. 5. Visualized in-cylinder phenomenon of LSPI (1st cycle)

At the last of compression stroke, objects with luminous flame flying from exhaust side to chamber center generated pre-ignition (Fig. 5(a)). Then it was observed that blue flame propagated from these luminous objects at exhaust side to chamber center direction (Fig. 5(b)). After a discharge ignition by spark plug (Fig. 5(c)), pressure oscillation which is expected to be a result of end-gas auto-ignition, was observed (Fig. 5(d)). After that, much amount of luminous flame was observed at whole viewing field (Fig. 5(e)). It is probably a result of rich fuel/air mixture generation with boundary layer destruction by the strong pressure oscillation and its rich combustion.

As a result of visualizing 1st abnormal cycle of LSPI many times, including different view angles, many of LSPI is triggered by flame propagation pre-ignited by flying objects with luminous flame like Fig. 5.

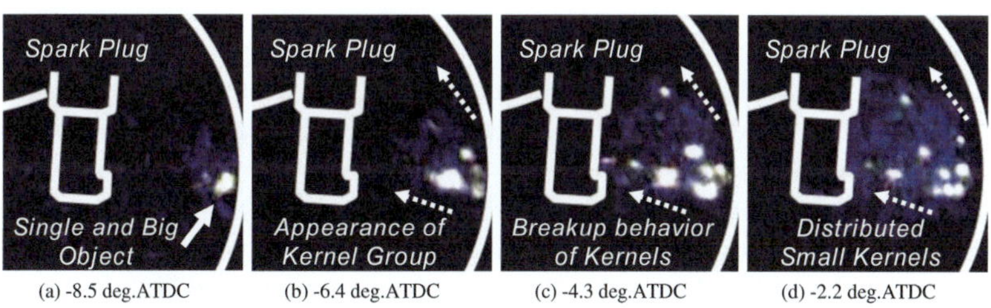

Fig. 6. Enlarged view of flying objects (flame kernels) triggering LSPI (1st cycle)

Figure 6 shows the enlarged view of flying object and its behavior on the same 1st abnormal cycle with Fig. 5. Firstly, a luminous object with relatively big diameter was emerged (Fig. 6(a)). Then, the number of luminous flame kernels was increased and simultaneously sizes of these kernels became smaller (Fig. 6(b), (c)) with flying distance (Fig. 6(d)). From these behaviors, we expected that these objects were droplets flying with making breakup and vaporization.

As a contrasting reference, Fig. 7 shows the visualized subsequent cycle of LSPI. The cycle just after the 1st abnormal combustion cycle same with Figs. 5 and 6 was visualized. Although fuel supply was cut off for this cycle with trigger controlled system in this case, a luminous flying particle was observed at compression stroke. The number of particle was only one and never increased. Also particle size was kept almost same (From image, its size was estimated to be approximately 0.4–0.7 mm). We can also easily assume that this particle is flying with riding on tumble flow from its trajectory (Fig. 7(a)–(f)). From these behaviors, this particle is expected to be solid, as already visualized in previous studies [9–11].

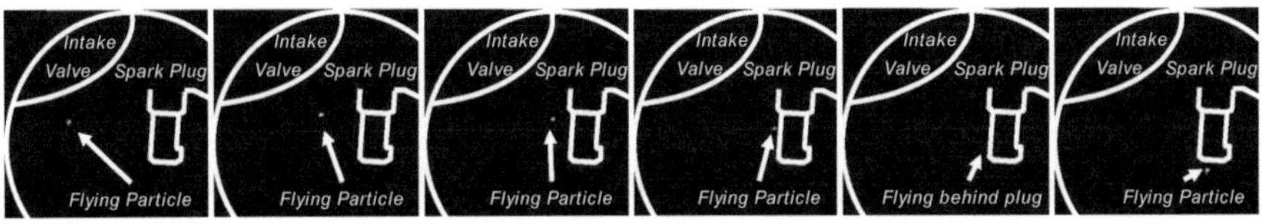

(a) -129 deg.ATDC (b) -119 deg.ATDC (c) -109 deg.ATDC (d) -99 deg.ATDC (e) -88 deg.ATDC (f) -78 deg.ATDC

Fig. 7. Behavior of flying particle at subsequent cycle of LSPI

From the observed behavior and comparison with sold-like particle in the subsequent cycle, we understood that the luminous objects at 1st LSPI cycle were flying droplets in liquid phase.

To supplement the understanding of LSPI phenomenon in 1st cycle, visualization in different viewing field, exhaust side, was also conducted. Figure 8 shows the result.

Same to the result in Fig. 5, objects with luminous flame flying from exhaust side (Fig. 8(a)) and flame propagation from these objects (Fig. 8(b)) were again observed. The combustion chamber became suddenly bright due to the end-gas autoignition (Fig. 8(c)). Soon after the flashlight by the autoignition, particles peeled off from piston surface near the edge (Fig. 8(d)) and enormous scattering of these particles (Fig. 8(e)) were observed.

To summarize these visualization results, we could thought that there are LSPI caused by flying droplets. Also, it can be understood that scattering of solid particles, which have been reported as a driver of subsequent LSPI cycles, are triggered by pressure oscillation of 1st LSPI cycle.

For further understanding on LSPI based on this "droplet triggered" mechanism hypothesis, we need to investigate two processes: The physical process of droplet flying and the chemical process of metallic additive's effect on droplet ignition.

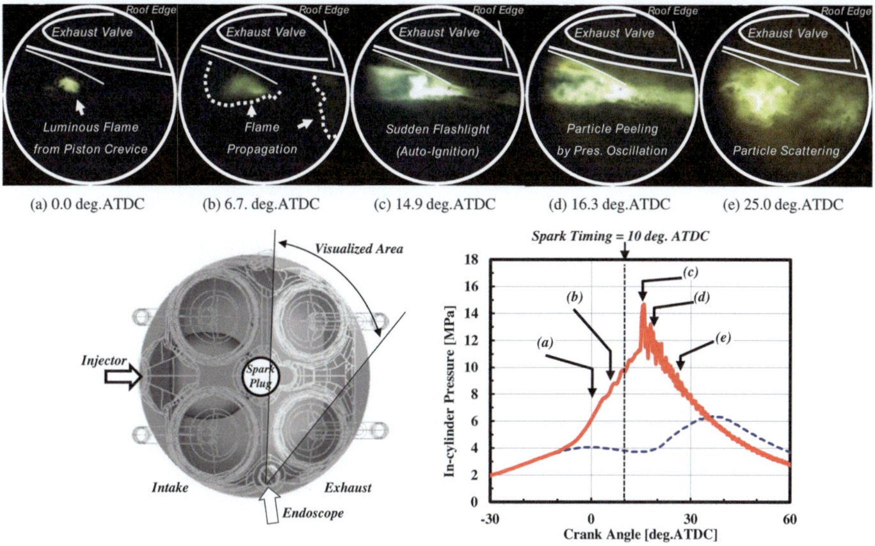

Fig. 8. In-cylinder phenomenon of LSPI (1st cycle) visualized in exhaust side

3 Analysis and Modeling on the Process of Droplet Flying

3.1 Experimental Methodology: Visualization Engine with Oil Lubrication

From understanding on the possibility of droplet triggered LSPI, analysis on detailed process and mechanism of droplet flying by using the visualization engine with oil lubrication was conducted.

Table 3. Specification of visualization engine

Engine specification	–	Visualization engine
Displacement	cc	400
Bore/Stroke	mm	78.0/84.0
Fuel preparation	–	GDI (Lateral)
Fuel pressure	MPa	20
Injection type	–	Multi-hole (6 holes)

Table 3 shows the specification of the visualization engine for this work. Experimental condition is shown in Table 4. Using same injection system with the single cylinder engine, fuel amount equivalent to the condition of highly boosted condition was injected. Then steady motoring operation was performed.

Table 4. Experimental conditions (Visualization)

Engine speed	rpm	1200
BMEP	MPa	0 (motoring)
Intake air temp.	K	333.15
Fuel	–	Market Gasoline (RON100)
Oil	–	Market Oil (5 W-30)

Figure 9 shows the apparatus of this visualization engine. A quart glass cylinder with dimension controlled to be same with production engine, a production piston, and production piston rings lubricated by oil were used for this visualization engine. With exposing light to the visualization area by metal halide lamp, behavior of oil and fuel at oil-lubricated sliding surface and its neighborhood (including piston crevice) can be visualized and recorded by high speed camera.

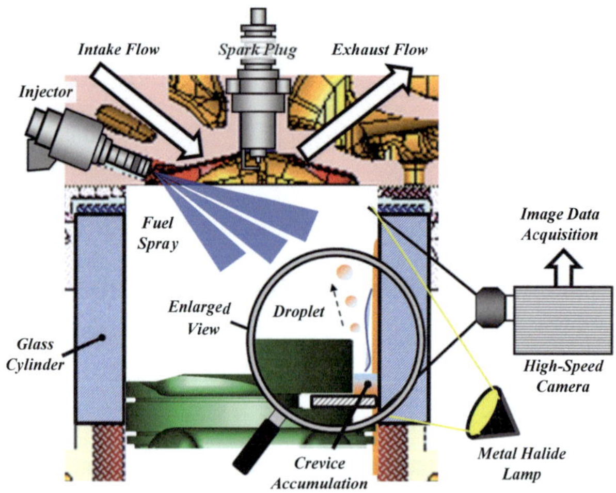

Fig. 9. System of visualization engine with oil lubrication

3.2 Visualized Droplet Flying Process

Figure 10 shows an example of visualized images by the visualization engine. In this case, start of injection was set to 240° BTDC. Series of phenomenon from intake stroke to compression stroke, written as below, was well observed.

- Adhesion of fuel spray on liner wall (Fig. 10(a))
- Scratch up of liner-adhered fuel by piston (Fig. 10(b))
- Droplets flying at the end of comp. stroke (Fig. 10(c))

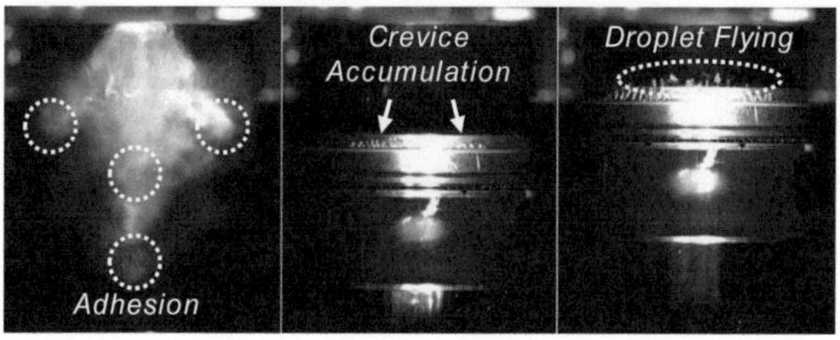

(a) -205 deg.ATDC (b) -40 deg.ATDC (c) -10 deg.ATDC

Fig. 10. Example of visualized intake & compression stroke

Figure 11 shows such typical behavior of droplets flying from piston land. These images were obtained by image procession of subtracting motoring images (background images without any fuel) from main images visualized at each crank angle. Obtained images are useful to understand the phenomenon of droplets flying in detail.

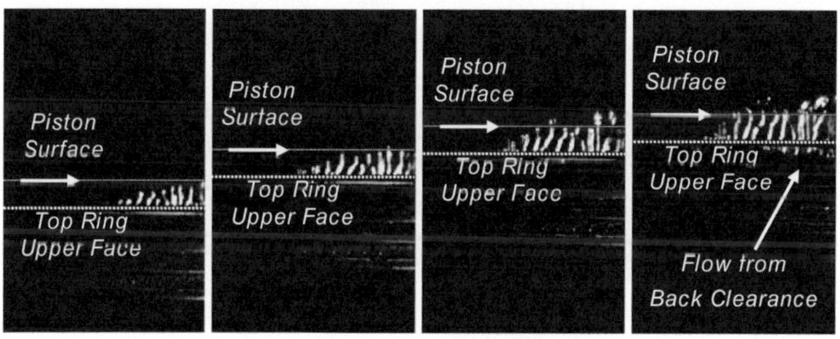

(a) -30 deg.ATDC (b) -20 deg.ATDC (c) -10 deg.ATDC (d) 0 deg.ATDC

Fig. 11. Visualized droplet flying process

Lubricant oil flying by piston inertia has been already reported by many literatures [13], mainly from the viewpoint of oil consumption. However, it was confirmed from this visualization that continuous operation at liner wetting condition and accumulation of fuel at piston crevice can easily leads to droplets flying, even at low engine speed (low piston inertia) which is focused on LSPI issues.

Behavior of droplets flying was almost same with the result of numerical analysis using VOF method by Moriyoshi et al. [14]. Mixture of oil and fuel go up on the piston land with generating liquid column and finally get on the piston crown surface and fly into combustion chamber. Such behavior is well explained by Rayleigh Taylor instability. Density difference between oil/fuel mixture (liquid phase) and fuel/air mixture in combustion chamber (gas phase) will develop disturbance and generate waves. Finally the heights of waves will exceed the piston land height then oil/fuel mixture get on piston crown surface and finally fly by piston inertia.

To investigate the effect of fuel adhesion or operating time on droplet flying, amount of flying droplets was calculated by image procession. Using the subtracted images, we defined control area as upper side of piston crown surface line. Then number of pixels brighter than criteria was counted up and used as an indicator for the amount of droplets.

Figure 12 shows the effect of injection timing and operation time on the amount of flying droplets at TDC. With more retarded injection timing, more amount of droplets was observed from earlier timing, due to more amount of fuel adhesion on liner wall. At the same time, after long enough operation time, amount of flying droplets looks saturated to some constant level. This is probably because of outflow from piston crevice. For example, evaporation from re-coated film and flow to the 2nd land across the ring gap/back clearance (Fig. 12(d)) are expected. Thus this system is thought to be a lag system with saturation behavior.

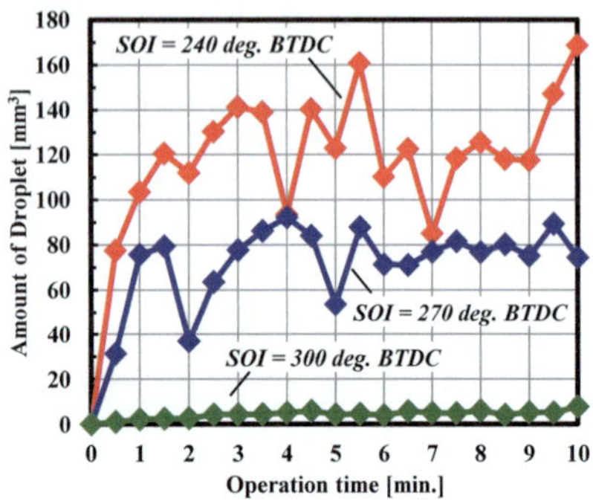

Fig. 12. Effect of injection timing on the amount of flying droplets

Figure 13 shows the amount of droplets flying at each crank angle with same injection timing (240° BTDC) and different operation time. With longer operation time, droplets flying at earlier timing (around −30° ATDC) was observed. It is because long operation time leads to more accumulation of fuel and oil/fuel mixture will have lower viscosity and surface tension.

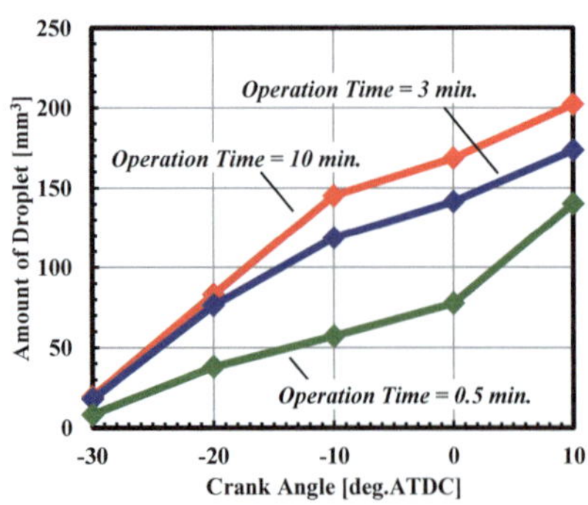

Fig. 13. Effect of operation time on the amount of droplets (SOI = 240° BTDC)

3.3 Modeling of Droplet Flying Process and Discussion

Experimentally visualized droplet flying process seems to be easily explained by Rayleigh-Taylor instability. We tried to model the phenomenon in simple way.

Table 5 shows the simplified process from wave generation to droplet flying based on Rayleigh-Taylor instability.

Table 5. Assumed droplet flying process based on Rayleigh-Tayloer instability

Phase	PhaseProcess of generation and growth of wave (Liquid column) based on Rayleigh-Taylor instability	
STEP0	Before Piston deceleration. Waveis not yet generated	
STEP1	Start of pistondeceleration. Liquidsurface start to generate wave based on Rayleigh–Taylor instability	
STEP2	Growth of wave continues. After the amplitude exceed the 0.4 times of wave length, wave growthtransition to non-linear zone and wave length is fixed	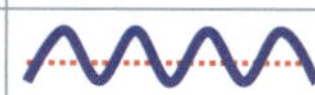
STEP3	Droplet will begenerated from grown wave with fixed wave length. Droplet flying timing can be assumed as the timing when the wave height exceed the piston top land height	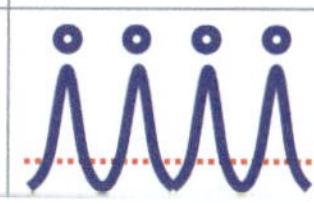

To estimate accurately the amount of oil/fuel mixture in piston top land, consideration of evaporation rate profile after fuel adhesion as well as fuel impingement and adhesion itself is necessary. However, for simple modeling, the amount of mixture expected to be in crevice at the end of compression stroke was assumed to be already in crevice at the calculation start timing. The amount of mixture was estimated from zero dimensional wetting simulation and experimental results.

Rayleigh-Tayler instability is a phenomenon that two types of fluids with different density will cause the instable boundary surface due to the effect of inertia. For the modeling of wave generation based on the instability, wave length and amplitude need to be estimated. In the simplified model, it was assumed that, after the amplitude exceed 0.4 times of wave length, wave growth transition to non-linear zone and wave length fixed. Model equations of Rayleigh-Taylor instability are shown from (1) to (3) [14–16]. Growth speed of generated waves are modeled as a function of kinetic viscosity, the amount of mixture and piston speed, based on the Eq. (4) proposed by Thirouard et al. [17, 18]. For the physical properties of mixture, iso-octane was used with assumption of enough local dilution of lubricant oil by fuel and saturation of the mixture properties [5].

Model equations of Rayleigh-Taylor instability:

$$\gamma_{peak}^2 = -\frac{kAta_{pst}}{2}\left(1 - \frac{k^2}{k_c^2}\right), k = k_{peak} \quad (1)$$

$$At = \left(\rho_{liq} - \rho_{gas}\right) / \left(\rho_{liq} + \rho_{gas}\right) \quad (2)$$

$$k_c = \left[\frac{-(\rho_{liq} - \rho_{gas}) a_{pst}}{Ts_{liq}} \right]^{1/2} \quad (3)$$

Model equation to estimate wave growth speed (proposed by Thirouard et al.):

$$H_{top} = H_0 + 1.188 \left(\frac{\left| u_p - u_{p(max)} \right| \sqrt{V_{liq}}}{2\nu} \right)^{0.394} \sqrt{V_{liq}} \quad (4)$$

Figure 14 shows the example of estimated wave height which is a representative parameter to judge the droplet flying. SOI = 300° BTDC, advanced injection timing with less liner wetting, didn't show the wave height higher than top land height. On the contrary, SOI = 200° BTDC, retarded injection timing with more liner wetting, showed higher wave and early droplet flying around −30° ATDC. These result qualitatively corresponds to the experimental results shown in Fig. 12 or Fig. 13. With SOI = 300° BTDC, no droplet flying was observed even with quite long operation time (Fig. 12). With retarded injection timing, the timing of droplet flying converged to around −30° ATDC. Estimated wave length (approximately 3.0 mm) and experimentally observed gap between liquid columns (approximately 2.0 mm) were also comparable.

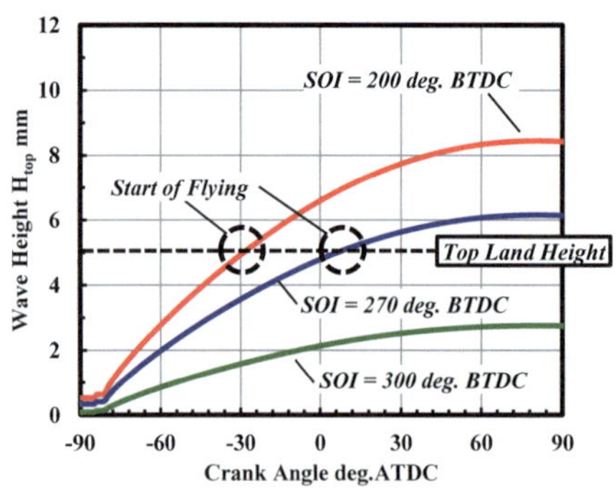

Fig. 14. Example of wave height estimation by RT instability model (1200 rpm)

These calculation is just rough estimation and further prediction accuracy on the amount and properties of mixture are required for the quantitative estimation. However, droplet flying mechanism based on Rayleigh-Taylor instability seems to be relevant.

4 Analysis and Modeling on the Process of Droplet Ignition

4.1 Experimental Methodology: Rapid Compression and Expansion Machine

From understanding on the possibility of droplet triggered LSPI, we focused on the chemical process of metallic additive's effect on droplet ignition.

To observe the ignition of fuel/oil droplets and subsequent flame propagation in the fuel/air mixture, a rapid compression and expansion machine (RCEM) are selected.

Figure 15 shows a schematic of the RCEM. This RCEM (Bore/Stroke = 125/140 mm) can simulate the piston motion of a reciprocating engine's compression and expansion strokes with its cam profile. The descending cam (total weight = 180 kg) moves the piston along the cam profile one time. Compression ratio settings can be changed easily by changing the cylinder head position.

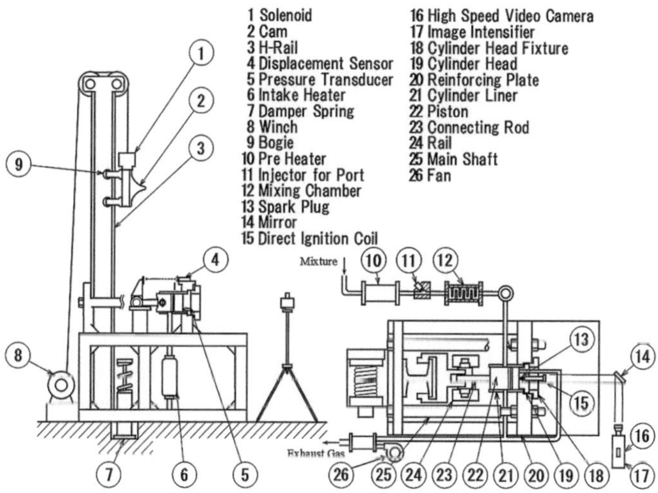

Fig. 15. Experimental schematic of RCEM

A transparent cylinder head made from polished acrylic was used to facilitate visualization of in-cylinder phenomena. The ignition kernels of fuel/oil droplets and subsequent flame propagation were observed using images obtained with the high-speed camera. Both the visualized images and pressure measurements allow quantitative evaluation of ignition delay times.

For the injection of the oil/fuel mixture, we prepared a special six-hole, solenoid-driven DI injector for gasoline engines. To avoid wetting of the spray stream on the piston crown surface, the injector was mounted at an inclined angle. Targets of the spray streams were also optimized. One stream for observation was designed to penetrate freely in the visualization area without wetting any wall, while the other five streams penetrated toward the outside of the visualization area. Figure 16 shows a schematic of the cylinder head, injector mounting design, and geometrical relation between the injector streams and the visualization area.

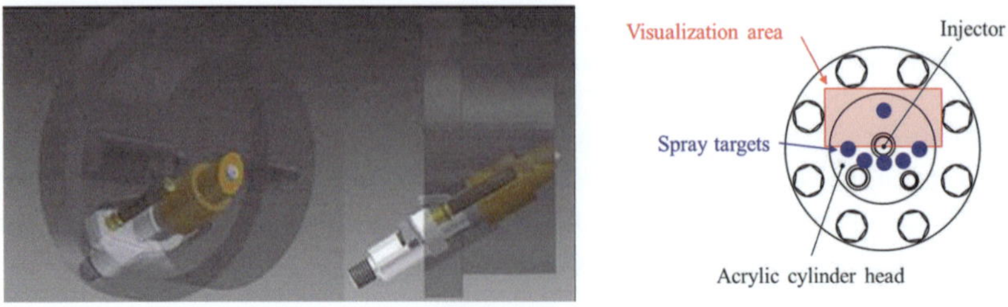

Fig. 16. Cylinder head design and geometrical layout of spray direction/visualization area

4.2 Analysis on the Droplet Ignition Phenomenon

Table 6 shows the experimental conditions with combinations of droplet composition with metallic additive variation and ambient mixture conditions with RON variation. Experiments were conducted 5 times per each condition.

Table 6. Experimental conditions of RCEM: ambient conditions and droplet composition

Case No.	Ambient Conditions		Droplet Composition			Droplets Injection Settings		
	Fuel Type	Comp. ratio	Fuel	Calcium	ZnDTP	Timing	Inj. Pres.	Amount
[-]	[-]	[-]	[wt %]	[wt %]	[wt %]	[deg. ATDC]	[MPa]	[mg]
Test 1	95 RON (PRF)	10.5 : 1	100 RON (Iso-octane)	0 - 1.57	N/A	-22	3.8	0.5
Test 2	90 - 100 RON (PRF)	10.5 : 1		0.784	N/A	-22	3.8	0.5
Test 3	90 RON (PRF)	10.5 : 1		0.784	0 - 0.3	-22	3.8	0.5

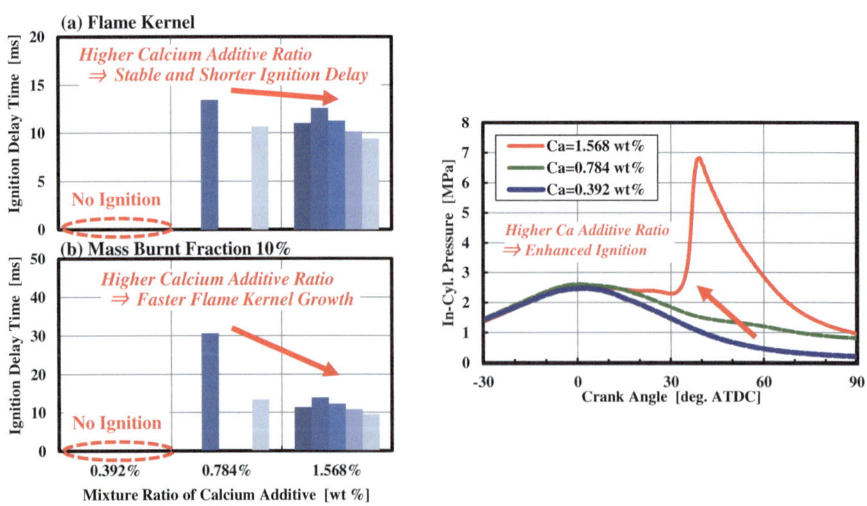

Fig. 17. Ignition delay times and pressure profiles with Ca additive variation

Figure 17 shows the obtained ignition delay times and pressure profiles of fuel droplets for various amounts of calcium additives. It definitely showed the enhanced droplet ignition and subsequent combustion due to calcium additive in droplet. Figures 18 shows the visualized enhancement of droplet ignition and combustion due to the different amount of calcium additives in droplets. Wider spatial spreading of the initial flame was clearly observed with higher amounts of calcium additives.

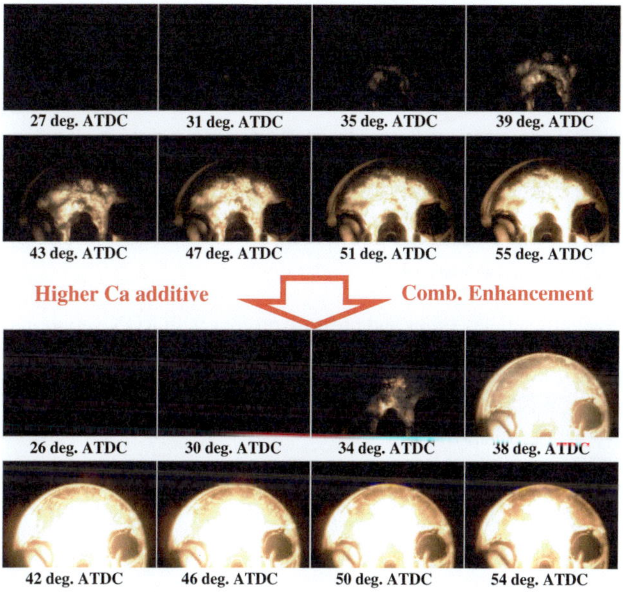

Fig. 18. Visualized combustion enhancement by Ca additive in droplet

Figure 19 shows the obtained ignition delay times and pressure profiles for various RON values of surrounding mixture. It definitely showed the enhanced droplet ignition and subsequent combustion due to lower RON value of surrounding mixture. Figure 20 shows the visualized enhancement of droplet ignition and combustion due to the lower RON value of mixture. Similar to calcium additive effect, wider spatial spreading of the initial flame was clearly observed with lower RON mixture.

Figure 21 shows the obtained ignition delay times and pressure profiles of fuel droplets for various amounts of zinc additives. It definitely showed the inhibited droplet ignition and subsequent combustion due to zinc additive in droplet. Figure 22 shows the visualized inhibition of droplet ignition and combustion due to the different amount of zinc additives in droplets. Zinc additive clearly show the opposite sensitivity to the calcium additive and to lower RON mixture.

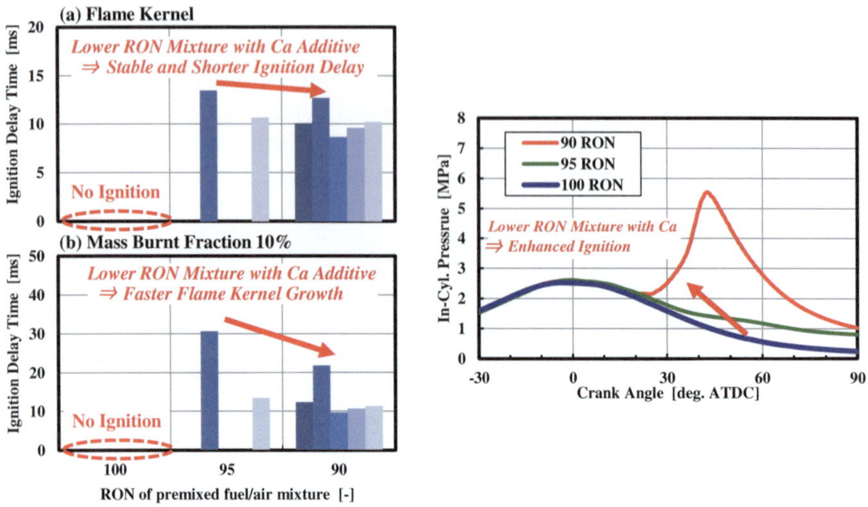

Fig. 19. Ignition delay times and pressure profiles with RON variation mixture

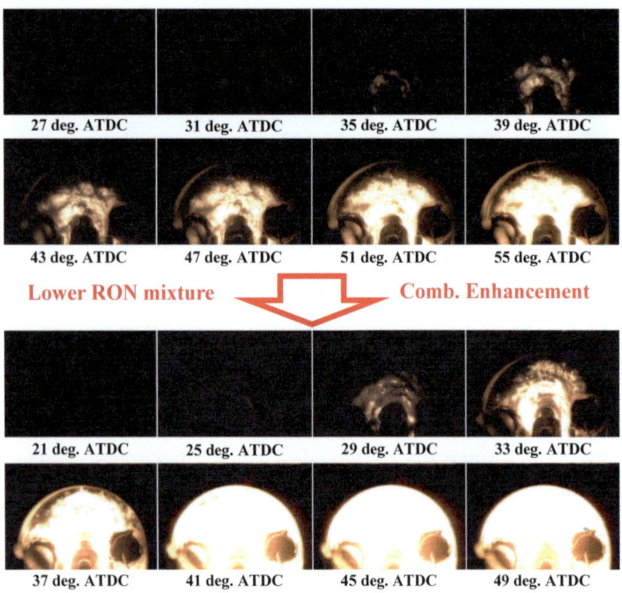

Fig. 20. Visualized combustion enhancement by low RON mixture (with Ca in droplet)

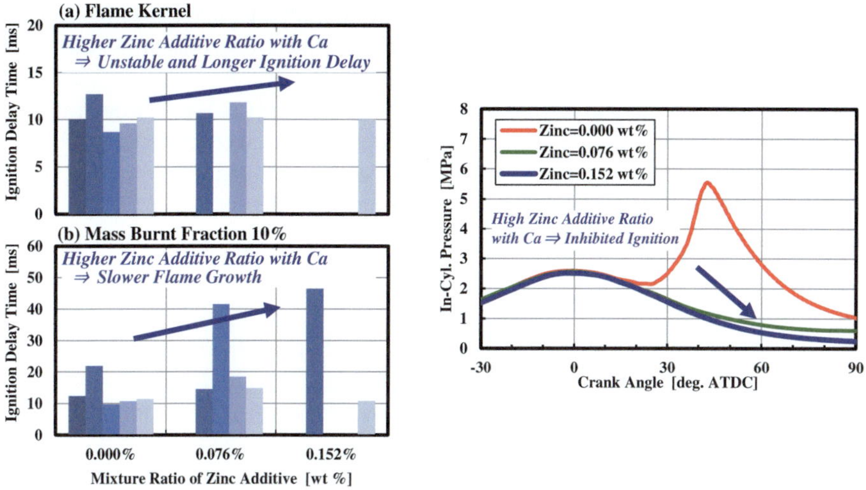

Fig. 21. Ignition delay times and pressure profiles with Zinc additive variation

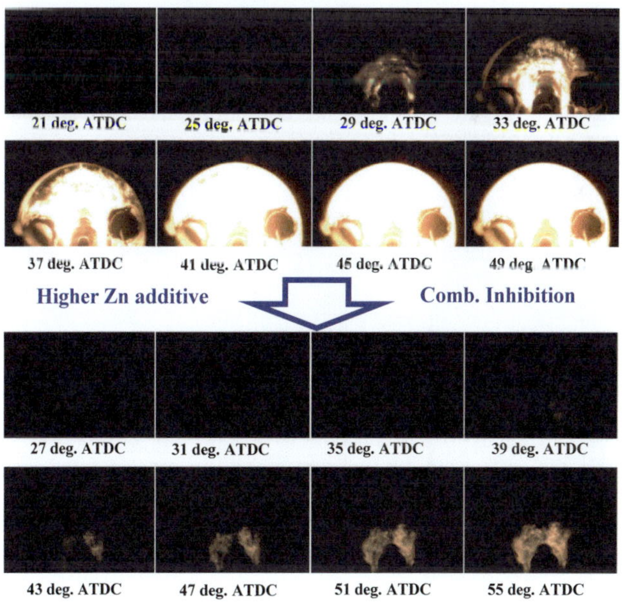

Fig. 22. Visualized combustion inhibition by Zinc additive in droplet

4.3 Modeling of Droplet Ignition

Based on the observation of the ignition delay times and the visualized images in these RCEM experiments, the gas phase reaction with the surrounding premixed mixture seems to be important for droplet ignition and subsequent combustion in the cylinder.

It is also expected that calcium additive have a promotion effect on gas-phase reaction and autoignition. On the contrary, zinc additive is expected to have an inhibition effect on gas phase reaction.

Historically, the catalytic promotion effects of $CaCO_3$ and CaO on hydrogen abstraction reactions have been well known in the research fields of thermal decomposition or gasification of coal [20, 21]. The hypothesized reactions (5) and (6) are shown below. If such reaction are applicable, the observed promotion effects of calcium additives can be explained.

$$RH + (CaO) = R + H + (CaO) \tag{5}$$

$$RH + \left(CaCO_3\right) = R + H + \left(CaCO_3\right) \tag{6}$$

It is well known that ZnDTP functions to prevent oxidation of the lubricant oil by changing radicals or peroxides into stable substances, as shown in (7) [22]. If such a reaction is effective and can be applied in the gas-phase reaction, the observed inhibition effects of zinc additives can be explained.

$$QOOH + ZnDTP = QOH + ZnO + DTP \tag{7}$$

We tried to model and simulate the chemical effects of these metallic additive with assuming such specific reactions. For the CFD calculation, KIVA-4 coupled with CHEMKIN was used. Figure 23 shows the computing mesh to simulate RCEM and Table 7 shows the sub-models used for this study.

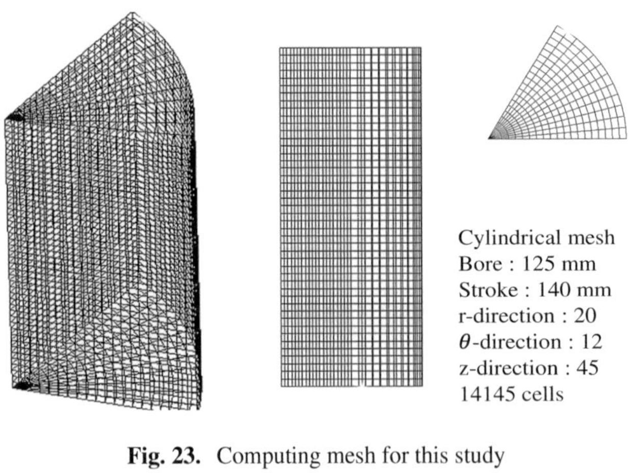

Fig. 23. Computing mesh for this study

Table 7. Sub-models used in the CFD code

Turbulence	RNG k-εmodel	
Breakup and atomization	TAB model	
Collision	O'Rourke model	
Chemical reaction	Partially changed Tsurushima scheme	
Chemical species and reaction scheme	Species	56
	Reactions	73
ODE solver	ERENA	
Chemistry and turbulence interaction	PaSR	
Heat transfer	Law-of-the-wall	
Parallel computation	OpenMP	

For the reaction scheme, skeletal PRF scheme developed by Tsurushima [19] was selected as a base scheme. Then promotion effect of calcium additive was simulated by enhancing hydrogen abstraction reaction and inhibition effect of zinc additive was simulated by inhibiting decomposition reaction of peroxides as shown in Fig. 24. To simulate ignition behavior of RCEM experiments, activation energy E_1 for hydrogen abstraction

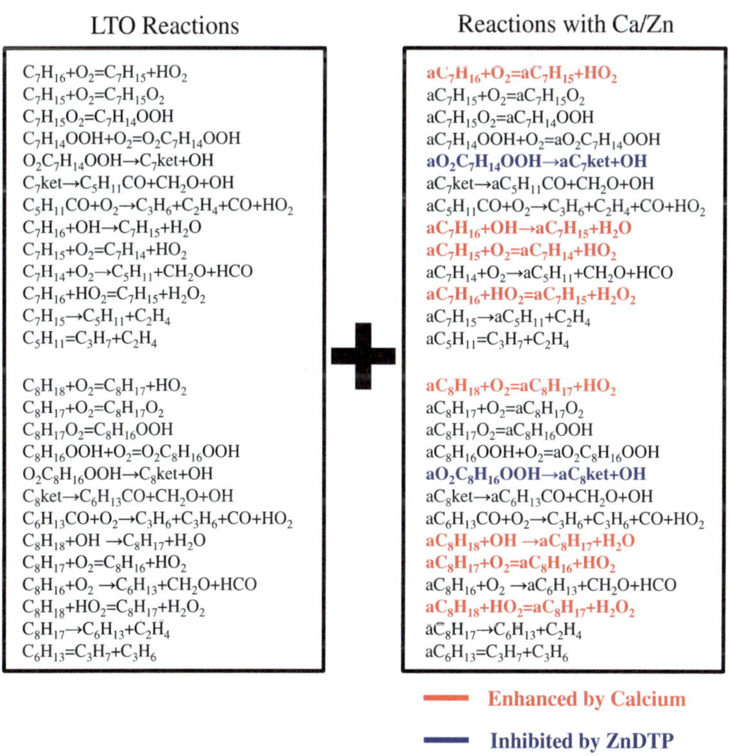

Fig. 24. Modified reaction scheme used for CFD calculation

reactions and activation energy E_2 for peroxide decomposition reaction are changed. To separate the droplet-effected reaction and bulk mixture reaction, related hydrocarbon species are treated as isomers. Initial conditions of calculation was summarized in Table 8.

Table 8. Calculation condition to simulate experimental results

Engine speed [rpm]		800
Compression ratio [-]		10.5
Intake gas	Pressure [kPa(gauge)]	25.0
	Temperature [K]	383.15
	RON of mixture [-]	90
	Equivalence ratio [-]	1.0
Injection	Amount [mg]	3.0
	Timing [deg. ATDC]	−22
	Duration [ms]	0.6
	Temperature [K]	293.15
Activation energies of reactions	Hydrogen Abstraction E_1 [%]	69.5/70/100
	Peroxide Decompositon E_2 [%]	100/115/120

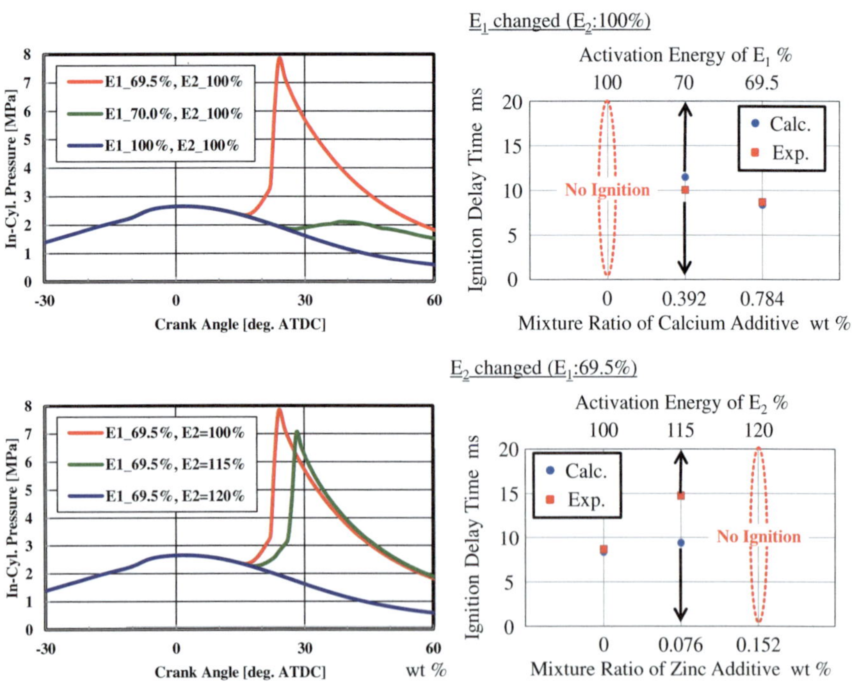

Fig. 25. Simulated pressure profiles and ignition delay times with E_1 and E_2 optimization

Figure 25 shows the results of calculation for both calcium and zinc additive effect simulating Test 1 (Fig. 17/18) and Test 3 (Fig. 21/22). With changing activation energy values, ignition timing was reproduced well. The relevance of absolute values of activation energy delta will need further discussion. However, required delta of these activation energies indicate that the effect of these metallic additive on droplet ignition in fuel/air mixture is really drastic.

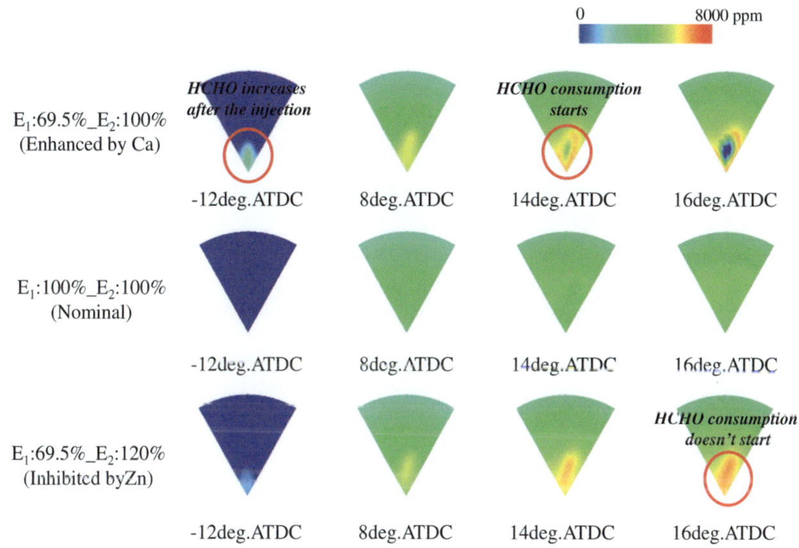

Fig. 26. Observation of LTO enhancement and inhibition via HCHO concentration

HCHO concentration in Fig. 26 and temperature distribution in Fig. 27 shows the effects of activation energy change on gas phase reaction process. Reduced E_1 for simulating enhancement by calcium additive shows early LTO represented by HCHO consumption and rapid temperature rise after LTO. Reduced E2 for simulating inhibition by zinc additive shows retarded HCHO consumption and temperature rising.

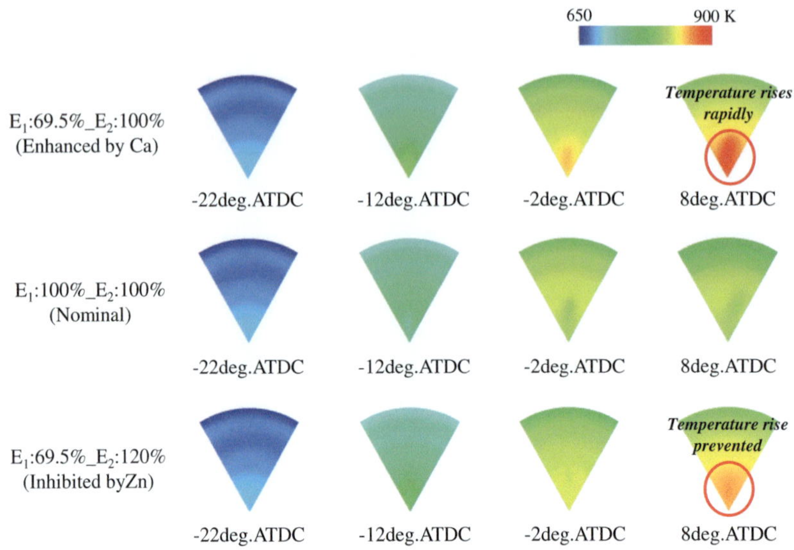

Fig. 27. Simulated temperature distribution in the cylinder

5 Summary

From the endoscopic visualization in single cylinder engine, we could assume that there are LSPI caused by flying droplets. Also, it can be understood that scattering of solid particles, which have been reported as a driver of subsequent LSPI cycles, are triggered by pressure oscillation of 1st LSPI cycle.

From the oil lubricated visualization engine experiment, it is confirmed that liner adhered fuel will accumulate in piston crevice then oil/fuel mixture can easily fly into chamber even at low engine speed condition.

Although further improvement on prediction of the amount and properties of mixture are required, droplet flying process and phenomenon can be well explained by Rayleigh-Taylor instability.

From the RCEM experiments on droplet ignition in air/fuel mixture, the gas phase reaction with the surrounding premixed mixture seems to be important for droplet ignition and subsequent combustion in the cylinder.

It is expected that calcium additive have a promotion effect on gas-phase reaction and autoignition. On the contrary, zinc additive is expected to have an inhibition effect on gas phase reaction. These chemical effects were tentatively modeled by enhancement of hydrogen abstraction reaction and inhibition of peroxide decomposition reaction.

References

1. Zahdeh, A., et al.: Fundamental approach to investigate pre-ignition in boosted SI engines, SAE2011-01-0340 (2011)

2. Inoue, T., et al.: Abnormal combustion in a highly boosted SI engine - the occurrence of super knock, SAE2012-01-1141 (2012)
3. Takeuchi, K., et al.: Investigation of engine oil effect on abnormal combustion in turbocharged direct injection - spark ignition engines, SAE2012-01-1615 (2012)
4. Dahnz, C., et al.: Investigations on pre-ignition in highly supercharged SI engines, SAE Paper 2010-01-0355 (2010)
5. Kassai, M., et al.: Mechanism analysis on LSPI occurrence in boosted SI engines, SAE Paper 2015-01-1867(2015)
6. Palaveev, S., et al.: Premature flame initiation in a turbocharged DISI engine-numerical and experimental investigations, SAE Paper 2013-01-0252(2013)
7. Okada, Y., et al.: Study of low-speed pre-ignition in boosted spark ignition engine, SAE Paper 2014-01-1218(2014)
8. Kuboyama, T., et al.: visualization and analysis of LSPI mechanism caused by oil droplet, particle and deposit in highly boosted si combustion in low speed range, SAE Paper 2015-01-0761(2015)
9. Palaveev, S., et al.: Simulations and experimental investigations of intermittent pre-ignition series in a turbocharged DISI engine. In: 4th International Conference on Knocking in Gasoline Engines (Berlin), pp. 414–442 (2013)
10. Döhler, A., et al.: A contribution to better understanding the pre-ignition phenomenon in highly charged internal combustion engines with direct fuel injection. In: 4th International Conference on Knocking in Gasoline Engines (Berlin), pp. 41–61 (2013)
11. Lauer, T., et al.: Model approach for pre-ignition mechanisms. MTZ Worldwide **75**(1), 44–49 (2014)
12. Schünemann, E., et al.: Pre-ignition analysis on a turbocharged gasoline engine with direct injection. In: 4th International Conference on Knocking in Gasoline Engines, pp. 380–393 (2013)
13. Thirouard, B., et al.: Oil transport in the piston ring pack (Part I): identification and characterization of the main oil transport routes and mechanisms, SAE Paper 2003-01-1953 (2003)
14. Sazhin, S.: Droplets and Spray. Springer, London (2014)
15. Drazin, P.G., Reid, W.H.: Hydrodynamic Stability. Cambridge Mathematical Library, 2nd edn. Cambridge University Press, Cambridge (2004)
16. Patterson, M.A., Reitz, R.D.: Modelling of the effects of fuel spray characteristics on diesel engine combustion and emission, SAE Paper 980131 (1998)
17. Thirouard, B., Tian, T.: Oil transport in the piston ring pack (Part I): identification and characterization of the main oil transport routes and mechanisms, SAE Paper 2003-01-1952 (2003)
18. Thirouard, B., Tian, T.: Oil transport in the piston ring pack (Part II): zone analysis and macro oil transport model, SAE Paper 2003-01-1953 (2003)
19. Tsurushima, T.: A new skeletal PRF kinetic model for HCCI combustion. Proc. Combust. Inst. **32**, 2835–2841 (2009)
20. Schachter, Y., et al.: Calcium-oxide-catalyzed reactions of hydrocarbons and of alcohols. J. Catal. **11**(2), 147–158 (1968)
21. Ellig, D.L., et al.: Pyrolysis of volatile aromatic hydrocarbons and n-heptane over calcium oxide and quartz. Ind. Eng. Chem. Process Des. Dev. **24**(4), 1080–1087 (1985)
22. Mortier, R.M., et al.: Chemistry and Technology of Lubricants, 3rd edn. Springer, New York (2010)

Chemical Analysis of Potential Initiating Fluid for Low-Speed Pre-ignition

Terence F. Alger and Thomas E. Briggs Jr.(✉)

Southwest Research Institute, San Antonio, TX, USA
thomas.briggs@swri.org

Abstract. Low-Speed Pre-Ignition (LSPI) has become a major challenge for the light duty automotive industry. The movement to downsize engines and add turbocharging has improved performance and fuel economy, but the in-cylinder conditions created have led to this stochastic and destructive knocking behavior. Many researchers studying the LSPI phenomenon have hypothesized that it is caused, at least in part, by fluid which is ejected from the piston crevice region.

If this hypothesis is true, it is critical to understand the nature of this fluid. It is well understood that there will be lubricating oil present in the crevice from the action of the rings upon the oil film on the liner. And, at the high load conditions where LSPI occurs, it may be assumed that some fuel will be present as well. However, neither of these fluids will auto-ignite in the manner seen during LSPI events. The hypothesis requires that some characteristics of the crevice fluid are altered from the original fluids.

To investigate this aspect of LSPI, a rapid-acting sampling valve was developed to permit sampling of the fluid in the piston crevice while avoiding sampling of cylinder contents when the piston crevice was not at the sampling location. The samples were analyzed with a two-dimensional gas chromatograph-mass spectrometer which allowed detailed speciation of the samples.

Sampling was performed with the engine running on fuels and lubricants which either increased or reduced LSPI occurrence. There were clear differences in the chemical content of the samples between these cases which indicate that some chemical reactions must take place in the crevice that impact the LSPI behavior of the engine.

1 Introduction

In the search for improved engine efficiency to meet current and future CO_2 regulations, the current industry trend is to increasingly downsize and downspeed engines. To maintain vehicle performance, turbochargers are used to match the torque output of larger engines. This has led to the engine operating under conditions where it is prone to experience a type of abnormal combustion that is distinct from classic hot-spot pre-ignition; this new combustion phenomenon is commonly referred to as Low-Speed Pre-Ignition (LSPI).

M. Günther and M. Sens (eds.), *Knocking in Gasoline Engines*,
https://doi.org/10.1007/978-3-319-69760-4_7

There have been many studies into the behavior and initiating mechanisms of LSPI. There are two predominant theories to explain the cause: droplet-caused ignition and deposit-caused ignition [1–3]. For the present work, the droplet causation hypothesis was considered. In this scenario, the droplets are released from liquid which is trapped in the piston-liner crevice. These droplets then move through the combustion chamber during the compression stroke, and ultimately auto-ignite before the spark initiates combustion. This droplet ignition then starts the main fuel-air mixture combustion process, but at an advanced timing. The result of this advanced combustion event is extremely strong knock which can easily damage the engine.

If the droplet hypothesis is true, then there must be chemical characteristics of the droplet which make it prone to auto-ignition. The fuel and lubricant both have properties which make them resistant to auto-ignition in the engine, so it is unlikely that they are the direct cause of the LSPI process. To investigate the chemistry of these drops, a rapid sampling valve was developed and installed on an LSPI test engine. The valve can pull fluid from just above the top ring when the piston is at top dead center, and can be closed when the piston crevice is not covering the valve location. This allows sampling of the fluid in the crevice while preventing other gases or fluids from the combustion chamber entering the sample. Once captured, the samples were analyzed with a two-dimensional gas chromatograph/mass spectrometer system to fully speciate the samples. The goal of this study was to determine if the chemistry of this crevice fluid was correlated to LSPI behavior when the engine was run on a range of fuels and lubricants that exhibited different LSPI occurrence rates.

The LSPI study was conducted using a 2.0 L GM Ecotec engine of the LHU variant. The engine is turbocharged with direct injection and was supported by GM with open ECU and operational guidance. Table 1 describes the general specifications of the engine. The engine was instrumented with temperature, pressure, and flow measurements. The locations of these measurements can be seen in Fig. 1. Cylinder pressure was recorded using AVL GU22CK pressure transducers and a high speed DAQ system that could continuously stream indicating data during LSPI testing.

Table 1. Specifications of LSPI test engine

Configuration	L4
Displacement	2.0 L
Bore and Stroke	86 mm × 86 mm
Compression Ratio	9.2:1
Maximum BMEP	22 bar

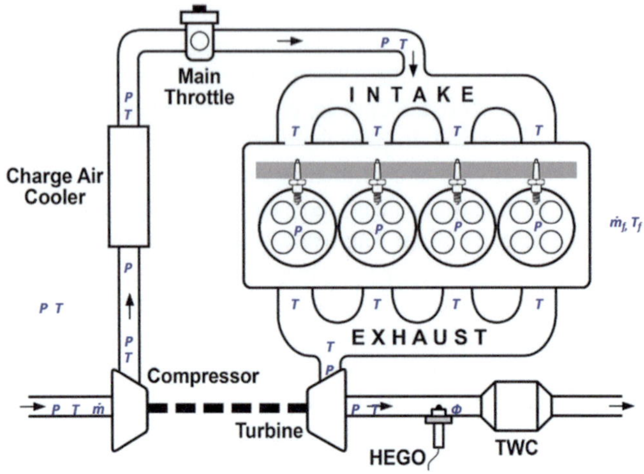

Fig. 1. Location of measurement points on the LSPI test engine

To determine LSPI behavior of lubricants and fuels, a cyclic test was developed that provided statistical relevance to the stochastic LSPI response of the engine to test fluids. There were two operating conditions of interest during the test; one was near the peak torque point of the engine and one was at the maximum torque at a lower engine speed. A portion of the test is shown graphically in Fig. 2. This represents a four-hour run cycle that was repeated four times for a given fuel/lubricant combination. Statistical analysis of the data was used to compare different fluids for their impact on LSPI.

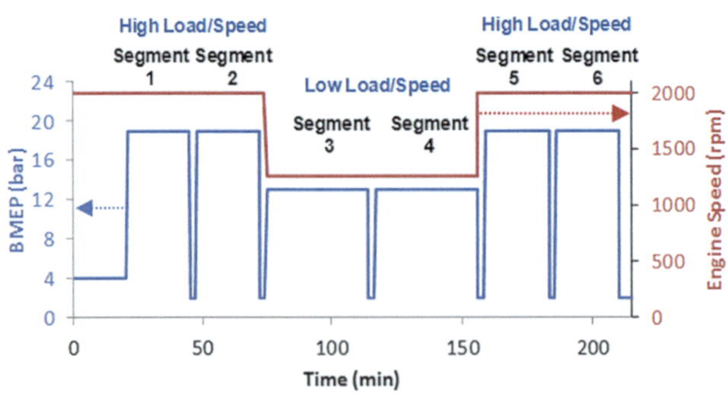

Fig. 2. Graphical depiction of LSPI test protocol

To investigate the composition of the crevice fluid, a similar engine was modified with a drilling through the block and liner into the cylinder just above the location of top ring reversal. A fast-acting sampling valve was then developed to open when the piston crown passed the drilling and to close when the piston reached the top of its stroke. This device can be seen in schematic form in Fig. 3. On each engine cycle a small amount of fluid from the crevice was forced through the drilling and into a sample collection area. After approximately 30 min of running the engine at an LSPI-generating condition the fluid sample was extracted for chemical analysis. For this work, a two-dimensional gas chromatograph coupled to a mass spectrometer was used to provide detailed speciation of the sample.

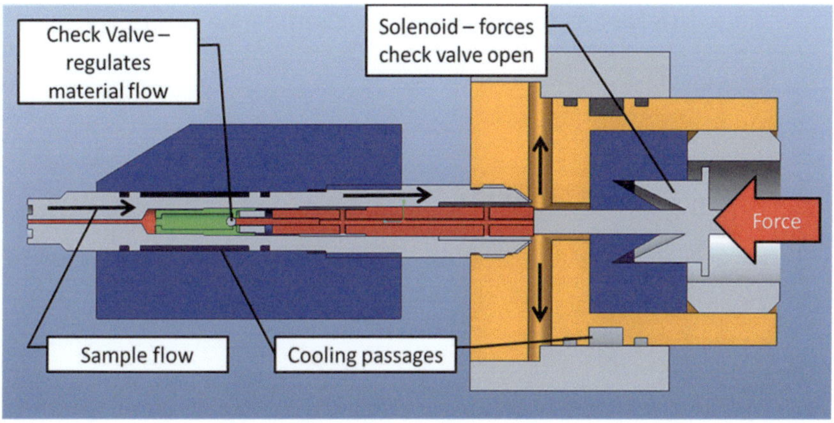

Fig. 3. Schematic of crevice sampling device

The following discussion will review the behavior of selected fuels and with respect to their LSPI response in the test engine and then will examine the crevice fluid speciation to examine characteristics which correlate with the observed LSPI behavior.

2 Engine Experiment Results

To evaluate the impact of fuel chemistry on LSPI, highly simplified fuels composed of only iso-octane, n-heptane, and toluene were blended and run under the LSPI testing conditions. A series of blends of these components were tested, corresponding to the primary fuel reference series from 87 through 96 octane and the toluene standard fuels at 93.4 and 96.9 octane. While highly simplified, these fuels offer a useful set of characteristics. Their knock behavior is well understood, being fuels that define the octane rating system. The boiling points of the three components are very close, eliminating distillation curve differences from the fuels. The addition of toluene offered the ability to examine fuels with matched RON values but with varying octane sensitivity (the difference between RON and MON).

When the LSPI tests are conducted, the number of events which occurred during the 16-hour run is recorded, with the counting broken down into the number of events per 25,000-cycle segment of the full test. For the higher load condition, which is of most interest, this provides 16 individual data points which can be analyzed using statistical methods. Figure 4 shows the least squares means of the LSPI counts for each of the test fuels, along with two samples of the baseline fuel used, which was a custom-blended 87 AKI version of the EPA Tier 2 certification fuel. Note that the data is reported as the square root of the counts plus 0.5. This is to transform the Poisson-distributed count data into a normal distribution to simplify the statistical analysis.

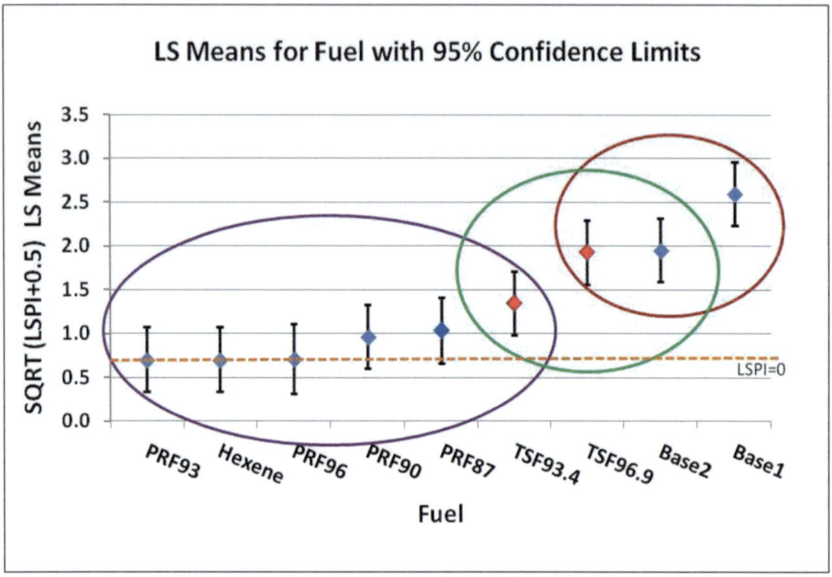

Fig. 4. Least squares mean LSPI results for simplified chemistry fuels

The fuels composed only of iso-octane and n-heptane exhibited no significant LSPI behavior. Two of the PRF blends had no LSPI events occur, and the other two blends only had one or two events. In contrast, the TSF blends had more LSPI events, enough to reach a statistically significant level. This is a significant finding, even for these simple fuels. The low boiling point of the components suggests that there is minimal liquid fuel penetration to the cylinder liner, which is one potential source of LSPI activity [4]. The fuels tested were composed primarily of iso-octane and toluene, both of which are high octane chemicals which do not easily auto-ignite. This suggested that the linkage between LSPI and the fuel chemistry was not observable in the fuel properties themselves. To identify the linkage, these same fuels were evaluated using the crevice fluid sampling system to observe the chemical activity occurring inside the engine.

3 Crevice Sampling Results

To simplify the chemistry as much as possible during the development of the analytical methods for crevice sampling, the PRF93 and TSF93.4 fuels were run in the engine while the crevice sampling system collected fluid from the piston crevice. As these were low boiling point fuels it was assumed that minimal collection of liquid fuel in the crevice would occur and that the samples would highlight any possible chemical changes which occurred inside the combustion chamber to create an ignitable fluid in the crevice. The graphical results from the two-dimensional gas chromatograph-mass spectrometer are shown in Figs. 5 and 6.

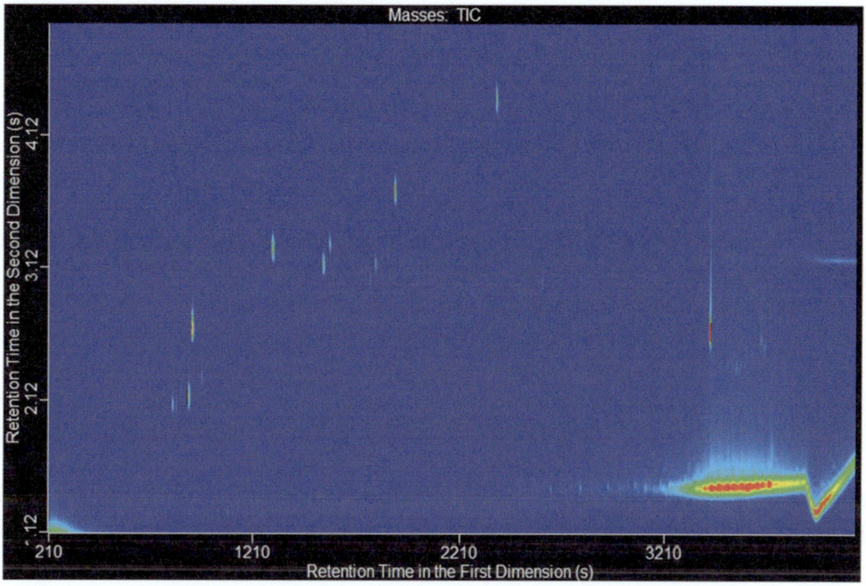

Fig. 5. 2D GCMS results for the PRF93 crevice sample

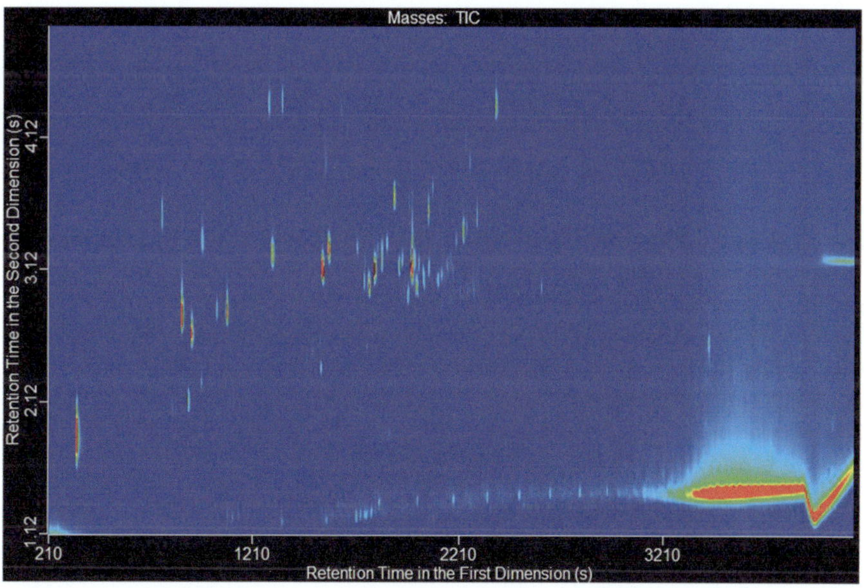

Fig. 6. 2D GCMS results for the TSF93.4 crevice sample

When interpreting the output of the gas chromatograph it is necessary to understand the system. There are two capillary columns within the device; the first column is very long and separates the sample by the number of carbon atoms in it. So C1 species emerge from the column before C2 species, and so on. At the exit of the first capillary there is a cutter device which will collect all the sample from a given carbon number and then admit it to the second capillary which is quite short. This second capillary separates the sample by molecular polarity. So, straight chain hydrocarbons emerge before branched chain, which emerge before simple ring structures, and so on. In this way, a highly-detailed map of the hydrocarbon content of a sample can be developed. In the graphical views presented here the color map corresponds to the quantity of material at a given carbon number and polarity. There is no absolute scale to the color map; it is a relative measure only.

Observing the results shown in Figs. 5 and 6 there are several key features. The large red zone above 3210 s of retention in the first dimension is indicative of the lubricating oil. The device can be operated at different temperatures; a low temperature method is used for getting detailed speciation of fuel-like samples and a high temperature method is used for detailed speciation of oil-like samples. These results were run with the low temperature method so the lubricant does not cleanly speciate but appears as a large continuous output at long retention times. The blue-green signal seen extending to the left of the lubricant zone is an artifact of the solvent that must be used to dilute the crevice sample.

In both samples, there is a significant toluene peak present at approximately 900 s retention time on the x-axis and 3 s on the y-axis. Immediately below the toluene peak, at approximately 2.12 s on the y-axis, is iso-octane, with n-heptane identified in one of the peaks to the left of iso-octane. If the fuels were found in pure form within the crevice these would be the only species detected by the gas chromatograph. Examination of the images shows that many other species are detected in the samples, though. It is in the difference between the samples and which species are detected that a potential link to LSPI may exist.

In Fig. 5, it is evident that there is some production of aromatic structures, either as combustion products which are adsorbed into the crevice fluid, or through reactions in the crevice itself. The peaks between 3.12 and 4.12 s on the y-axis are larger aromatic species with one ring and long chain tails or with two rings. The magnitude of the peaks is relatively low and there are not many identified species, however. By comparison, the TSF93.4 sample shown in Fig. 6 shows many more aromatic peaks, and these are of much greater magnitude. The analysis presented here does not identify the chemical process by which these species are formed. However, it is possible to make a reasonable inference about the source based on existing knowledge of combustion. Particulate matter formation in engines is understood to begin with formation of poly-aromatic hydrocarbons (PAH). And it has been well documented that the aromatic content of fuel has a strong correlation with the formation of PAH species which lead to particulates. Since the two fuels under consideration are primarily differentiated by the presence or absence of an aromatic specie, it is reasonable to assume that the large aromatics detected in the crevice are generated as part of the combustion process, and are adsorbed into the crevice fluid.

Another difference is present in the samples, which is less obvious than the aromatic structures. The paraffin species will be found along the 2.12 s y-axis location. Comparing Figs. 5 and 6 shows that there are more mid-range paraffin species present in the TSF93.4 sample than in the PRF93 sample. These species are of great interest; in general, larger aromatic species have lower RON values as there are more opportunities for hydrogen abstraction to start the combustion reactions in a larger molecule. As the fuels supplied to the engine do not contain any of these larger paraffin species, they must have been created in-cylinder. There are two possible pathways for this to occur: the fuel compounds can polymerize in the crevice under the high pressure, moderate temperature conditions present or the lubricant compounds can break down under the same conditions. It is not clear from the present study which of these pathways is more likely. The crevice does provide a suitable environment for such reactions to occur. It is maintained at a temperature much lower than the main combustion chamber through the heat transfer to the piston and liner; there is water vapor present which can assist in some chemical reactions; there is a relatively long time where the fluid may be retained in the crevice, providing time for slower reactions to occur. It is also possible that some metals, either as wear products from the engine or those present in the lubricant additive package, act as a catalyst to enhance the reactions which appear to be occurring in the crevice fluid. At present, the authors believe that it is these mid-range paraffin species which are a likely contributor to LSPI through the crevice fluid.

4 Conclusions

Low-speed pre-ignition is a complex phenomenon occurring within modern GDI engines. There are several reasonable hypotheses for the causes of LSPI, and it is quite possible that more than one of them may be correct. The hypothesis of crevice fluid-initiated LSPI was investigated in the present work using simplified fuels to reduce the number of chemical variables present.

Statistically-guided engine tests to determine LSPI response with these fuels showed that fuels which only contained paraffins exhibited little or no LSPI, while adding toluene increased the occurrence of LSPI by a statistically-significant amount. These tests could not provide any insight into why the aromatic chemistry was significant for LSPI, so additional investigation was performed using a sampling system to draw fluid from the piston crevice for chemical analysis.

The sampling study identified significant differences in the crevice fluid composition between a paraffin-only fuel and one containing toluene as well as paraffins. There was a significant increase in the concentration and number of aromatic species in the crevice when toluene was present in the fuel. This appears to be correlated to the known link between fuel aromatics and PAH formation. More surprising was the presence of larger paraffin species in the crevice sample, and that the toluene fuel had more of these species present. It is not clear what pathway creates these species, but there is a potential link to LSPI; larger paraffins tend to have lower autoignition temperatures and are more likely to autoignite under compression conditions in an GDI engine. This may be the link between fuel chemistry and LSPI occurrence.

Further studies are ongoing to investigate the chemical processes which may lead to the presence of these paraffin species and to determine if any of the known lubricant interactions with LSPI also link to these intermediate paraffins. If these additional studies support the findings presented here, then a more fundamental understanding of LSPI can be used to eliminate this undesired feature of modern engine combustion.

References

1. Dahnz, C., Has, K.-M., Spicher, U., Magar, M., Schiessl, R., Maas, U.: Investigations on pre-ignition in highly supercharged SI engines. SAE Int. J. Engines **3**(1), 214–224 (2010)
2. Kuboyama, T., Moriyoshi, Y., Morikawa, K.: Visualization and analysis of LSPI mechanism caused by oil droplet, particle and deposit in highly boosted si combustion in low speed range. SAE Int. J. Engines **8**(2), 519–537 (2015)
3. Kassai, M., Hashimoto, H., Shiraishi, T., Teraji, A., Noda, T.: Mechanism analysis on LSPI occurrence in boosted S. I. engines (2015)
4. Zadeh, A., Rothenberger, P., Nyugen, W., Anbarasu, M., Schunk-Soldan, S., Schaefer, J., Goebel, T.: Fundamental approach to investigate pre-ignition in boosted engines. SAE Paper 2011-01-0340 (2011)

Simulation of the Effects of Spark Timing and External EGR on Gasoline Combustion Under Knock-Limited Operation at High Speed and Load

Michal Pasternak[1(✉)], Corinna Netzer[1], Fabian Mauss[1], Michael Fischer[2], Marc Sens[3], and Michael Riess[3]

[1] Brandenburg University of Technology, Cottbus, Germany
{michal.pasternak, corinna.netzer, maussf}@b-tu.de
[2] Tenneco GmbH, Edenkoben, Germany
mfischerl@tenneco.com
[3] IAV GmbH, Berlin, Germany
{marc.sens, michael.riess}@iav.de

Abstract. Combustion in a spark ignition engine operated at high speed and load is investigated numerically with regard to knock behavior. The study focuses on the concurrent impact of spark timing and exhaust gas recirculation (EGR) on the severity of knock. Specifically, the possibility of knock reduction through the lowering of nitrogen oxide (NO) content in the rest-gas is examined. Simulations are carried out using a stochastic reactor model of engine in-cylinder processes along with a quasi-dimensional turbulent flame propagation model and multicomponent gas-phase chemistry as gasoline surrogate. The knock-limited conditions are detected using the detonation diagram. By lowering the NO content in the external EGR the end-gas auto-ignition is suppressed. This prevents a transition to knocking combustion and enables advancing of spark timing that yields better combustion phasing. As a result, fuel economy is improved and the potential benefits of cleaning the EGR are indicated.

1 Introduction

Meeting the ever tightening emission standards and future fuel economy demands are the driving forces behind the development of high-efficiency combustion engines. In the context of spark ignition (SI) gasoline engines, the development efforts are frequently targeted at improving the resistance to knocking combustion that is one of the major cause of rather poor thermal efficiency of SI engines. The tendency for knock prevents an optimum combustion phasing and high compression ratio that along with mixture specific ratio govern thermal efficiency of gasoline engines. Besides deteriorating thermal efficiency and consequently also increasing fuel consumption, the knocking combustion increases exhaust emissions and may cause a damage of engine hardware [1, 2].

M. Günther and M. Sens (eds.), *Knocking in Gasoline Engines*,
https://doi.org/10.1007/978-3-319-69760-4_8

Commonly accepted theories assume engine knock to be an abnormal combustion (see, e.g. [2–8] that underlie the description in this section). It results from a spontaneous auto-ignition of a portion or portions of the end-gas air-fuel mixture that occur ahead of the propagating flame front and if the associated release of chemical energy is sufficient to cause the formation of pressure waves of substantial amplitude and frequency. The occurrence of pressure waves is revealed by a characteristic metallic noise with a frequency between 10–30 kHz [3] referred to as knock. Auto-ignition is the primary process for the development towards knocking combustion. It does not happen well distributed within the end-gas region, but in localized spots entrapped inside the heterogeneous area in front of the propagating flame. These so called exothermic centers with specific induction time and specific excitation characteristic (spatial and temporal heat release) emerge from incomplete mixing and temperature inhomogeneity. Whether a self-ignition in the end-gas region will develop into what is referred to as engine knock is dependent on the properties of the exothermic centers, such as size gradient and spatial distribution [4]. These properties determine the magnitude of the heat released by the exothermic center and hence, also the magnitude of the resulting pressure oscillations.

Most commonly, the detection and analysis of knock are carried out based on the measured in-cylinder pressure data or acoustic signals by means of acceleration sensors mounted to the engine crankcase. These methods indicate the occurrence of knock, but provide no information about its phenomenology. This drawback is avoided in the complex and expensive optical methods. These methods enable the detection of intermediate species and radicals that are markers of auto-ignition. They may also provide the time and space resolved information about the flame development. Hence, they contribute to better understanding of the transition from auto-ignition to knocking combustion. Further support of the analysis of knock and its detection is offered by numerical simulations. Here, one can list simple zero-dimensional (0D) or one-dimensional (1D) approaches and complex three-dimensional (3D) computational fluid dynamics (CFD) (see, e.g. [8, 9]). All these methods become particularly robust if coupled with chemical kinetic models that enable the prediction of end-gas auto-ignition that initiates knocking combustion.

Because of the negative effects of knock on the combustion process of SI engines, significant efforts are being made to develop methods of its suppression. Classically, the tendency for knock has been avoided by retarding the spark timing towards the expansion stroke or by reducing the compression ratio. However, both these methods deteriorate thermal efficiency. Furthermore, late combustion phasing increases exhaust gas temperature that at high loads may significantly increase thermal loading and cause engine damage. To avoid this drawback, mixture enrichment is frequently applied that in turn influences negatively fuel economy. Among other methods of knock suppression one can list [2, 5, 8, 10] increasing turbulence and combustion speed, application of cooled exhaust gas recirculation (EGR) and dedicated EGR composition, direct fuel injection, water injection, variable valve timing, developing of fuels with anti-knocking performance, such as liquefied petroleum gas fuels, increasing fuel octane index by using variety of antiknock additives or the reduction of nitrogen oxide (NO) content in the rest-gas.

Among the listed above methods, so far relatively less work has been devoted to the possibility of knock suppression by the reduction of NO contained in the rest-gas. The increased reactivity of the air-fuel mixture in the presence of even trace amounts of NO is known from the literature [11, 12]. The negative impact of NO on auto-ignition tendency and knock in SI engines has already been indicated in [13–15] and more recently in [16, 17]. To our best knowledge, the work reported in [14] is the most thorough numerical analysis of the chemical impact of NO on the end-gas auto-ignition in SI engines. In that work, calculations were carried out using a simple two-zone and homogeneous model of SI combustion along with a reaction mechanism for primary reference fuel (PRF) as gasoline surrogate. In some extent, the present work can be considered as a further development of that in [14]. However, here simulations are carried out with consideration for mixture inhomogeneity due to the impact of fuel injection, heat transfer and turbulent mixing. Thus, the effects of temperature fluctuations and inhomogeneity, which are present in the gas mixture in actual engines [18], are respected in the model. Furthermore, a multicomponent reaction mechanism is employed. In contrast to the commonly used PRF models, here we respect also the impact of ethanol and aromatic contents that enable better representation of fuels used in actual SI engines. Overall, the present approach provides a modeling framework that is tailored for the simulation of modern SI engines with direct fuel injection.

The specific purpose of this article is to provide a parametric analysis of combustion to explore the potential of knock suppression throughout the reduction of NO in the external EGR. The reduction of NO is carried out with respect to the possibility to optimize combustion phasing and hence, to improve fuel economy while operating the engine at the knock limit. The analysis deals with a gasoline engine that is operated at high load and speed and with cooled EGR. The rest part of the article is structured as follows. Firstly, a brief description of the numerical model is given. Next, the applied experimental and numerical methods of knock detection are presented. Subsequently, the modeling framework and assumptions are discussed. Finally, exemplary simulation results are presented and the work is summarized.

2 Modeling Approach

The modeling approach adopted for the present study features a stochastic reactor model (SRM) of in-cylinder processes in SI engines along with a turbulent flame propagation model and multicomponent model of gas-phase chemistry as gasoline surrogate. The models belong to the LOGEengine [19] simulation platform that is specifically tailored to the simulation of physical and chemical processes in combustion engines. Below, a basic description of each model is presented for readability of the article. The description has been freely adopted from our previous works [19–21] that contain thorough description of the sub-models.

2.1 Stochastic Reactor Model

The SRM is a 0D model of physical and chemical processes occurring in combustion engines during the closed part of the engine cycle. It is formulated within a probability density function (PDF) approach for the simulation of turbulent flows with chemical reactions. For SI engines, the SRM is formulated as two-zone model; burned and unburned zones are distinguished that result from the flame propagation across the combustion chamber. There is no interaction between the zones except for the transfer of mass as the combustion proceeds. The gas inside each zone is treated as an ensemble of notional particles. Each zone is assumed to be statistically homogenous, and particle-particle interaction is purely stochastic. Particles can mix with each other and exchange heat with the cylinder walls. Each particle has a chemical composition, temperature and mass and hence, each particle represents a point in phase-space for species mass fraction and temperature. These scalars are treated as random variables and can vary within the cylinder. They determine the composition of the gas mixture and are described with probabilities using PDFs. The frequency of scalar mixing between particles is governed by the mixing time that is explained in the next section. Mixing time is the main modeled parameter for the SRM. By its modeling, the inhomogeneity is introduced into the gas-phase for species concentration and temperature. Besides the impact of mixing, the mixture inhomogeneity is further enhanced by the heat transfer to the wall and fuel injection. Overall, the modeled inhomogeneity of the mixture mimics, in some extent, the turbulence effects in actual engines.

2.2 Turbulent Flame Propagation and Mixing Time

The SRM employs a quasi-three dimensional turbulent flame propagation model that governs the mass transfer from the unburned zone to the burned zone. A spherically expanding flame is assumed that is limited by the cylinder walls. The turbulent flame speed (S_{T}) is calculated after [22] based on the laminar flame speed (S_{L}) and turbulent intensity that is represented by the velocity fluctuations (u').

$$S_T/S_L = 1 + C(u'/S_L)^m \tag{1}$$

Here, C and m are constants and equal to 2.5 and 0.9, respectively. The S_L is taken from a pre-generated library for pure iso-octane and for different pressures, temperatures, equivalence ratios and EGR rates. The value of u' is calculated as the ratio between the integral length scale (l_I) of the flow and the turbulent mixing time (τ_{t}) that represents the time taken to transfer the turbulent kinetic energy (k) from the large scales to small scales, where due to viscosity it is dissipated (ε) into heat. Hence, the turbulent flame propagation is calibrated by only the turbulent mixing time.

$$\tau_t = l_I/u' \cong k/\varepsilon \tag{2}$$

The turbulent flame speed is assumed to be constant in all directions unless the flame encounters a barrier. The geometric effects of near-wall flame quenching on the flame shape are not modelled.

With respect to the SRM simulations, rather than the turbulent mixing time the scalar mixing time is used. For simplicity, it is called in the later part of this article just as mixing time and is denoted by τ. The mixing time used in the SRM is related to the turbulent mixing time, τ_{t}, by the mixing time constant C_ϕ (see, e.g., [23]).

$$\tau = C_{\phi}^{-1}\tau_t \tag{3}$$

The constant C_ϕ is interpreted as velocity-scalar decay time scale. Knowing an overall trend of τ_t and calibrating C_ϕ, the right-hand side of Eq. (3) can be determined and it models the mixing time that is used as an input for the SRM.

2.3 Reaction Kinetics for Gasoline Surrogate

A four component reaction kinetics mechanism from LOGEfuel [19] has been used as gasoline surrogate. The mechanism has been successfully used in our recent work [24] devoted to the analysis of combustion in SI engines. It contains sub-models describing the oxidation process of ethanol, toluene, iso-octane and n-heptane and is referred to as ethanol-toluene reference fuel (ETRF). The mechanism contains also the thermal model of NO formation and growth pathways for polyaromantic hydrocarbons that are compatible with the soot model used in the LOGEengine platform. The version of the mechanism used in the present study contains 193 species and 2088 reactions. The content of ethanol, toluene, iso-octane and n-heptane was set to 0.15, 0.39, 0.29 and 0.17, respectively and on a mole fraction basis. For such defined mixture the research octane number (RON) index is 91 and lower heating value (LHV) is 41.6 MJ/kg. These values match exactly with the values reported for the fuel used during the experimental work; the exact composition of the fuel used during the experimental work was unknown.

3 Detection of Knocking Combustion

The detection of engine knocking combustion and subsequently the determination of engine operating conditions at the knock limit underlie all investigations presented in this article. They were carried out separately for both simulated and measured data.

3.1 Experimental Method for Engine Knock Detection and Control

In the development of combustion engines, the precise measurement of the in-cylinder pressure with a high temporal and signal resolution via quartz-crystal pressure transducers is a standard. In this work, the knock indicator system from IAV (KIS4) was employed in order to identify knocking combustion [25, 26]. The system uses the in-cylinder pressure signal from a quartz-crystal sensor that is analyzed in the time domain. In Fig. 1 the signal chain inside KIS4 is presented. The cylinder and cycle individual maximum peak of the high pass filtered and rectified pressure, which is denoted by PS, is used to describe the severity of engine knock.

In order access whether each individually recorded and analyzed cycle is knocking or not, the peak pressure PS is set in relation to a normative. This normative or threshold might be different for each engine and is application tailored. Frequently, the knock threshold is set to consider the background noise in the in-cylinder pressure

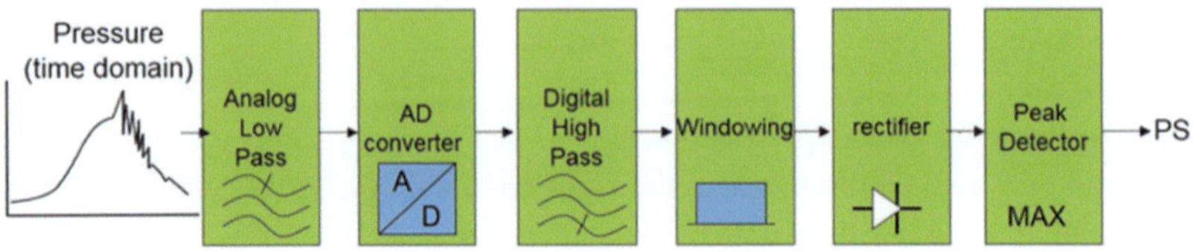

Fig. 1. Overview of pressure signal conversion in KIS4

signal coming along with an increase of engine speed. Thus, a typical empirical definition of the maximum pressure PS allowed for a cycle to be judged not knocking is to undercut the logic: PS < engine speed/2000, where the right hand side of the inequality defines the knock threshold. This definition was also employed in the present work. In order to account for the stochastic nature of engine knock, a statistical validation of the analyzed PS data is performed. Typically, the spark timing on the engine test bench is controlled to allow a certain percentage of cycles to be the knocking cycles according to the defined knock threshold. It is assumed that the knock limit is achieved if for 2% of the analyzed cycles the PS is greater than the assumed knock threshold.

3.2 Knock Detection During Simulations

As stated in the introduction section, according to the commonly accepted theories the occurrence of auto-ignition of the end-gas mixture is a necessary but not sufficient condition to give rise to knocking combustion [27, 28]. To cope with this issue, in the present work we adopt a concept of developing detonation to determine the occurrence of knock and to define engine operating conditions at the knock limit. Overall, the concept assumes that the developing detonation, which is associated with the presence of pressure waves and follows the auto-ignition, may occur if the chemical and acoustic waves reinforce each other and if the gas mixture is sufficiently reactive [28]. The foundation of the concept is dated back to the work reported in [29], where different modes of a propagating chemical reaction in the reactive mixture are distinguished. This concept was extended in [28, 30] and resulted in a theory of wave reinforcement. In [31], the occurrence of hot-spots causing temperature gradients was added to the auto-ignition concept. These concepts were further investigated and extended in [32, 33] with reference to the analysis of super-knock.

For the analysis of different modes of developing detonation that follows the auto-ignition with hot-spots, detonation diagrams are frequently applied. The concept of a detonation diagram was introduced in [28] and was also used in our recent work [24]. In the detonation diagram the evolution of an ignition to a developing detonation, which is understood as knocking combustion, is described using two dimensionless parameters ξ and ε that after [34] are referred to as resonance parameter and reactivity parameter, respectively. The parameter ξ is calculated as the ratio between the speed of sound, a, and the reaction velocity, u. This ratio determines if the two waves are in the same order of magnitude and therefore, may combine and grow to a developing detonation or if the reaction front will travel at subsonic or supersonic speed. The reaction velocity can be determined from the temperature, T, and the ignition delay time, τ_i [31].

$$\xi = u/a = a(\partial T/\partial x)(\partial \tau_i/\partial T) \quad (4)$$

The reactivity parameter ε is determined by the ratio of the constant temperature kernel length l and the length that the speed of sound travels during the ignition event. Thus, it is a measure of the reactivity and the size of the ignition kernel [31].

$$\varepsilon = l/(a \cdot \tau_e) \quad (5)$$

Here, τ_e is the excitation time that denotes time between 5% of heat release until the maximum heat release during the ignition. As ε increases, the range of ξ, which can lead to a developing detonation, also increases that results in a characteristic peninsula bounded by the solid lines in Fig. 2 (results elaborated from [28, 35, 36]). The upper limit of the peninsula is denoted by ξ_u and the lower limit by ξ_l. Overall, the ε-ξ solution indicates about the pressure development mode. Besides the peninsula of developing detonation, the remaining three regimes based on ξ and ε values are deflagration, subsonic auto-ignition and thermal explosion [36]. The transition from the subsonic auto-ignition to the developing detonation is most interesting from the point of view finding operating conditions at knock limit.

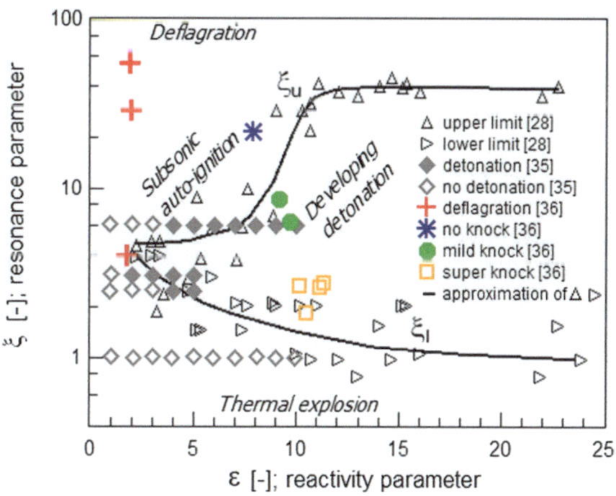

Fig. 2. The parameters ε-ξ defining the detonation peninsula (solid lines, [28]) and indicating different modes of reaction front propagation (elaborated from [28, 35, 36])

The detonation diagram was first developed for syngas mixtures [28, 30]. In [35], it was verified that the boundaries for detonation (Fig. 2) are also valid for other fuels, such as n-heptane and iso-octane, since the developing detonation depends on the high temperature chemistry. In [36] and references therein reported, engine-relevant-regimes were defined on the detonation diagram (Fig. 2) and based on the 3D large eddy simulations.

The methodology reported in [28, 30–33] for the calculation of parameters ξ and ε has been transferred to the SRM and is discussed in detail in [24]. Here we summarize this methodology for readability of the article. The methodology is based on the analysis of heat release and the local conditions in the unburned zone and is applied in a post-process. Since there is no spatial resolution for the particles, all particles that contribute to the investigated high temperature peak are collected and treated as one ignition kernel that is assumed to be a sphere. To determine the maximum knock intensity of the current cycle, all particles are selected that have their maximum heat release in the same time step as the maximum gradient in the rate of heat release that is calculated for the entire in-cylinder mixture. Particles that ignite in an earlier or later time step are accounted to a different auto-ignition kernel. In this way, the most severe auto-ignition event (steepest heat release gradient) of the current cycle is investigated and evaluated. Due to the auto-ignition, the temperature of the kernel is increased and hence, the volume also increases accordingly. The difference between the radius of the ignition kernel before and after the ignition gives the length x that is needed for Eq. (4). The gradient of ignition delay time, τ_i, is calculated using the maximum particle temperature that is included in the ignition kernel before the ignition occurs and using the mean temperature of the unburned zone. The speed of sound is calculated as a function of the mean properties of the end gas. The excitation time is calculated analogously to the calculation of ignition delay time. Fuel characteristics, such as ignition delay time and excitation time, are calculated using a constant volume reactor from [19].

4 Computational Set-up and Method

4.1 Engine Layout and Operating Points

The computational analysis is carried out for a 1.4 L, four-cylinder, direct injection, 16-valve, turbocharged SI engine that is operated with external EGR. The engine is fueled with standard gasoline fuel that is characterized by RON 91. The basic specification of the engine is summarized in Table 1.

Table 1. Engine basic specification

Parameter	Value
Bore [mm]	74.5
Stroke [mm]	80
Compression ratio [-]	9.7
Connecting rod [mm]	140

Two operating points, which are defined by engine speed and load, are used in this study. Their basic characteristic is given in Table 2.

The operating point OP1 is used to verify the predictive capability of the SRM in the context of simulating spark timing (ST) sweep. For this operating point the measured in-cylinder pressure was available for three different spark timing positions, namely −6, −3 and 0 crank angle degrees (CAD) after top dead center (ATDC). The

Table 2. Engine operating points

Parameter	OP1	OP2
Engine speed [1/min]	1750	4750
IMEP [bar]	16	17
Equivalence ratio [-]	1.0	1.0

operating point OP2 is the main point for the analysis of the impact of NO reduction in the rest-gas on the suppression of knock. At this operating point the engine is operated at knock limit. According to the measurement methodology described in Sect. 3.1, it means that knocking combustion occurred in 2% out of 500 measured pressure cycles. Furthermore, at OP2 the engine is operated with 7.9% of external EGR.

4.2 SRM Set-up and Procedure

The SRM is applied to simulate combustion and tendency for knocking during the closed part of the engine cycle, between inlet valve closure (IVC) and exhaust valve opening (EVO). The in-cylinder mixture is discretized into 1000 particles. Calculations are carried out with 0.5 CAD time step. The detailed ETRF model outlined in Sect. 2.3 is used as gasoline surrogate.

The initial conditions for the SRM are calculated using a method implemented in the LOGEengine platform. The method employs a genetic algorithm to find the parameters at IVC, such as initial mass and temperature, mixture composition, internal EGR and absolute pressure. The apparent rate of heat release is used as a reference and target parameter during the analysis. The algorithm employs detailed chemistry through the calculation that enables accurate representation of the initial gas composition.

The SRM is calibrated using a crank angle dependent mixing time (τ) that is optimized until the simulated engine parameters, such as in-cylinder pressure and rate of heat release, match the experimental counterparts. The optimization is carried out using the LOGEengine. The history of mixing time is modeled as a combination of linear and exponential function [19]. The modeling is targeted at mimicking the changes of mixing time that usually are seen in 3D CFD solutions (see, e.g. [21]) and based on Eq. (2). The modeled mixing time maps the impact of the 3D character of actual in-cylinder processes into the 0D domain of the SRM.

Several assumptions have been made to isolate the impact of NO contained in the rest-gas on the auto-ignition tendency and knock. The rest-gas is assumed to be composed of nitrogen (N_2), carbon monoxide (CO), carbon dioxide (CO_2), water steam (H_2O) and NO. Such a composition was obtained during the preliminary calculations as an average composition from the multi-cycle simulations using the SRM. The concentration of CO, CO2 and H_2O in the rest-gas is kept constant for all calculations carried out in this work. The changes of NO content are suitably compensated by an increase or decrease of N_2 content. The impact of cycle to cycle variation on the in-cylinder pressure and knock tendency is not considered here. The possible impact of stochastic mixing, heat transfer and fuel injection on the combustion rate and knocking tendency is eliminated by representing these parameters as means calculated over thirty consecutive cycles; the rest-gas composition still remains unchanged for each cycle. Overall, all computed results are understood as coming from a single representative cycle for a given set of initial and operating conditions.

4.3 Model Extrapolation for Varied Spark Timing

The possibility to predict engine in-cylinder parameters, such as rate of heat release and pressure, for varied spark timing is crucial for the overall performance of the model. It enables an estimation of the concurrent impact of reducing NO in the rest-gas and advancing of spark timing on the combustion process. In the present work, the predictive capability of the SRM is obtained through the calibration of mixing time history at a given operating point and subsequently, its extrapolation to different ST positions at that operating point.

Overall, mixing time varies according to the variation of turbulent kinetic energy that in turn is affected by engine geometry and depends strongly on engine speed (see, e.g. [21] and references therein). Hence, for a varied spark timing, and if the engine speed does not change, it is assumed that the history of mixing time also remains unchanged. Only the anchorage of mixing time history (see the profiles in Fig. 3) on a crank angle basis has to be extrapolated according to the ST changes.

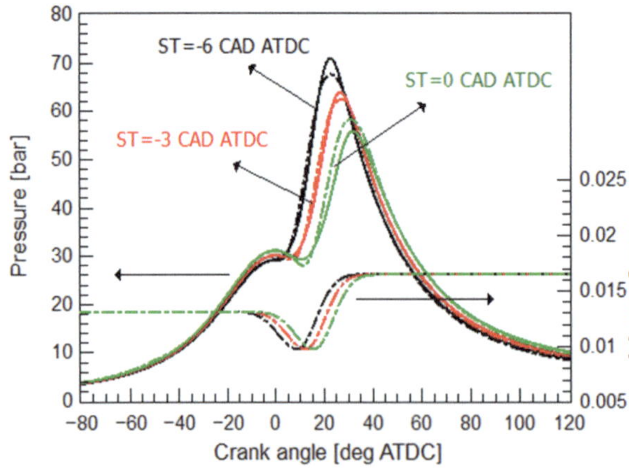

Fig. 3. Experimental (solid lines) and simulated (dash-dotted lines) pressure histories for different STs along with the calibrated mixing time history for ST = −3 CAD ATDC and the extrapolated histories for ST = −6 and 0 CAD ATDC (results for OP1: n = 1750 rpm and IMEP = 16 bar)

The performance of the proposed modeling of mixing time extrapolation is demonstrated based on the calculations carried out for OP1 (Table 2). For this OP the engine speed (n) is 1750 rpm and indicated mean effective pressure (IMEP) is 16 bar. Thus it is rather low speed but high load operating point. The experimental pressure data for this OP are available for three ST positions, namely −6, −3 and 0 CAD ATDC.

The SRM was calibrated for ST = −3 CAD ATDC. To predict in-cylinder performances at ST equal to −6 and 0 CAD ATDC, the mixing time history is moved by 3 CAD receptively to the earlier and later positions during the cycle. More specifically, only the nonlinear part of the mixing time history shown in Fig. 3, in between −10 and 40 CAD ATDC is moved; this regime corresponds approximately to the duration of combustion. The results from the calibrated and extrapolated SRM are presented in Fig. 3 for in-cylinder pressure and in Fig. 4 for combustion progress. The combustion progress is defined by crank angles corresponding to 10%, 50% and 90% of the total energy release during combustion. These characteristic CA positions are denoted as CA10, CA50 and CA90, respectively.

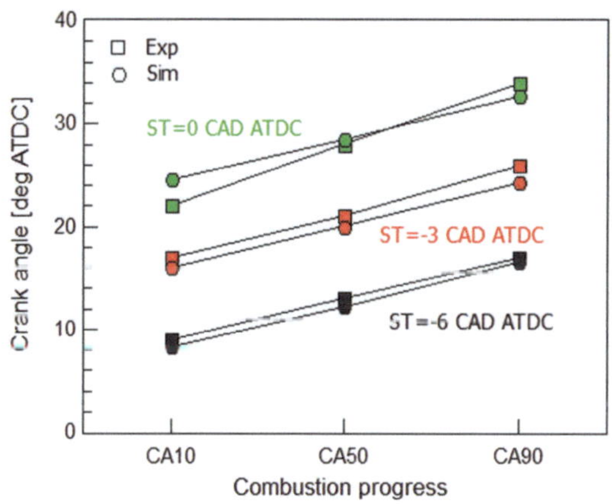

Fig. 4. Computed (circles) and experimentally obtained (squares) CA10, CA50 and CA90 for different ST positions for the engine operated at OP1 with n = 1750 rpm and IMEP = 16 bar

Using mixing time as the only calibration parameter, the simulated pressure histories and combustion progress match well the experimental counterparts. Some small deviations observed at ST = 0 CAD ATDC for CA10 and CA90 indicate that besides extrapolating mixing time with respect to ST position it may also be beneficial to extrapolate its values. This may help to improve the accuracy of the model for very early and late phases of combustion. Overall, the accuracy of the obtained results is sufficient with respects to the objectives of the present work. For this reason, the outlined methodology of mixing time extrapolation is also applied in Sect. 5 that discusses the impact of NO contained in the rest-gas on the combustion process.

5 Results and Discussion

The analysis of the interaction between ST and NO content in the rest-gas is carried out for OP2 (Table 1) that is operated at the knock limit and at n = 4750 rpm and IMEP = 17 bar. The ST for this operating point equals −9 CAD ATDC and is denoted as the reference position (ST_{ref}). The initial NO content in the in-cylinder mixture, which was found during the preliminary tests, is 330 ppm. This combination of NO and ST is referred to in this work as the baseline configuration. First, the quality of the model calibration is discussed. Then, the determination of knock limit is presented. Subsequently the impact of ST changes and NO content in the rest-gas on the combustion progress and knocking tendency are presented. Finally, the possibility to reduce fuel consumption through the concurrent reduction of NO content in the rest-gas and advancing ST is demonstrated.

5.1 Numerical Model Calibration

The combustion process simulated by the SRM is verified quantitatively by the comparison of the computed in-cylinder pressure and rate of heat release (RoHR) histories to the measured data (Fig. 5). The calibration process is based on mixing time calibration and it follows the methodology outlined in Sect. 4.2. Overall, there is a good agreement between the simulated and measured quantities and hence, the SRM is assumed to be calibrated with respect to the combustion process.

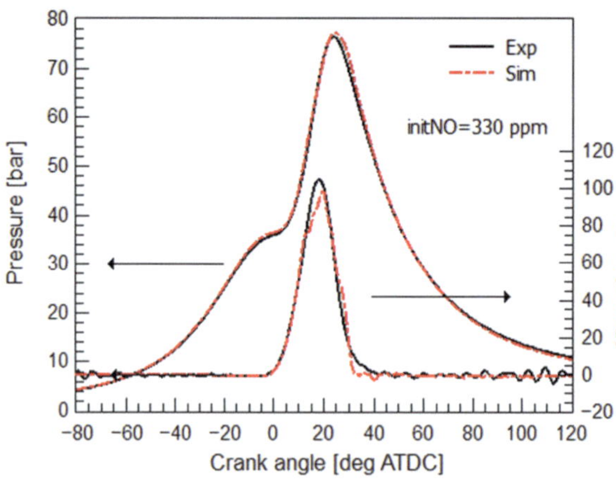

Fig. 5. Experimental (solid line) and computed (dash-dotted line) histories of in-cylinder pressure and rate of heat release for OP2 (n = 4750 rpm, IMEP = 17 bar, 330 ppm of initial NO)

5.2 Engine Operation at Knock Limit

To verify if the simulated pressure and RoHR data in Fig. 5 correspond to the operation at knock limit, the rate of heat release in the unburned zone is analyzed. Furthermore, the solution from the SRM, in the form of non-dimensional resonance parameter, ε, and reactivity parameter, ξ, is mapped over the detonation limits as presented in Sect. 3.2. In the baseline configuration (ST_{ref} and results in Fig. 5), the history of RoHR-u (solid line in Fig. 6a) shows two characteristic peaks due to low and high temperature chemistry, respectively. Both peaks are of the same order of magnitude. This indicates that the end-gas has mainly a low temperature that may not give rise to knock. A confirmation of it is given by the ε-ξ solution (circle in Fig. 6b) that is found shortly before the upper limit of detonation (ξ_u).

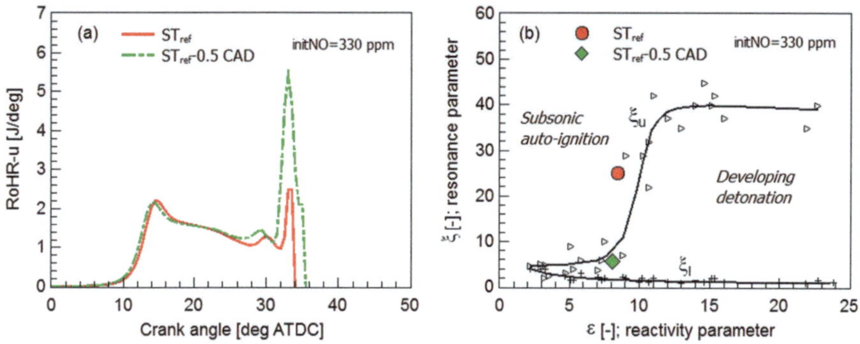

Fig. 6. Computed RoHR-u histories (a) and ε-ξ solutions (b) in the detonation diagram from [28] for the engine operated at OP2 with 330 ppm of initial NO in the mixture for not-knocking conditions at ST_{ref} and for knocking conditions at ST_{ref} −5 CAD

In order to check if the obtained solution corresponds indeed to the knock limit, the spark timing is advanced by 0.5 deg (ST_{ref} − 0.5 deg) from the baseline position. The ST has been selected since it is one of the most sensitive parameters for the occurrence of knock. The RoHR-u history and ε-ξ solution after the advancing of ST are represented by dash-dotted line in Fig. 6(a) and filled diamond in Fig. 6(b), respectively. The advancing of ST just by 0.5 deg changes the conditions in the end-gas significantly. The second peak in the RoHR u history, which is due to the high temperature chemistry, is now significantly higher than the peak resulting from the low temperature chemistry. This behavior is reflected in the detonation diagram by the ε-ξ solution being within the peninsula of developing detonation, between the ξ_l and ξ_u. Thus, knocking combustion is indicated for these conditions. Therefore, we consider the simulated results for ST_{ref} as corresponding to the knock limit. This evaluation agrees with the evaluation obtained from the measured in-cylinder pressure signal and using the method presented in Sect. 3.1.

5.3 Combustion Phasing Effects

To further demonstrate the impact of ST changes on the auto-ignition and knocking tendency, the calculations are carried out for six different ST values that are listed in Fig. 7. The calculations refer to 287 ppm of NO in the initial mixture. Starting from ST = −9 CAD ATDC, the mixing time histories for other ST positions are obtained by following the methodology described in Sect. 4.3. By advancing the ST, the start of combustion, combustion center (CA50) and maximum peak pressure occur earlier during the cycle. This is an expected tendency and hence, the predicted results are considered plausible.

Through the analysis of RoHR-u and the ε-ξ solution on the detonation diagram, the results presented in Fig. 7 are quantitatively diagnosed with respect to the occurrence of knock. The ε-ξ results are presented in Fig. 8.

The dependency of RoHR-u on ST changes and for the initial NO = 287 ppm is similar to the one presented in Fig. 6. The advancing of ST enhances the tendency for

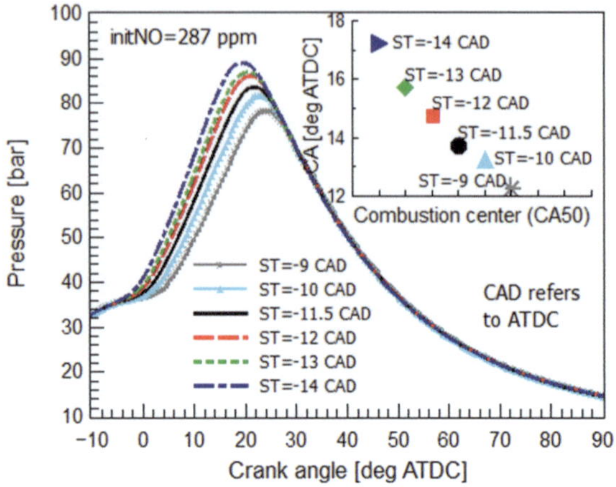

Fig. 7. Simulated in-cylinder pressure history and combustion center (CA50) for different ST values and constant NO content in the initial mixture (287 ppm) for OP2

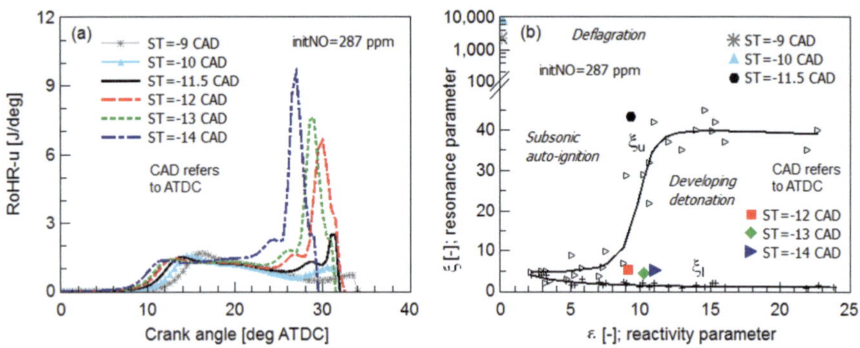

Fig. 8. RoHR-u histories (a) and the detonation peninsula from [28] along with the computed ε-ξ data (b) for different ST at constant initial NO content (287 ppm) in the mixture for OP2

auto-ignition, whereas the delaying of ST mitigates it. The change of ST has no significant impact on the maximum value of the first peak in the RoHR-u history. Only the position of maxima on a crank angle basis is different for different ST. Here, for the initial NO = 287 ppm the ST = −11.5 CAD ATDC corresponds to the operating of the engine at knock limit that is confirmed by the results in Fig. 8(b). The advancing of ST from −11.5 CAD ATDC by 0.5 CAD results in a transition from the regime of subsonic auto-ignition to the regime of developing detonation. Further advancing of ST moves the ξ-ε solution towards the lower limit of the resonance parameter (ξ_l) and slightly higher values of the reactivity parameter (ε). On the contrary to that, the delaying of ST by 1.5 CAD or 2.5 CAD with reference to the value at knock limit (ST = −11.5 CAD ATDC) moves the ξ-ε solution towards very small values of ε and very high values of ξ that are characteristic for deflagration (see Fig. 2).

5.4 Influence of the Initial NO Content

The reduction of NO in the rest-gas is adopted in this work to avoid entering the detonation mode. The calculations refer to ST = −9 CAD ATDC and are presented in Figs. 9 and 10.

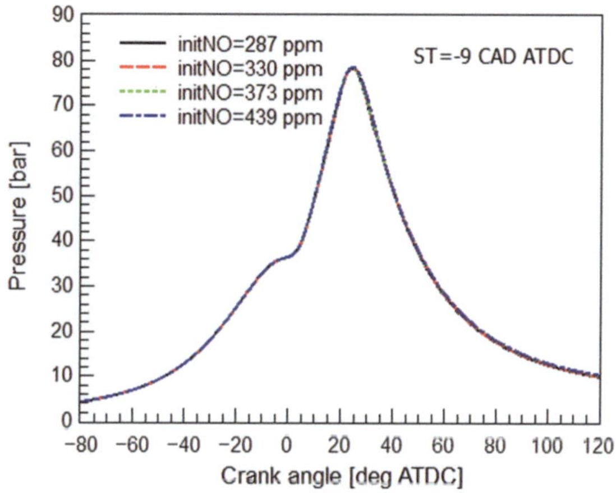

Fig. 9. Simulated pressure histories for different initial NO content in the in-cylinder mixture and at constant ST = −9 CAD ATDC for OP2

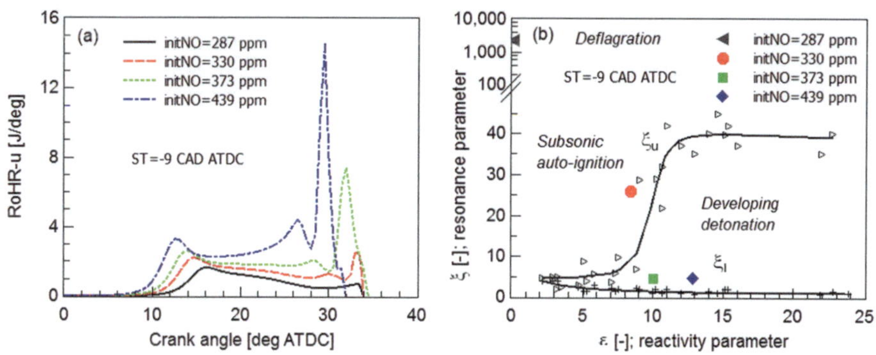

Fig. 10. Computed RoHR-u histories (a) and the detonation peninsula [28] along with the computed ε-ξ data (b) for varied initial NO content in the mixture at ST = −9 CAD ATDC for OP2

Varying only the initial NO content in the mixture between 287 ppm and 439 ppm has no impact on the combustion rate. The computed in-cylinder pressure histories overlap each other. A contrary situation is observed with respect to knocking tendency that is gained from the analysis of RoHR-u and the ξ-ε solution in Fig. 10.

The rate of heat release in the unburned zone is very sensitive to the changes of NO in the rest-gas. Overall, by increasing NO, the reactivity of the mixture is enhanced that is revealed by significantly higher maximum values of the second peak of RoHR-u histories. Also the timing of the second peak of RoHR-u moves to earlier crank angle positions as the NO content increases, though the spark timing is unchanged. Furthermore, we also see the sensitivity of the first peak in the RoHR-u history. This is on the contrary to the impact of ST variation for a given constant NO content that is discussed in the previous section.

For the operating point investigated and for NO = 287 ppm there is only one peak recognized in the RoHR-u history that is attributed to low temperature chemistry. The lack of the second peak corresponds to very high ξ value (Fig. 10b) that for this combination of NO and ST indicates deflagration. Increasing NO to 330 ppm gives a rise the second peak in the RoHR-u. As a matter of fact, NO = 330 ppm and ST = −9 CAD ATDC correspond to the operation at knock limit that is discussed in Sect. 5.2. Further increasing of the NO content to 373 ppm and then to 439 ppm, result in high values of the second peak in the RoHR-u histories (Fig. 10a). These results are reflected in entering to the regime of developing detonation as it is shown in Fig. 10(b). Here, very small ξ and rather high ε values for these NO concentrations indicate knock of high severity.

5.5 Reduction of NO in the External EGR and ST Advancing

In the previous two sections we discuss separately from each other the impact of spark timing and the impact of reducing NO in the rest-gas on the combustion progress and knocking tendency. In order to avoid knocking combustion, spark timing must be delayed from its optimum position that results in less efficient combustion. The negative impact of the delayed spark timing can be compensated by reducing the NO content in the rest-gas that suppresses the end-gas auto-ignition and prevents a transition to knocking combustion.

The calculations in this section refer to OP2 that is also used for the analysis in previous sections. In the baseline configuration, which corresponds to the SRM calibrated to the experimental data, the ST is −9 CAD ATDC and the initial and total content of NO inside the cylinder is 330 ppm. This value corresponds to 12% of the rest-gas from which 7.9% is attributed to the external EGR. Such a distribution of the rest-gas was obtained during the preliminary analysis involving numerical simulations and measurements. Here, it is assumed that only NO that is contained in the external EGR can be reduced that corresponds to NO = 217 ppm. Overall, this can be understood as a cleaning of the recirculated gases as it has been recently shown in [16].

Starting from the baseline configuration, i.e. no reduction of NO in the external EGR and ST = −9 CAD ATDC, calculations are carried out for NO being reduced by 10%, 30%, 50%, 80% and 100%. This gives respectively, 330 ppm, 287 ppm, 221 ppm, 178 ppm, 134 ppm and 113 ppm of the total initial NO content in the mixture. For each content of NO, the ST is iteratively calibrated until the position is found that leads to the operation of the engine at knock limit. The knock limit is determined according to the method presented in Sect. 5.2. The results of the analysis for each pair of NO and ST are presented in Fig. 11 with distinguishing between the engine operation at the knock limit (a, b) and just after it (c, d).

The results in Fig. 11 are qualitatively similar to those discussed in Sect. 5.2 – the advancing of ST by just 0.5 CAD causes a transition from a regular combustion mode to the mode of developing detonation. We note also a high sensitivity of the model for (1) NO = 134 ppm and ST = −17.5 CAD ATDC and for (2) NO = 178 ppm and ST = −16.5 CAD ATDC. For these two configurations and based on the ε-ξ solutions in Fig. 11, the advancing of ST by 0.5 CAD causes a direct transition from the deflagration mode (Fig. 11a) to the detonation mode (Fig. 11b) and no solution is found in the subsonic regime. Furthermore, a shorter RoHR-u duration is recognized as the NO content decreases and ST is advanced (Fig. 11a, c).

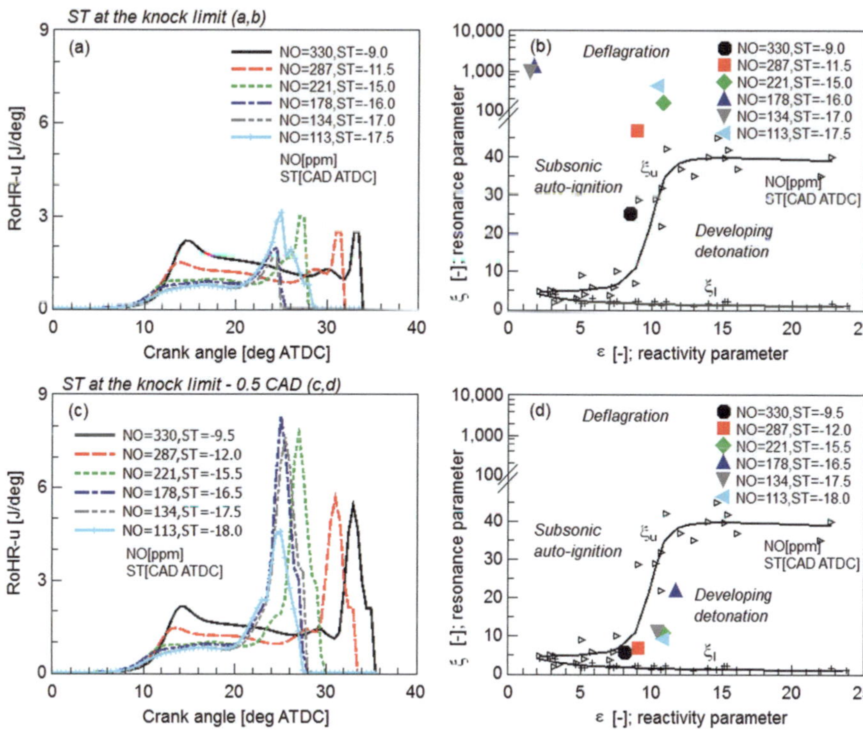

Fig. 11. Computed RoHR-u (a, c) and ε-ξ solutions (b, d) in the detonation diagram from [28] for different pairs of initial NO content and ST positions at the knock limit (top raw) and just after passing into the detonation mode (bottom raw) due to the advancing of ST by 0.5 CAD

With respect to the combustion process (Fig. 12), the strongest impact of NO reduction for a given ST position is observed between 178 ppm and 330 ppm of the total NO content in the initial mixture. This corresponds to the 70% reduction of NO in the external EGR from 217 ppm to 65 ppm; the content of NO associated with the internal rest-gas (4%) remains constant. The impact of NO reduction is revealed by fast

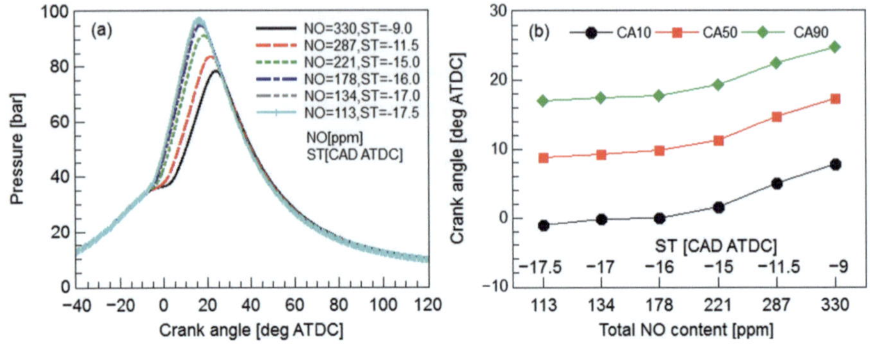

Fig. 12. The impact of NO reduction in the rest-gas and ST advancing on the in-cylinder pressure (a) and combustion progress (b) for the engine operated at OP2 and knock limit

advancing of CA10, CA50 and CA90 as well as higher maximum values of the in-cylinder pressure. Further reduction of NO does not have such a strong impact. The in-cylinder pressure histories for NO = 178 ppm, NO = 134 ppm and for 113 ppm are close to each other. The same tendency is observed for CA10, CA50 and CA90.

The results presented in Fig. 12 find reflection in the results plotted in Fig. 13 that contains the predicted IMEP, indicated specific fuel consumption (ISFC) and knock margin improvement.

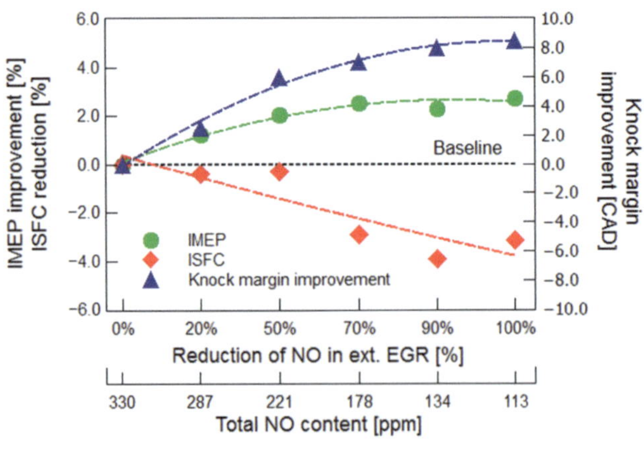

Fig. 13. The impact of NO reduction in the initial in-cylinder mixture on improving knock margin and fuel economy for OP2 (n = 4750 rpm, IMEP = 17 bar)

The dotted line in Fig. 13 represents the baseline solution that was obtained for ST = −9 CAD ATDC and NO = 330 ppm. The performance results in Fig. 13, such as IMEP, ISFC and knock margin improvement, are presented versus the total content of NO in the in-cylinder mixture and versus the percentage reduction of NO from the external EGR. For the operating point investigated, the complete reduction of NO in the external EGR reduces the total NO content in the rest-gas by 66%. This reduction enables to advance the ST from the baseline value (ST = −9 CAD ATDC) to ST = −17.5 CAD ATDC and without the risk of knock occurrence (see Fig. 11a, b). The knock margin, which is calculated as a difference between the current ST and the baseline ST = −9 CAD ATDC, is improved by 8.5 CAD and results in 3.2% reduction of ISFC. Similarly as in the results in Fig. 12, here the strongest impact of lowering NO on the knock margin improvement is observed up to 70% reduction of NO in the external EGR.

The results in Fig. 13 are an outcome of the numerical analysis that has not been verified directly by measurements. Therefore, it is very important to note here that the computed improvement of knock margin by 8.5 CAD would be very difficult, if not possible at all, to realize in practical applications. It would require a very precise control of the gas composition as well as very precise control and determination of knock-limited operation. As a consequence of this, the computed IMEP and ISFC are also expected to be lower in the actual engine operation than these here predicted. Nevertheless, an overall tendency of the presented computations looks sound. Confirmation of it is given by the existing experimental works. In [16], it is presented that the complete reduction of NO_x from the external EGR moves the combustion center towards an earlier position by 2 CAD. In [17], a complete reduction of NO_x from the initial mixture resulted in 6 CAD improvement of knock margin for 20% EGR and 3 CAD improvements for 10% EGR. These values correspond to 4.5% and 2.5% reduction of fuel consumption, respectively. The results in this article represent the same order of magnitude and therefore are considered plausible. As a consequence, the overall simulation methodology is also sound.

6 Summary

This paper reports a computational study about knocking tendency for an engine operated at high speed and load and with predefined EGR composition. Specifically, the impact of lowering the NO content in the initial in-cylinder mixture on knock onset along with the impact of advancing the spark timing to improve fuel economy are discussed. The investigations refer to the engine operated at n = 4750 rpm and IMEP = 17 bar. They are carried out using the SRM of engine in-cylinder processes that employs a detailed reaction kinetic model for the analysis of the end-gas auto-ignition process.

A combination of the SRM solution along with the evaluation of auto-ignition in the unburned zone and based on the detonation diagram provides a modeling framework that enables an efficient analysis of knocking tendency in SI engines. Not only the occurrence of auto-ignition is determined but also its severity that in consequence may give rise to the development of knock. Similarly as in our previous work [24], here it is also confirmed that this methodology can be used to determine the knock-limited engine operating conditions.

For the conditions investigated in this work, a comparison between the RoHR-u histories and the ξ-ε solutions on the detonation diagrams indicate that the knock-limited conditions correspond usually to the occurrence of two peaks in the RoHR-u that are of the same height. Thus, the RoHR-u results can be used for very first and rough estimation of the end-gas conditions with respect to knock occurrence. However, the assessment whether the auto-ignition may or may not give rise to knock can only be gained from the ξ-ε solution.

Similarly as found in experimental works (see, e.g. [16, 17]) here it is also proved that the presence of NO in the mixture has a strong impact on knock onset. Specifically, by lowering the NO content the knock margin can be improved. For the conditions investigated in this work, a complete reduction of NO in the external EGR enabled the advancing of spark timing by 8.5 CAD with reference to the baseline position and without violating knock limit. The improvement of the knock margin results in 3.2% reduction of ISFC. The strongest impact of lowering NO on the knock margin improvement and ISFC reduction is observed up to 70% reduction of NO in the external EGR.

Although the obtained improvement of knock margin and ISFC have not been confirmed by measurements yet, they agree qualitatively to existing publications. Therefore, they are considered plausible and indicate the potential of cleaned EGR to improve fuel economy of gasoline engines. Furthermore, also the overall simulation methodology presented in this work is sound.

Notation

ATDC	after top dead center
CA/CAD	crank angle/crank angle degree
CA10,50,90	crank angle of 10%, 50% and 90% of the cumulative heat release
CFD	computational fluid dynamics
EGR	exhaust gas recirculation
EVO	exhaust valve opening
IMEP	indicated mean effective pressure
ISFC	indicated specific fuel consumption
IVC	intake valve closure
OP	operating point
PS	pressure signal in knock detection using KIS4 from IAV
RoHR	rate of heat release
RoHR-u	rate of heat release in the unburned zone
SI	spark ignition
ST	spark timing
$C\phi$	mixing time constant
l_I	integral length scale of the flow
n	engine speed
u'	velocity fluctuation
S_L	laminar flame speed
S_T	turbulent flame speed
ε	reactivity parameter

τ	scalar mixing time
τ_i	ignition delay time
τ_e	excitation time
τ_t	turbulent mixing time
ξ	resonance parameter

References

1. Heywood, J.B.: Internal Combustion Engine Fundamentals. McGraw-Hill, New York (1988)
2. Zhen, X., Wang, Y., Xu, S., Zhu, Y., Tao, C., Xu, T., Song, M.: The engine knock analysis – an overview. Appl. Energ. **92**, 628–636 (2012)
3. Rechs, M.: Untersuchungen von Zylinderdruck- und Motorstrukturschwingungen zur Auslegung von Antiklopf-Regelsystemen, Dissertation, RWTH Aachen (1990)
4. König, G.: Auto-ignition and Knock Aerodynamics in Engine Combustion, Dissertation, Leeds University (1993)
5. Chen, L., Li, T., Yin, T., Zheng, B.: A predictive model for knock onset in spark-ignition engines with cooled EGR. Energy Convers. Manage. **87**, 946–955 (2014)
6. Wang, Z., Liu, H., Reitz, R.D.: Knocking combustion in spark-ignition engines. Prog. Energ. Combust. **61**, 78–112 (2017)
7. Merola, S.S., Vaglieco, B.M.: Knock investigation by flame and radical species detection in spark ignition engine for different fuels. Energy Convers. Manage. **48**(11), 2897–2910 (2007)
8. Bozza, F., De Bellis, V., Teodosio, L.: Potentials of cooled EGR and water injection for knock resistance and fuel consumption improvements of gasoline engines. Appl. Energ. **169**, 112–125 (2016)
9. Netzer, C., Seidel, L., Pasternak, M., Mauss, F., Lehtiniemi, H., Ravet, F.: 3D CFD engine knock prediction and evaluation based on detailed chemistry and detonation theory. Ber. Energie Verfahrenstechnik (BEV) **17**(1), 185–196 (2017). ISBN 978-3-945806-08-1
10. Alger, T., Mangold, B.: Dedicated EGR: a new concept in high efficiency engines. SAE Int. J. Engines **2**(1), 620–631 (2009)
11. Prabhu, S.K., Li, H., Miller, D.L., Cernansky, N.P.: The effect of nitric oxide on autoignition of a primary reference fuel blend in a motored engine. SAE Technical Paper 932757 (1993)
12. Amano, T., Dryer, F.L.: Effect of dimethyl ether, NOx and ethane on CH4 oxidation: high pressure, intermediate temperature experiments and modeling. Proc. Comb. Inst. **27**, 397–404 (1998)
13. Kawabata, Y., Sakonji, T., Amano, T.: The effect of NOx on knock in spark-ignition engines. SAE Technical Paper 1999-01-0572 (1999)
14. Stenlåås, O., Gogan, A., Egnell, R., Sundén, B., Mauss F.: The influence of nitric oxide on the occurrence of autoignition in the end gas of spark ignition engines. SAE Technical Paper 2002-01-2699 (2002)
15. Hoffmeyer, H., Montefrancesco, E., Beck, L., Willand, J., Ziebert, F., Mauss, F.: CARE – catalytic reformated exhaust gases in turbocharged DISI-engines. SAE Int. J. Fuels Lubr. **2** (1), 139–148 (2009)
16. Fischer, M., Kreutziger, P., Sun, Y., Kotrba, A.: Clean EGR for gasoline engines – innovative approach to efficiency improvement and emissions reduction simultaneously. SAE Technical Paper 2017-01-0683 (2017)

17. Takaki, D., Tsuchida, H., Kobara, T., Akagi, M. et al.: Study of an EGR system for downsizing turbocharged gasoline engine to improve fuel economy. SAE Technical Paper 2014-01-1199 (2014)
18. Poschl, M., Sattelmayer, T.: Influence of temperature inhomogeneities on knocking combustion. Combust Flame **153**, 562–573 (2008)
19. LOGE AB. LOGEfuel gasoline v1.04, LOGEsoft v1.0, LOGEengine v1.0, software manuals (2017). http://www.logesoft.com
20. Pasternak, M., Mauss, F., Xavier, F., Riess, M., Sens, M., Benz, A.: 0D/3D simulations of combustion in gasoline engines operated with multi spark plug technology. SAE Technical Paper 2015-01-1243 (2015)
21. Pasternak, M., Mauss, F.: Sens., M., Riess, M., Benz, A., Stapf, K.G.: Gasoline engine simulations using zero-dimensional spark ignition stochastic reactor model and three-dimensional computational fluid dynamics engine model. Int. J. Engine Res. **17**(1), 76–85 (2016)
22. Peters, N.: Turbulent Combustion. Cambridge University Press, Cambridge (2000)
23. Pope, S.: Pdf methods for turbulent reactive flows. Prog. Energ. Combust **11**(2), 119–192 (1985)
24. Netzer, C., Seidel, L., Pasternak, M., Klauer, C., et al.: Engine knock prediction and evaluation based on detonation theory using a quasi-dimensional stochastic reactor model. SAE Technical Paper 2017-01-0538 (2017). https://doi.org/10.4271/2017-01-0538
25. Fischer, M., Günther, M., Röpke, K., Lindemann, M., Placzek, R.: Knock Detection in Spark-Ignition Engines. MTZ Worldwide 3/2003, vol. 64 (2003)
26. Stahr, A., Langfritz, P., Günther, M., Kratzsch, M.: The DELTA knocking control – the necessary paradigm shift for engines with high power density. In: Proceedings 3rd Conference on “SI Engine Knock – Irregular Combustion”, Berlin (2013)
27. Pilling, M.J.: Low-temperature combustion and autoignition. In: Hancock, G., Compton, R. G. (eds.) Comprehensive Chemical Kinetics, vol. 35, pp. 1–794. Elsevier, Amsterdam (1997)
28. Bradley, D., Morley, C., Gu, X., Emerson, D.: Amplified pressure waves during autoignition: relevance to CAI engines. SAE Technical Paper 2002-01-2868 (2002)
29. Zeldovich, Y.: Regime classification of an exothermic reaction with nonuniform initial conditions. Combust. Flame **39**, 211–214 (1980)
30. Bradley, D., Kalghatgi, G.T.: Influence of autoignition delay time characteristics of different fuels on pressure waves and knock in reciprocating engines. Combust. Flame **156**, 2307–2318 (2009)
31. Gu, X.J., Emerson, D.R., Bradley, D.: Modes of reaction front propagation from hot spots. Combust. Flame **133**, 63–74 (2003)
32. Kalghatgi, G.T., Bradley, D.: Pre-ignition and ‘super-knock’ in turbocharged spark-ignition engines. SAE Int. J. Engines **13**(4), 399–414 (2012)
33. Peters, N., Kerschgens, B., Paczko, G.: Super-knock prediction using a refined theory of turbulence. SAE Int. J. Engines **6**(2), 953–967 (2013)
34. Pan, J., Shu, G., Wei, H.: Interaction of flame propagation and pressure waves during knocking combustion in spark-ignition engines. Combust. Sci. Techn. **186**(2), 192–209 (2014)
35. Peters, N., Kerschgens, B., Jochim, B., Paczko, G.: Mega knock in super-charged gasoline engines interpreted as a localized developing detonation. In: Kratzsch, M., Guenther, M. (eds.) Knocking in Gasoline Engines. IAV Automotive Engineering, Berlin (2013)
36. Bates, L., Bradley, D., Paczko, G., Peters, N.: Engine hotpots: modes of auto-ignition and reaction propagation. Combust. Flame **166**, 80–85 (2016)

Development of a Model for Predicting the Knock Boundary in Consideration of Cooled Exhaust Gas Recirculation at Full Load

Alexander Fandakov[1(✉)], Michael Grill[5], Michael Bargende[1], Max Mally[2], Marco Günther[2], Stefan Pischinger[2], Liming Cai[3], Heinz Pitsch[3], Karl Alexander Heufer[4], Ajoy Ramalingam[4], Heiko Minwegen[4], and André Casal Kulzer[6]

[1] Institute of Internal Combustion Engines and Automotive Engineering, University of Stuttgart, Stuttgart, Germany
alexander.fandakov@ivk.uni-stuttgart.de
[2] Institute for Combustion Engines, RWTH Aachen University, Aachen, Germany
[3] Institute for Technical Combustion, RWTH Aachen University, Aachen, Germany
[4] Physico-Chemical Fundamentals of Combustion, RWTH Aachen University, Aachen, Germany
[5] FKFS – Research Institute of Automotive Engineering and Vehicle Engines Stuttgart, Stuttgart, Germany
[6] Porsche AG, Weissach, Germany

Abstract. Engine knock limits the efficiency of turbocharged SI engines at high loads. The occurrence of this phenomenon can be inhibited by deploying recirculation of cooled exhaust gas (EGR) at full load. However, the development of full load EGR combustion systems cannot be per-formed in the 0D/1D engine simulation, as no meaningful models for the reliable prediction of the knock limit under the influence of EGR exist.

Measurements of ignition delay times in a shock tube and a rapid compression machine under the influence of exhaust gas have been carried out. The addition of 25% EGR prolonged the ignition delay time up to 100%. A detailed reaction mechanism for gasoline surrogates was defined and validated against the measurement results. Furthermore, the effects of EGR on combustion, knock behavior and emissions were investigated on a single-cylinder re-search engine. The center of combustion could be advanced by up to 9° CA with the addition of EGR leading to a four percenter higher indicated efficiency. The influence of catalytically treated exhaust gas was examined as well. At high EGR rates of 25% catalytically treated exhaust gas allowed a 2° CA earlier center of combustion. Furthermore, the influence of nitric oxide on the knocking tendency was investigated. It has been found that a total cylinder NO concentration of about 100 ppm leads to the highest knocking tendency. At NO concentrations below 40 ppm NO the knocking tendency was decreased. Higher concentration than 100 ppm of NO in the cylinder decreased knocking tendency as well.

M. Günther and M. Sens (eds.), *Knocking in Gasoline Engines*,
https://doi.org/10.1007/978-3-319-69760-4_9

The pressure trace analysis of the measured single working cycles shows that the pre-reaction state of the unburned mixture at knock onset calculated with commonly used knock models decreases with rising EGR rate and engine speed, although by definition it must be constant at the time of auto-ignition. Consequently, reaction kinetics simulations at in-cylinder conditions proved that, under specific boundary conditions, the auto-ignition of the unburnt mixture resulting in knock happens in two stages. In this case, low-temperature ignition occurs in the unburnt mixture while the combustion is taking place. This phenomenon significantly influences the ignition delay of the mixture, which severely impairs the prediction capabilities of commonly used knock models.

Based on these findings, a new knock modeling approach capable of predicting the low-temperature ignition occurrence as well as reproducing its influence on the mixture's auto-ignition was developed. The results from 3D-CFD simulations accompanying the model development supported all model assumptions made. The developed knock model was successfully validated against measurement data at various boundary conditions, such as different inlet temperatures and mixture compositions as well as EGR rate and engine speed variations. It can predict the knock limit very accurately with errors in center of combustion below 2° CA and thus contributes to an efficient development process of full load EGR combustion systems in the 0D/1D engine simulation.

1 Introduction

The knock limitation at higher loads is one of the major challenges for increasing the efficiency of gasoline engines. For low part load fuel consumption a high compression ratio is desirable but a compromise with full load efficiency must be made. Another challenge is the component protection at higher engine loads and speeds due to exhaust gas temperature limitations. Currently enrichment is used but the upcoming "Real Driving Emissions" legislation will not permit this strategy anymore. A suitable countermeasure is the application of high load exhaust gas recirculation to overcome both challenges.

Yet, the development of a high load EGR combustion system is complex. The engine needs a specific layout of the charging system, the charge air and EGR cooling system and various other parameters, like compression ratio and valve timing. Due to the complexity of the layout phase only 0D/1D engine simulation can be applied. Thanks to the high prediction quality of the models and the low computational times, this is a powerful tool used to reduce development costs by partially eliminating the need for cost-intensive test bench investigations. However, the existing models for predicting knocking combustion in 0D/1D SI engine simulations are known for their poor performance and the great effort needed for their calibration. Hence, the reliable prediction of knock occurrence is a challenging task, especially if the entire engine map has to be considered. This fact results in significant restrictions on the development of full load EGR combustion systems with the help of 0D/1D simulations. Therefore, except the fundamental understanding of the effects of EGR, the goal of this work is the development of a reliable knock model for the application in EGR combustion system development.

2 Fundamental Combustion Kinetic Investigation

The development of the chemical kinetic mechanism for the combustion of gasoline fuels and their surrogates is described in this section. Due to the limited ignition delay time data for model validation, especially at low temperature conditions with EGR, fundamental kinetic measurements of ignition delay times were performed in a rapid compression machine and a shock tube for an ethanol-doped research grade gasoline (RON95E10). The obtained data covers a wide range of conditions including initial temperatures, equivalence ratios, pressures, and EGR ratios. The model was derived based on the state-to-art kinetic knowledge as well as advanced mathematical methods and was validated successfully against the data obtained as part of this study and those available in the literature.

2.1 Experimental Investigation

Rapid compression machine
The ignition delay times in the low-to-intermediate temperature regimes were measured in a rapid compression machine (RCM). A complete description of the facility can be found in Ref. [Lee12]. Briefly, the reaction chamber of the RCM, the mixing vessels along with the tubing's are electrically heated in order to accommodate the high boiling components of the provided gasoline. The initial temperature is monitored by type 'T' thermocouples mounted on the reaction chamber. The mixtures were prepared with the aid of two static pressure sensors which monitor the partial pressures required for the mixtures. The dynamic pressure changes during a reactive experiment were recorded with the help of PCB 113B24 sensor. The Kistler 6125C pressure sensor was used during a non-reactive experiment in order to account for the thermal heat shock resistance. The compressed temperature was calculated using the adiabatic compression and expansion routine of Gaseq [Mor05]. The experimental uncertainty on compressed temperature using error propagation analysis was estimated to be ± 5 K and the observed variation in ignition delay time is within 10%. In the RCM, heat loss and radical pool generation during compression phase are the two non-idealities that affects the reactivity behavior at longer and shorter ignition delay times, respectively, and it is essential to include these effects during the zero-dimensional simulations by using non-reactive pressure profiles [Sun14].

Shock tube
The ignition delay times in the high temperature regime were measured in a shock tube (ST) facility. Its detailed description is available in Ref. [Zha16]. The complete driven section of the ST, mixing vessel, and the connecting tubes are electrically heated and monitored by type 'T' thermocouples. The partial pressures of the mixture preparation process were monitored with two static pressure sensors. The dynamic pressure changes in the driven section of the ST were recorded with 5 PCB 113B22 pressure sensors which are used for the determination of shock velocity. The reflected shock conditions were calculated using an in-house code based on the shock and detonation toolbox [She14] routines implemented for Cantera [Goo15]. The uncertainty in the

reflected shock temperature and pressure is estimated to be 1.1% and 3.5%, respectively, while that of the ignition delay time is 20%. At longer ignition delay times, the effect of boundary layers constitutes to a constant rate of pressure rise, thereby affecting the reactivity behavior. This non-ideal behavior can be included during the simulations by assuming an average pressure rise of 8%/ms.

Measurement of ignition delay times
The ignition delay times for a fully characterized research grade gasoline (RON95E10) were measured at the conditions defined in Table 1. The composition of exhaust gas was specified with respect to lean and rich conditions. As mentioned above, the experimental work aids in providing a comprehensive data set for the validation of chemical models.

Table 1. Measurement of ignition delay times in the RCM and ST

Pressure/bar	Rel. AFR/1	EGR rate/%	Exhaust gas composition/mass %				
			N_2	CO_2	CO	H_2O	H_2
20–40	0.85	0	–	–	–	–	–
20–40	1.30	0	–	–	–	–	–
20–40	0.85	25	71.21	14.15	5.16	9.35	0.11
20–40	1.30	25	72.29	18.87	–	8.83	–

2.2 Fuel Properties and Surrogate Formulation

The gasoline, RON95E10, was supplied by Shell global solutions GmbH. Its key properties are summarized in Table 2. In addition, its temperature-dependent specific heats and distillation curve were also characterized for the calculation of compressed

Table 2. Properties of RON95E10 and the proposed surrogate mixture.

Property	RON95E10	Surrogate[a]	Surrogate[b]	Surrogate[c]
RON/-	96.5	95.0	91.3	96.2
MON/-	85.2	86.6	87.2	88.7
RON-MON/-	11.3	8.4	4.1	7.5
H/C ratio/-	1.94	1.93	1.93	
Density/kg/l	0.7417	0.7443	0.7443	0.7493
Composition/mass%	10.3% Ethanol	13.6% n-Heptane 45.7% iso-Octane 30.5% Toluene 10.3% Ethanol		

[a]RON and MON are calculated using the proposed blending rules.
[b]RON and MON are determined based on the volumetric-blending rule.
[c]Properties are measured.

mixture temperatures in RCM measurements and for the specification of initial temperatures in the reactor and mixing vessel, respectively.

Since real petroleum fuels are composed of a large variety of hydrocarbon components, surrogate mixtures of only few representative species are typically employed in computational research and in engine development. For ethanol-doped gasoline fuels, mixtures of n-heptane, iso-octane, toluene, and ethanol are commonly used as the surrogate components and their composition is optimized to mimic target properties of real fuels. In general, the properties, research octane number (RON), motor octane number (MON), H/C ratio, and liquid density, are of interest due to their importance. To calculate the H/C ratio and liquid density of surrogate mixtures, mole- and volume fraction-based blending rules are used, respectively. For the determination of RON and MON, volumetric blending approach is applied in general. However, recent studies [Mor10,And10] reveal that this approach reproduces the RON and MON of surrogate fuels with large uncertainties. Therefore, a revised blending rule for RON and MON of ethanol-doped gasolines is proposed here. For the base gasoline fuel, which is emulated with the mixtures of n-heptane, iso-octane, and toluene, non-linear volumetric blending approach [Mor10] is applied to determine their octane numbers. Mole fraction-based blending rule is then employed to determine the RON and MON for the mixtures of the base gasoline fuel and ethanol, following the suggestion of Andersen et al. [And10]. The MON and RON can be calculated as follows:

$$\begin{aligned} RON = (1 - X_{Eth})\Bigg(100\frac{X_{Iso}}{X_{Iso}+X_{Hep}} &+ \\ 142.79\frac{X_{Tol}}{X_{Iso}+X_{Hep}+X_{Tol}} - 22.65\Big(&\frac{X_{Tol}}{X_{Iso}+X_{Hep}+X_{Tol}}\Big)^2 \\ - 111.95\frac{X_{Iso}}{X_{Iso}+X_{Hep}}&\frac{X_{Tol}}{X_{Iso}+X_{Hep}+X_{Tol}}\Bigg) + X_{Eth}RON_{Eth} \end{aligned} \tag{1}$$

$$\begin{aligned} MON = (1 - X_{Eth})\Bigg(100\frac{X_{Iso}}{X_{Iso}+X_{Hep}} &+ \\ 128\frac{X_{Tol}}{X_{Iso}+X_{Hep}+X_{Tol}} - 19.21\Big(&\frac{X_{Tol}}{X_{Iso}+X_{Hep}+X_{Tol}}\Big)^2 \\ - 119.24\frac{X_{Iso}}{X_{Iso}+X_{Hep}}&\frac{X_{Tol}}{X_{Iso}+X_{Hep}+X_{Tol}}\Bigg) + X_{Eth}MON_{Eth} \end{aligned} \tag{2}$$

with X_{Iso}, X_{Hep}, X_{Tol}, and X_{Eth} indicating the mole fraction of iso-octane, n-heptane, toluene and ethanol in the surrogate mixtures, respectively. RON_{Eth} and MON_{Eth} are the octane numbers of ethanol.

The surrogate composition is optimized numerically by minimizing the property differences between real and surrogate fuels. The compositions and the calculated surrogate properties are presented in Table 2, in comparison with those of the RON95E10. As the validation of the suggested blending rule, the key properties of the proposed surrogates were measured and shown in Table 2, together with the RON and MON calculated with the conventional volumetric-blending rules. It can be seen that application of the revised blending rules leads to improved prediction accuracy of the RON and octane sensitivity (RON-MON).

2.3 Development of a Kinetic Mechanism for Gasoline Surrogate Fuels

The chemical mechanism was developed based on the published mechanism for gasoline surrogate fuels from Cai et al. [Cai16]. The reference mechanism [Cai16] includes the oxidation chemistry of various C_0–C_8 hydrocarbon species and substituted aromatic species, including n-heptane, iso-octane, toluene, and ethanol, allowing the formulation of multi-component surrogate fuels with capability of predicting Polycyclic Aromatic Hydrocarbon (PAH) formation in gasoline engines. Recent studies [Cai16_2] suggested the modification of oxidation chemistry of alkanes in terms of thermochemistry, reaction rates, and pathways. The aim is to avoid the error compensation in chemical models and to improve the model prediction ability of ignition delay times, which is of particular importance in this work for the prediction of knocking behavior. Therefore, the sub-models for n-heptane and iso-octane in the reference mechanism [Cai16] are revised according to the state-to-art kinetic knowledge.

The combustion chemistry of n-heptane was extracted from the recently published mechanism [Cai16_2] of normal alkanes, which includes the latest combustion kinetic knowledge in terms of thermochemistry, rate rules, and reaction pathways. Following the model development concept introduced in Ref. [Cai16_2], a revised iso-octane mechanism was proposed. The detailed mechanism for iso-octane from Curran et al. [Cur02] was first updated by incorporating recalculated thermochemical data of species involved in the oxidation of iso-octane using the group additivity method [Ben76] with the revised group values [Bur15]. Subsequently, the alternative reaction pathways of peroxy hydroperoxide species were included in the mechanism. Following this, the rate rules for the primary and secondary carbon sites were updated by using the rate rules recommended in Ref. [Cai16_2]. The rate rules for the tertiary sites were calibrated for good model performance using an uncertainty quantification framework [Cai16_2]. In order to reduce the computational cost, the detailed mechanisms for n-heptane and iso-octane were reduced to a skeletal level using a multi-stage reduction strategy developed by Pepiot-Desjardins and Pitsch [Pep07]. The reduced models were further built as additional modules upon the reference model [Cai16]. The mechanism consists of 489 chemical species and 3369 elementary reactions.

2.4 Model Validation

The measured ignition delay times of RON95E10 are shown in Fig. 1., compared to those calculated. Note that, facility effects of both setups are taken into account in the simulation. By means of the RCM and ST, the measured ignition data covers a very wide range of initial temperature, extending the validation database available in the literature. Reasonable agreement is observed between data and simulation results. More importantly, the model reflects the influence of pressure, equivalence ratio and EGR ratio very well, as illustrated in Fig. 2. The model was also validated successfully against the data available in the literature, such as ignition delay times and burning velocities. For the sake of brevity, these cases are not shown here.

This model will be coupled with 3D-CFD simulations in the following to study the knock behavior of gasoline with EGR at full load operating conditions reduction.

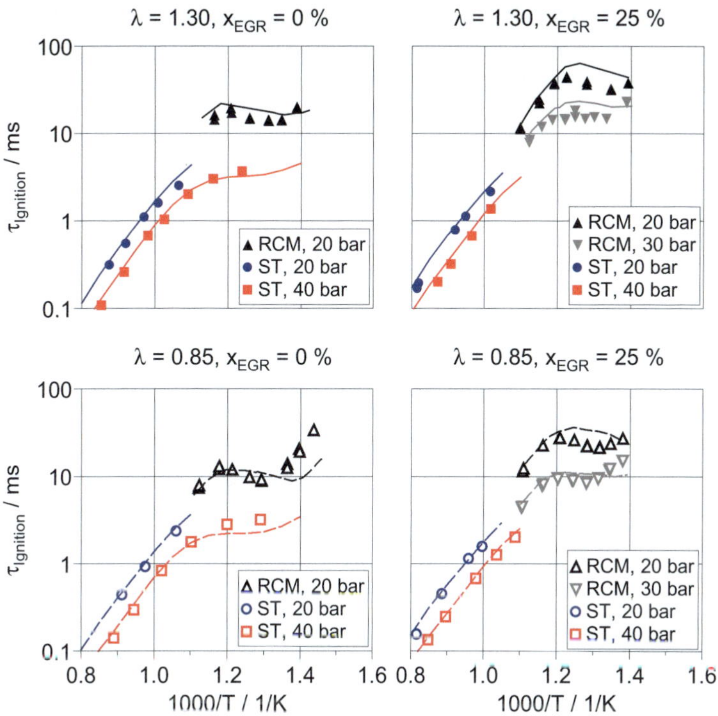

Fig. 1. Ignition delay times of RON95E10. Symbols and lines denote the data and the numerical results, respectively.

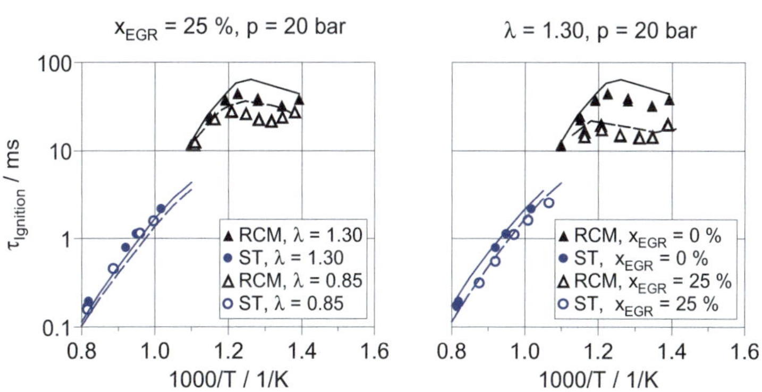

Fig. 2. Ignition delay times of RON95E10. Symbols and lines denote the data and the numerical results, respectively.

In order to reduce the CFD computational effort, a reduced model with 327 species and 1730 reactions was derived using the method of isomer lumping introduced in Ref. [Pep07]. Marginal difference (<1%) is observed between the reduced model and the detailed one with 428 species at conditions of interest.

3 Thermodynamic Testing

3.1 Experimental Setup

The experimental investigations were performed on a homogeneously operated direct injection spark ignition single cylinder research engine. The engine features external boosting, low pressure exhaust gas recirculation and a tumble generation device. The tumble generation device has been used at 50% actuation during the experimental investigations. Information about the level of charge motion and the flow performance can be found in [Jak11]. Further technical data of the engine is summarized in Table 3.

The layout of the cylinder head and combustion chamber dome is shown in Fig. 3. The engine features a central injector, which is located between the intake valves. A spark plug with a heat value of 8 according to NGK is mounted between the exhaust valves.

In order to perform testing under realistic boundary conditions the engine is equipped with an exhaust throttle to simulate the back pressure of the turbine. During boosted operation the average exhaust back pressure is controlled to be equal to the intake pressure. The extraction point of the low pressure EGR system is downstream of the throttle. The EGR inlet is located upstream of the external charging system (three-stage roots blower). In order to avoid condensation of water in the EGR cooler and the intake system the outlet temperature of the EGR cooler was set to 70 °C.

Table 3. Engine specifications

Engine specifications					
Stroke (s)	90.5	mm	Valve train	DOHC (4 V)	
Bore (D)	75	mm	Exhaust valve opening*	13 (bBDC)	° CA
s/D	1.207	1	Exhaust valve closing*	10 (bTDC)	° CA
Engine displacement	399	cm^3	Intake valve opening*	15 (aTDC)	° CA
Connection rod length	152	mm	Intake valve closing*	30 (aBDC)	° CA
Piston-pin offset (thrust side)	0.5	mm	Injection system	Bosch HDEV 5.2 DI solenoid multi-hole	
Compression ratio	10.9	1	Fuel pressure	200	bar

*: Referred to 1 mm lift

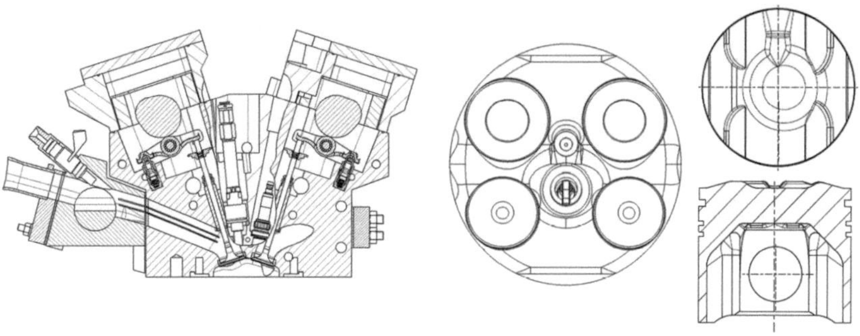

Fig. 3. Research engine layout: (a) Sectional view of cylinder head, (b) combustion chamber dome and (c) piston geometry for compression ratio of 10.9

The EGR rate is measured via CO_2 measurement in the exhaust and the intake system. Based on the measured concentrations the EGR rate can be calculated:

$$x_{EGR} = \frac{1}{1 + \frac{Y_{CO_2,Exh} - Y_{CO_2,Int}}{Y_{CO_2,Int} - Y_{CO_2,Air}} \cdot \frac{\dot{m}_{Exh}}{\dot{m}_{Air}} \cdot \frac{M_{Air}}{M_{Exh}}} \quad (3)$$

The exhaust gas composition was measured via a flame ionization detector (HC), a paramagnetic oxygen analyzer (O_2), an infrared gas analyzer (CO and CO_2), a chemiluminescence analyzer (NO_x) and a TSI EEPS™ (PN). Furthermore, the engine has been equipped with a conventional catalyst from a gasoline passenger car in order to run the engine with catalyzed exhaust gas. Due to the high volume of the catalyst and thus low space velocities a high conversion rate could be achieved. During catalyst operation the components in the exhaust gas were measured upstream and downstream of the catalyst. To investigate the influence of single reactive components of EGR on the combustion an external dosing system has been added prior to the charging system. High accuracy mass flow controllers were used to dose the single components into the intake system.

For all measurement points spark sweeps in the area of the knock limited spark advance have been carried out. The characteristic number of knocking combustion was set to be the knocking peak to peak (KPP) value of the high-pass filtered pressure trace. The limit for KPP was defined as engine speed divided by 1000 in bar. Since knocking is a statistic phenomenon the knock frequency is the criterion which has been used to define the knock limited spark advance (KLSA). If 3–10% of the cycles reveal a knocking combustion, it is defined that the engine is operating at the knock limit (the spark timing closest to 3% is chosen). Based on KLSA a spark timing sweep from 1° CA retarded to 1° CA advanced spark timing was conducted. Thus the measurement data contains operating points from slight to severe knock occurrence.

An overview of the test matrix is given in Table 4. Overview of experimental test matrix (p_{mi} = 16 bar). Measurements with rich and lean exhaust gas recirculation as well as the addition of CO and H_2 to the fresh charge have been carried out as well, but are not included for the sake of brevity. The EGR rate has been increased by increments

Table 4. Overview of experimental test matrix (p_{mi} = 16 bar)

EGR type	EGR rate/%	T_{Intake}/°C	Rel. AFR/1	Engine speed/1/min	Synthetic component
Bef. Cat.	0–25	35–55	1.0	1500	–
Bef. Cat.	0–25	35–55	1.0	2500	–
Bef. Cat.	0–20	45–65	1.0	4000	–
Aft. Cat.	0–25	35–45	1.0	1500	–
Aft. Cat.	0–20	35–45	1.0	1500	NO

of 5% and the intake temperature by increments of 10 °C. The load has been fixed at 16 bar indicated mean effective pressure (pmi).

For validation datasets with a different cylinder head (tumble runner design) and an increased compression ratio of 11.84 have been generated. Besides the standard camshaft, data with a Miller camshaft (230° CA event length) have been generated as well. The load ranges from 12 to 20 bar p_{mi}. Thus, the validation database covers a wide range of the engine operation map.

3.2 Experimental Results

During baseline testing the engine was operated with non-catalyzed exhaust gas. In order to cover a wide range of thermodynamic states the intake temperature, the engine speed and the EGR rate have been varied.

A summary of the baseline results with EGR is given in Fig. 4. Only measurements at the KLSA are depicted. EGR proved to inhibit knock at all investigated points when the intake temperature is kept on a constant level. With increased EGR rate the spark timing had to be advanced in order to compensate the longer heat delay and duration. In all cases the spark timing could be advanced in that way, that an earlier center of combustion could be achieved. Subsequently the engine efficiency was increased with EGR. Another contributor to the increased efficiency is the reduced heat loss to the cylinder wall. The effect of reduced temperatures during combustion can be seen as well in the decrease of NO emissions. Due to the lower flame temperatures and thus increased flame quenching the HC emissions are increased.

For knock model development, the identification of the knock limited spark advance is of major importance. To this end, spark timing sweeps in steps of 0.5° CA were conducted. The results at 1500 1/min are depicted in Fig. 5. For the investigated cases, the number of knocking cycles is slightly increasing with advanced spark timing at first. Starting at the point of about 5% knock frequency, the number of knocking cycles steeply rises, indicating that the number of knocking cycles is following an exponential trend. For the cases with EGR, no significant differences in this behavior can be observed. The maximum peak to peak amplitude of the pressure oscillation is below 8 bar for all cases, except for the EGR rate of 20%. The point closest to 3% knocking cycles has been chosen as the KLSA. Regarding the dependence of the knock tendency on the spark timing relative to KLSA it can be concluded that the addition of EGR does not have a high impact on the knocking behavior of the engine itself.

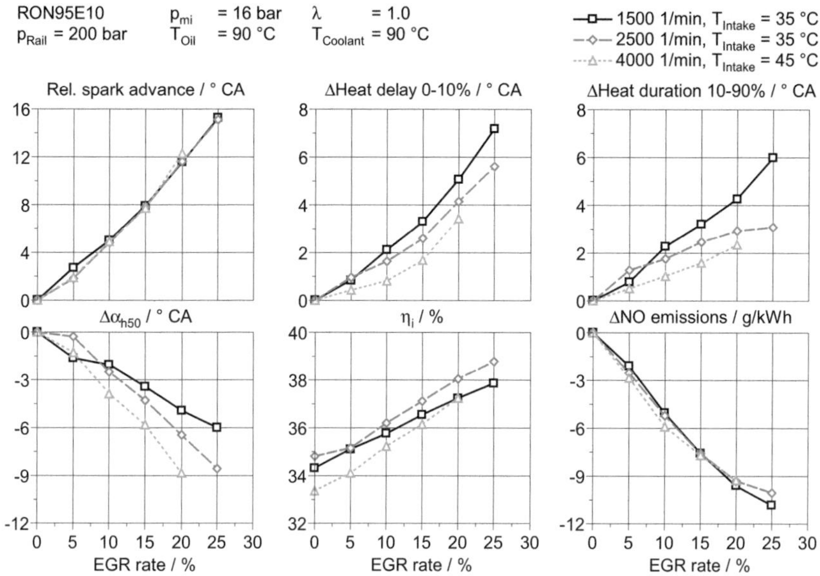

Fig. 4. Results of baseline investigation

The general trend in knock frequency was found for higher engine speeds as well. For 10% EGR, the comparison in dependence of the engine speed for knock frequency and maximum KPP are depicted in Fig. 6. The trend in knock frequency is very similar for all engine speeds. Nevertheless, the severity of the knock events is increasing with engine speed. At 2500 1/min an about 4.5 bar higher maximum KPP is observed compared to 1500 1/min. Taking the reduced duration of the pressure fluctuations at higher engine speeds into account, the stress due to the pressure fluctuations is similar for the points with the maximum KPP values.

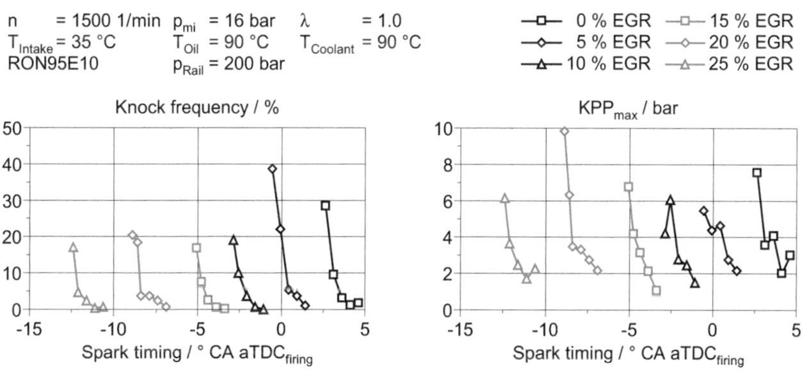

Fig. 5. Knock frequency and maximum KPP at 1500 1/min

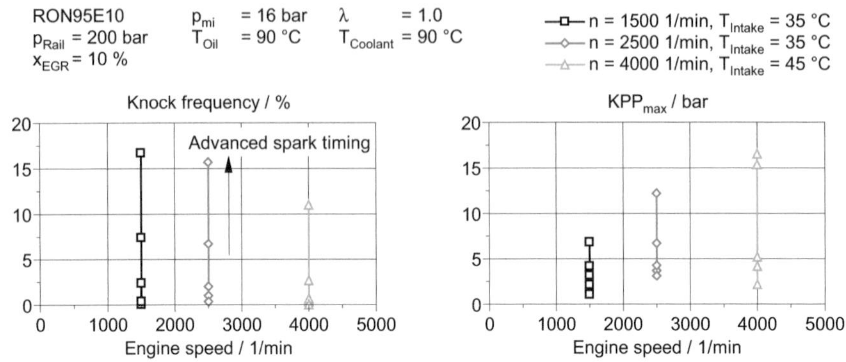

Fig. 6. Dependence of knock frequency and maximum KPP on engine speed

For the knock model development, spark timing sweeps at increased intake temperatures have been carried out as well. Thus, the measurement data covers a wide range of thermodynamic conditions in the end-gas, which allows the development of a reliable knock model for full load EGR combustion systems.

3.3 Investigation of the Influence of Nitric Oxide on Engine Knock

In dependency of the rel. AFR in the combustion chamber the exhaust gas composition is varying. About three-quarter of the volumetric composition is nitrogen, which does not have a strong chemical effect on the combustion. During rich combustion high amounts of CO and small amounts of H_2 are formed. For stoichiometric combustion higher concentrations of NO are formed. When the exhaust gas is recirculated these species can take part in the combustion process. For CO no influence on the auto-ignition in a HCCI engine was found [Deb06]. Hydrogen is known to increase the combustion speed and stability [Alg07]. Especially NO is considered to be a species which can strongly affect the auto-ignition in the cylinder. Various researches have investigated the influence of NO on knocking combustion but the results are often contradicting, as summarized in [Par15].

Measurements with NO addition to the intake charge were performed while running the engine with catalyzed exhaust gas. Due to the size of the catalyst the NO concentration could be reduced to less than 10 ppm in the recirculated exhaust gas. NO was added prior to the external boosting system in order to achieve a homogeneous mixing with the EGR and air. Measurements were performed with three different EGR rates and an intake NO concentration from 0 to 300 ppm (100 ppm steps). In order to take the cylinder NO concentration into account CFD simulations of the corresponding operating points were carried out to determine the internal residual gas mass fraction. With the exhaust gas measurement prior to the catalyst the internal NO concentration can be calculated. Finally, the total NO concentration is derived with the addition of external NO. With rising external EGR rate the internal NO rate is decreasing. The calculated NO concentration in the cylinder due to internal EGR is varying from 100 ppm at 0% EGR to 20 ppm at 20% EGR. The results are shown in Fig. 7.

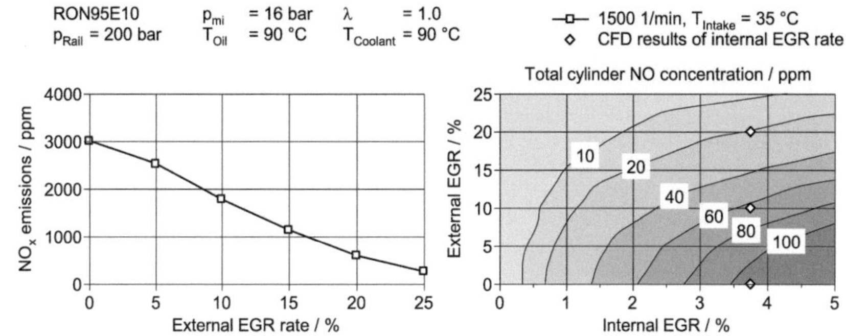

Fig. 7. Calculation of cylinder NO concentration based on pre-catalyst measurement and the internal EGR rate (calculated with CFD simulations)

The experimental results of the NO addition in the intake port are depicted in Fig. 8. The average combustion and the combustion stability are not affected by the addition of NO. However, the number of knocking cycles changes with the addition of NO. At 0% EGR the number of knocking cycles is reduced with the addition of NO. This gives an indication that the knock promoting effect of NO is already saturated at low NO concentrations in the cylinder. At 10% EGR no significant difference between 0 ppm and 100 ppm external NO is observed. With further increase of NO the number of knocking cycles decreases again. Surprisingly no knock promoting effect is observed at these two operating points. At 20% EGR the number of knocking cycles is drastically increasing at 100 ppm external NO. Again a further increase of external NO reduces the number of knocking cycles. This behavior was confirmed at a second intake temperature. Without knowledge of the internal EGR rate and the cylinder NO concentration this behavior cannot be explained. Taking the cylinder NO concentration into account it can be observed that the maximum knock promoting effect of NO is observed at about 100 ppm cylinder NO concentration. Higher NO concentrations as observed for the 0% EGR case lead to a reduced knock tendency. At 20% EGR the internal NO concentration is about 20 ppm NO. Here external addition of NO leads to a strong increase of knocking cycles. The effect is most pronounced at 100 ppm external NO. Further increase of cylinder NO concentration leads to a reduction of knocking cycles. These findings give a good indication about the different results obtained from literature. Often the internal residual gas mass fraction is unknown and thus the internal cylinder NO concentration cannot be derived from measurement results.

This phenomenon was further investigated with the detailed mechanism developed in Sect. 2.3. The underlying chemical reactions of the effect of NO on auto-ignition are shown in Fig. 9. On the one hand NO can promote auto-ignition at low temperatures via reaction with HO_2 which ultimately yields two OH radicals. At higher temperatures NO can react with OH radicals and form rather unreactive HONO. The second path ultimately leads to a chain termination. In Fig. 9 the effect of NO addition on ignition delay times is depicted. The auto-ignition delay time in the low temperature regime (here evaluated at 700 K) is decreased with addition of 100 ppm NO, but a further addition of NO does not have any effect. In the high temperature regime (evaluated

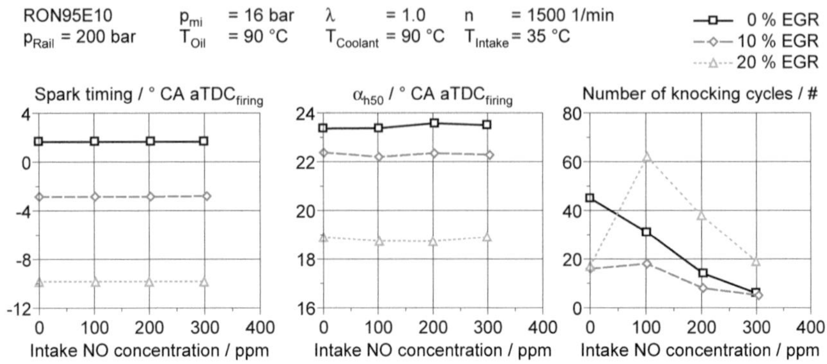

Fig. 8. Effect of NO additional on combustion and knock

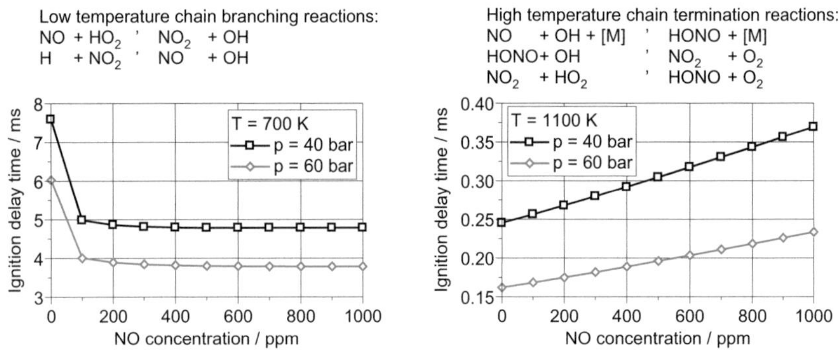

Fig. 9. Isochoric reactor simulation with gasoline surrogate and NO addition at rel. AFR = 1

at 1100 K) the auto-ignition delay time is increased with addition of NO. No saturation of this effect is observed with higher amounts of NO.

3.4 Comparison of Pre- and Post-catalyst Extraction of EGR

One key motivation of this work is to identify whether pre or post catalyst extraction is favorable. Despite many published papers on the comparison of pre and post catalyst extraction, no well accepted opinion has been established yet [Par15]. A dedicated measurement approach was used to compare pre and post catalyst extraction. The engine was operated with identical boundary conditions (as long as possible) between both EGR extraction points. Due to the size of the catalyst complete conversion of NO_x could be achieved for all conducted measurements. To ensure a correct calculation of the EGR rate with the CO_2 measurements in the exhaust system the post catalyst extraction point was used for calculation of the EGR rate.

To analyze how the concentration of NO in the cylinder is varying between pre and post catalyst extraction the concentration was calculated for pre catalyst extraction based on internal and external EGR rate. Figure 10 shows the results. For pre catalyst

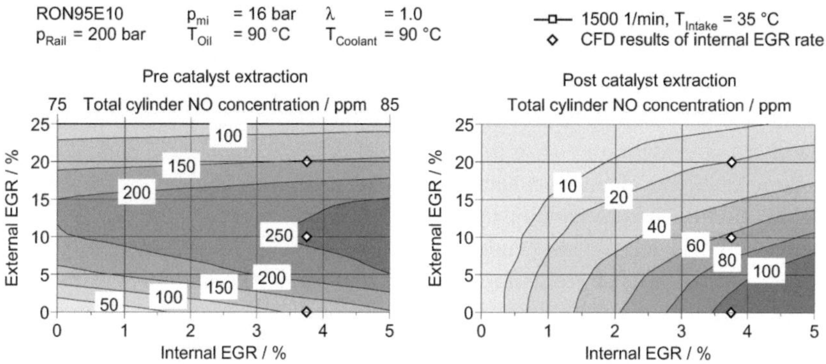

Fig. 10. Comparison of cylinder NO concentration between pre and post catalyst extraction

extraction the cylinder NO concentration cannot be reduced below 100 ppm for the investigated operating points. Peak concentration with pre catalyst extraction is slightly above 250 ppm. For post catalyst extraction the cylinder NO concentration can be decreased steadily with external EGR rate.

The experimental results for pre and post catalyst extraction at the KLSA are shown in Fig. 11. For low EGR rates no differences between pre and post catalyst extraction were apparent from the measurement data. First small differences appeared at 15% EGR rate which indicated a slightly less knock limitation for post catalyst extraction. With even higher EGR rates the spark timing could be further advanced with post catalyst extraction leading to a gain of about 2° CA in center of combustion. These findings are in-line with the measurement results of the single component investigation of NO. Similar trends were found at an inlet temperature of 45 °C.

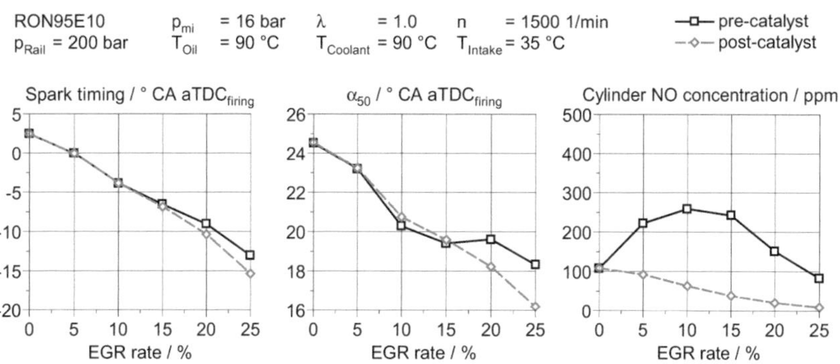

Fig. 11. Comparison of pre- and post-catalyst extraction

4 Numerical Investigation

For further understanding of the test bench results numerical investigations were carried out. The scope of the numerical investigation was to gain comprehension of the thermodynamic and mixture state in the cylinder during compression and combustion. Furthermore, the developed mechanism from Sect. 2.3 was coupled to the simulation model in order to simulate combustion and the occurrence of auto-ignition events. The numerical domain for the in-cylinder simulation includes the geometry of the single cylinder research engine from intake to exhaust pressure transducer. Crank angle resolved boundary conditions were used at the intake and exhaust boundary. The composition of the exhaust gas was calculated based on test bench measurements. Propane was used as chemical representative for HC emissions. The composition of the fresh charge was calculated as a perfect mixture of air and EGR depending on the operating points EGR rate. A comprehensive overview of the simulation model is given in Fig. 12.

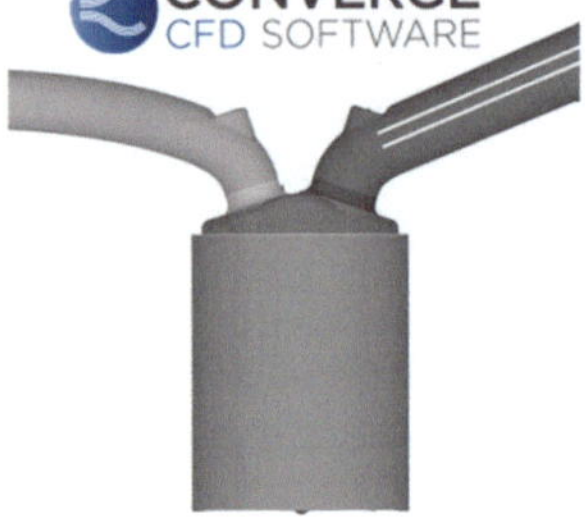

Numerical models and settings
Cylinder cell size < 1 mm
CFL_{Mach} < 3 to solve pressure fluctuations
Kelvin-Helmholtz/Rayleigh-Taylor spray breakup
Kuhnke wall film modeling
Standard two-equation k-μ turbulence model
Amsden Heat Transfer Model
Software Converge CFD v2.3.14

Fig. 12. Numerical models and settings for the in-cylinder 3D-CFD simulation

4.1 Analysis of Thermodynamic State

The thermodynamic and mixture state in the cylinder (temperature, pressure and composition) is the most important input parameter of the knock model. This state was assessed in detail with help of 3D-CFD simulations. Furthermore, simulation data was evaluated for local inhomogeneities (e.g. hot spots or rich mixture pockets). Thus 3D-CFD simulations including gas exchange and injection were carried out for all investigated engine speeds and three EGR rates (0%, 10% and 20%). The intake temperature was set at 35 °C for the speeds 1500 and 2500 1/min and 45 °C for 4000 1/min.

For an EGR rate variation at constant speed it was found that the end of compression temperature is rising with increased EGR rate. At first glance an opposite trend is expected since the three-atomic gases of the exhaust gas have a higher heat capacity than air. However, the heat capacity of the fuel is much higher and since the dilution with EGR reduces the fuel mass fraction the effect of the EGR is overpowered and a slight decrease of the heat capacity is observed, Fig. 13. However, during combustion

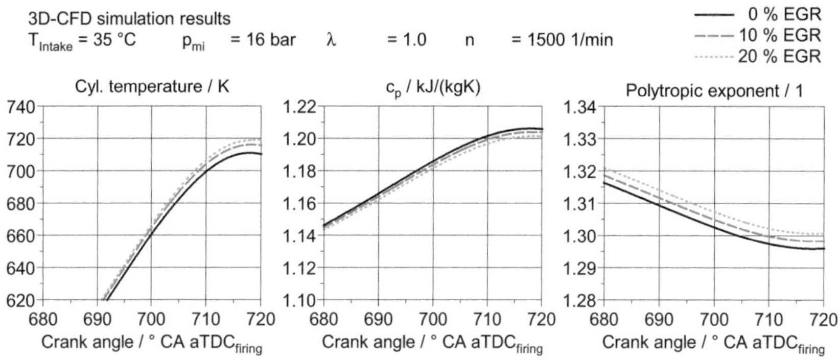

Fig. 13. Impact of EGR on the end of compression temperature at 1500 1/min

the unburned zone is compressed nearly adiabatically and the higher pressures for cases with EGR lead to a reduced temperature increase in the unburned zone [Höp12].

In a next step the stratification at TDC position was analyzed. The simulation results were classified for the quantities residual gas mass fraction (RGF), temperature and rel. AFR. The RGF is distributed very homogeneously since an ideal mix of air and EGR is introduced into the induction system and the high level of charge motion leads to a good mixing with the internal EGR. Small inhomogeneities were found for the rel. AFR in the range of ± 0.15. For the temperature it was found that usually hot spots with approximately 15 K higher temperature than the bulk gas temperature exist. With the help of the simulation results these hot spots were found to be close to the hot exhaust valves. The effect was strongest when a relatively rich zone is located at one exhaust valve. Due to the less evaporation cooling of the fuel the mixture is found to have the highest temperature in the combustion chamber.

4.2 Combustion Simulation

To analyze the effect of EGR on engine knock, combustion simulations have been carried out. Instead of a conventional combustion model the reduced reaction mechanism from Sect. 2.3 was directly coupled into the CFD simulation. To reduce the numeric effort a multi-zone approach was used. Cells with similar temperature and reaction progress are combined in one chemical reactor [Bap05]. The combustion is initiated with an energy source between the spark plug electrodes and the energy input is set to be half to energy of the primary circuit of the ignition system (45 mJ). For the simulations the KPP value can be derived in the same manner as at the test bench with the help of virtual pressure transducers. Thus the simulation results can be directly compared with the test bench results. For the experimental results cycle-to-cycle variations are observed, which cannot be resolved with the Reynolds-averaged Navier-Stokes (RANS) approach. In order to mimic the test bench behavior spark timing variations in the 3D-CFD simulation were performed to simulate cycles with early, intermediate and late center of combustion.

In Fig. 14 the simulation results at 2500 1/min are depicted. In total five spark timings were simulated for the 0% EGR case of which the last two are within the area of the experimental results. Further advanced spark timings were simulated to verify if the simulation correctly catches the trend of increasing knocking peak to peak values. As shown in the bottom left hand side of Fig. 14 KPP is increasing as expected when the combustion is advanced. The simulation approach is able to correctly reproduce the combustion and knocking behavior at this operating point.

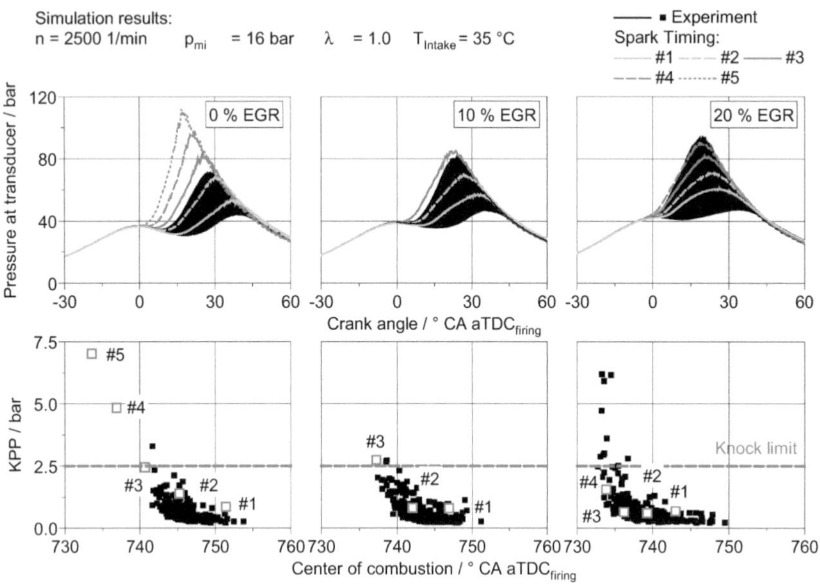

Fig. 14. Comparison of experimental and simulation results at 2500 1/min 16 bar p_{mi}

Figure 14 depicts the results of the EGR variation as well. At 10% EGR the combustion and the knock tendency is accurately reproduced by the simulation results. At 20% the earliest combustion is showing less KPP than measured at the test bench. The higher cycle-to-cycle variations at this operating point make it more difficult to reproduce the knock limit with a RANS approach. However, a similar trend is observed.

In a next step the thermodynamic state of the end gas was analyzed in detail for the different EGR rates. Monitoring positions at a distance of 2.5 mm from the liner surface were set up. During the combustion process the local temperature in the end-gas is increasing by approximately 100 K before the gas at the monitoring positions is consumed by the flame, Fig. 15. When evaluating the local pressure at the time of the first temperature increase no pressure oscillations are observed. Thus this first energy release is not directly connected with the occurrence of knock. The simulation results show that first stage ignition is playing an important role for the relevant

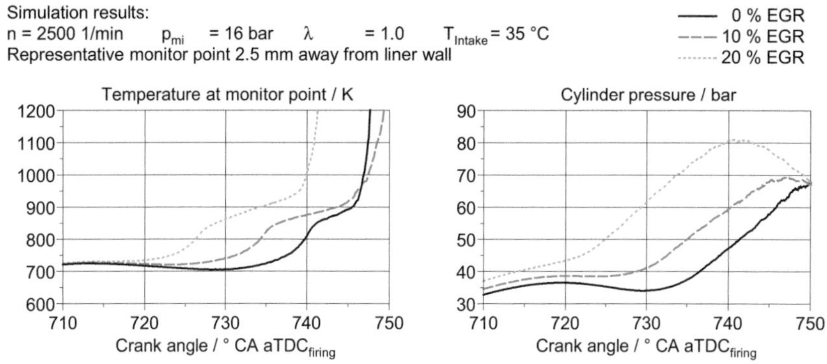

Fig. 15. Representative temperature and pressure in the end-gas zone at 2500 1/min

temperature for engine knock. In all investigated cases the whole end-gas is showing first stage ignition at relatively low mass fraction burned ($x_b < 0.5$).

With increased EGR rate first stage ignition is observed at earlier timings. To understand this behavior constant volume reactor simulations under the influence of EGR were carried out, Fig. 16. The main effect of EGR on the ignition delay time is visible in the prolonged time for the 2nd stage ignition, while first stage ignition is occurring at a similar time. Increasing pressure with EGR leads to similar timings of first stage ignition for simulation with and without EGR. Thus the effect of the first stage ignition is more pronounced with the application of EGR. It can be concluded that considering the first stage ignition is of major importance for the knock modelling (Sect. 5) to correctly predict the occurrence of knock with EGR.

The advantage of using CFD is that the underlying mechanisms of engine knock can be analyzed in a very detailed way. In Fig. 17 the pressure fluctuations for the 0% EGR case with spark timing #5 of Fig. 14 during occurrence of knock are depicted. At 736° CA, a steep pressure rise at the intake side of the combustion chamber is observed. Subsequently, a pressure wave is formed, which travels through the combustion chamber. At 737.2° CA the pressure wave is then reflected at the opposite side

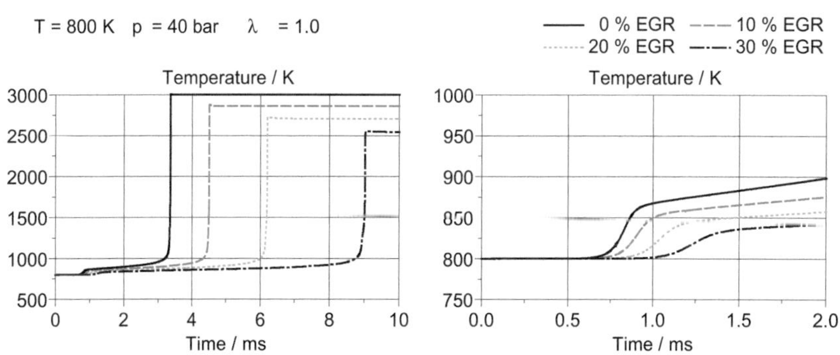

Fig. 16. Temperature trace of auto-ignition under the influence of EGR

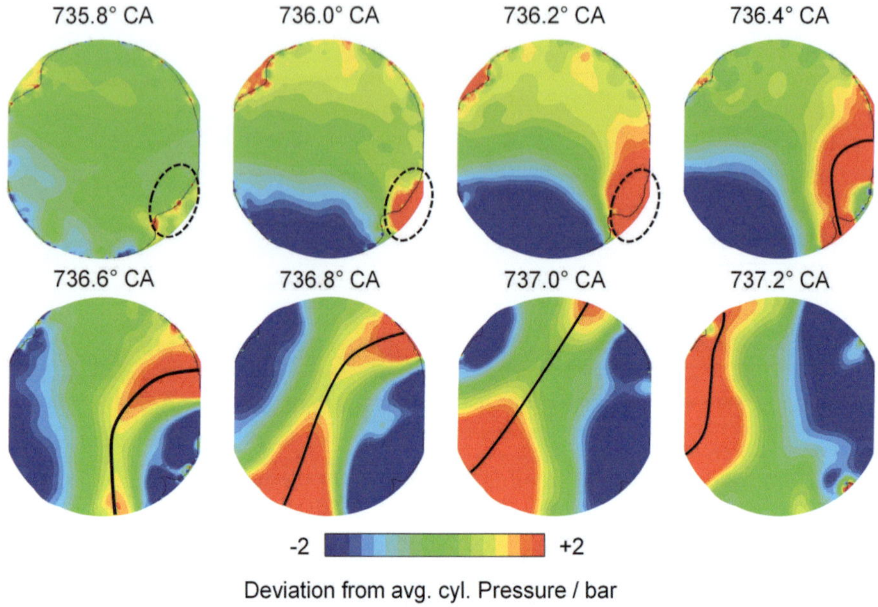

Fig. 17. Visualization of pressure fluctuations during knocking combustion at 2500 1/min, 16 bar p_{mi} and 0% EGR

of the combustion chamber. Finally, a standing wave is created, which leads to the typical pressure oscillations in the simulated pressure at the transducer location.

The coupling of the kinetic mechanism with the CFD model allows analyzing the species concentrations during the occurrence of knock. In Fig. 18 the concentrations of formaldehyde (CH_2O) and hydrogen peroxide (HOOH) are shown for the same operating point. Formaldehyde is a tracer of low temperature combustion and is formed

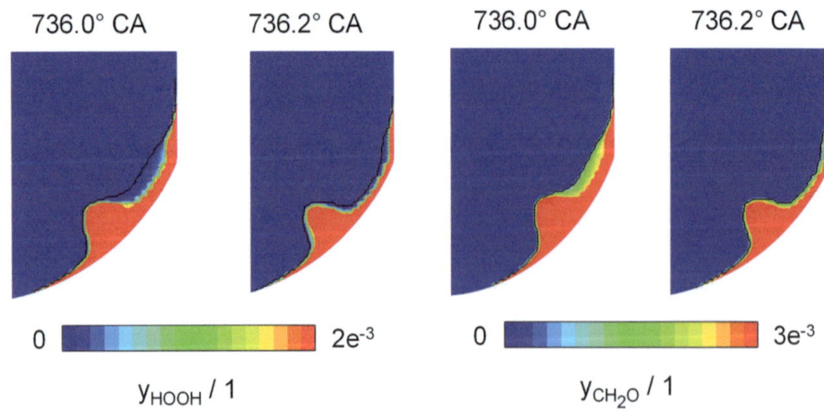

Fig. 18. Consumption of HOOH and CH_2O during the occurrence of knock

during the first stage ignition of the unburned mixture. The flame front is indicated by the black line. At 736° CA the concentration of HOOH ahead of the flame front is decreasing and two OH radicals are formed by the corresponding chemical reaction. Due to the increase in radical concentration CH_2O is consumed and the second stage of ignition is taking place. The fast heat release leads to an increase in temperature and a pressure wave is triggered. With these findings of the 3D-CFD simulations it can be concluded that the new knock model must be capable of predicting both stages of the auto-ignition correctly.

5 Knock Model Development

5.1 General Engine Knock Modeling Approach

The 0D/1D knock models that have been developed in the past 50 years are generally based on the evaluation of an integral representing the pre-reaction state of the unburnt mixture (also known as knock, pre-reaction state and Livengood-Wu integral). Livengood and Wu originally formulated this correlation in 1955 [Liv55]. They proved that the ignition delay time of an air-fuel mixture in motored and firing SI engines can be estimated by evaluating an integral representing the degree of chemical reaction progress and thus the pre-reaction state of the mixture, Eq. (4). Hence, it is possible to predict when auto-ignition of a mixture in an engine will occur (end of integration t_e), if the ignition delay times τ are known at every integration step. These can be measured in a RCM or a shock tube, Sect. 2.1.

$$1 = \int_{t=0}^{t=t_e} \frac{1}{\tau} dt \tag{4}$$

$$\tau \approx C \cdot p^{C_1} \cdot e^{C_2/T} \tag{5}$$

Subsequently, an Arrhenius-type correlation, Eq. (5), can be calibrated by estimating the empirical constants C, C_1 and C_2 and then used for their calculation. The integration process performed in Eq. (4) represents the process of concentration of chain carriers building up, until a pre-defined critical concentration has been reached and exceeded, resulting in an auto-ignition of the air-fuel mixture. The value of the pre-defined critical concentration is constant and thus independent of changes in the boundary conditions, such as e.g. an engine speed increase. Knock modeling approaches commonly used in the 0D/1D engine simulation today are all based on the correlation in Eq. (4). They evaluate the change in the pre-reaction state I_k of an air-fuel mixture between inlet valve close (IVC)/90°CA before Firing Top Dead Center (FTDC) and knock onset (KO) or a constant Mass Fraction-Burnt-point (usually between MFB75 and MFB85) respectively, as shown in Eq. (6) [Fra91].

$$I_k = \frac{1}{n} \cdot \frac{1}{C} \int_{\alpha=\alpha_{IVC}/90^\circ CA\ bFTDC}^{\alpha=\alpha_{KO/const.MFB}} \frac{1}{\tau} d\alpha \tag{6}$$

As soon as the pre-reaction state calculated in Eq. (6) exceeds a constant pre-defined value, the so-called critical pre-reaction state $I_{k,crit}$, knock onset has been reached. The $I_{k,crit}$ value at knock onset corresponds to the critical concentration of chain carriers in the original Livengood-Wu correlation that results in auto-ignition. Hence, the only difference between Eqs. (4) and (6) is the integration domain. Again, the ignition delay times needed for the integration process can be calculated with Eq. (5), after the empirical constants have been estimated. This is commonly done by fitting to measurement data, so that the critical pre-reaction state $I_{k,crit}$ is reached at the measured knock onset. Generally, the mean values of the unburnt zone's parameters estimated with an appropriate two-zone SI combustion modeling approach, e.g. the Entrainment model [Gri06], are used as knock model inputs.

Many different variations of the commonly used knock approach in Eq. (6) have been developed based on measurement data. By adding corresponding terms to Eq. (5), they consider the influence of various boundary conditions on the ignition delay of the air-fuel mixture, such as the AFR and exhaust gas fraction. Alternatively, detailed kinetic reaction mechanisms can be used to calculate the ignition delay times of air-fuel mixtures at various boundary conditions and hence to estimate the values of the empirical constants and to expand Eq. (5). This method has the main advantage of being able to cover a much wider range of boundary conditions than a parameter estimation performed with measurement data from a test bench. Furthermore, the use of detailed mechanisms ensures that no engine-specific effects affect the calculated values. For these reasons, the ignition delay times integrated in the new knock model were estimated with the mechanism developed in Sect. 2.3. However, detailed kinetic reaction mechanisms cannot be integrated in 0D/1D simulations, as this would have a significant negative impact on the typically short computational times. Hence, suitable models for the ignition delay times have to be developed, Sect. 5.4.

5.2 Modeling Approach Performance Investigation

Firstly, the performance of the commonly used knock modeling approach was evaluated by estimating the pre-reaction state of the unburnt mixture of measured knocking single cycles at knock onset. To this end, the knock onset of each knocking cycle, Sect. 3.1, was detected as the location of the minimum of the third derivative of the filtered cylinder pressure. 200 000 single cycles were evaluated in total. In order to obtain the data needed for the performance evaluation, Pressure Trace Analysis (PTA) of the single working cycles was performed. These calculations yield the mean temperature of the unburnt mixture. Subsequently, the pre-reaction state of the unburnt mixture at the measured knock onsets was obtained, Fig. 19. The ignition delay times at each integration step were estimated with a three-domain modeling approach presented in Sect. 5.4.

By neglecting the time span between auto-ignition and the occurrence of oscillations in the measured cylinder pressure resulting from knock (or assuming that it does not change significantly with the operating conditions), it can be concluded that hat the measured knock onset is also the time of auto-ignition in the unburnt mixture. As discussed in Sect. 5.1, this assumption implies that the calculated pre-reaction states at knock onset have to be equal to the critical pre-reaction state for the investigated engine

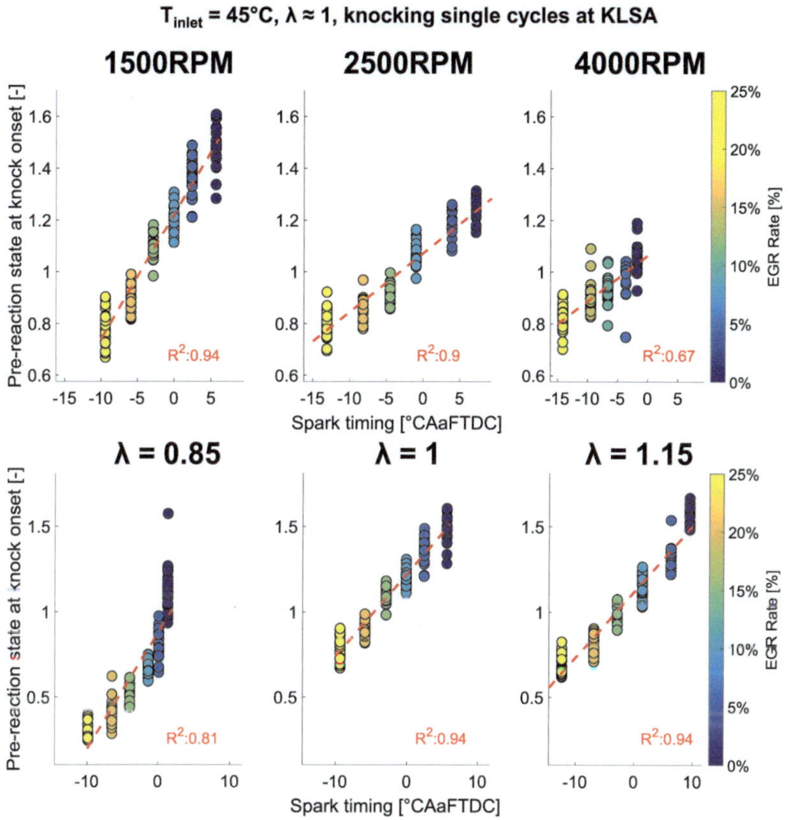

Fig. 19. Pre-reaction state of the unburnt mixture and heat release at measured knock onset of knocking single working cycles at various boundary conditions.

and hence all of the same magnitude. However, Fig. 19 clearly shows significant changes of the values over all investigated operating condition variations. Hence, the knock integral is not capable of accurately estimating the progress of the chemical reactions leading to auto-ignition at in-cylinder conditions, resulting in poor knock prediction performance.

Furthermore, the evaluation of the knock integral is generally performed to an always constant MFB-point, as it is assumed that auto-ignition after this point does not result in knock because of the small unburnt mass and volume fractions left. However, the measured MFB at knock onset changes significantly with the boundary conditions, Fig. 20. Hence, the assumed constant end-of-integration point will also result in a prediction error. Overall, this performance investigation clearly shows that neither the measured knock onset nor the knock boundary can be reliably predicted with the commonly used knock integral, Eq. (6).

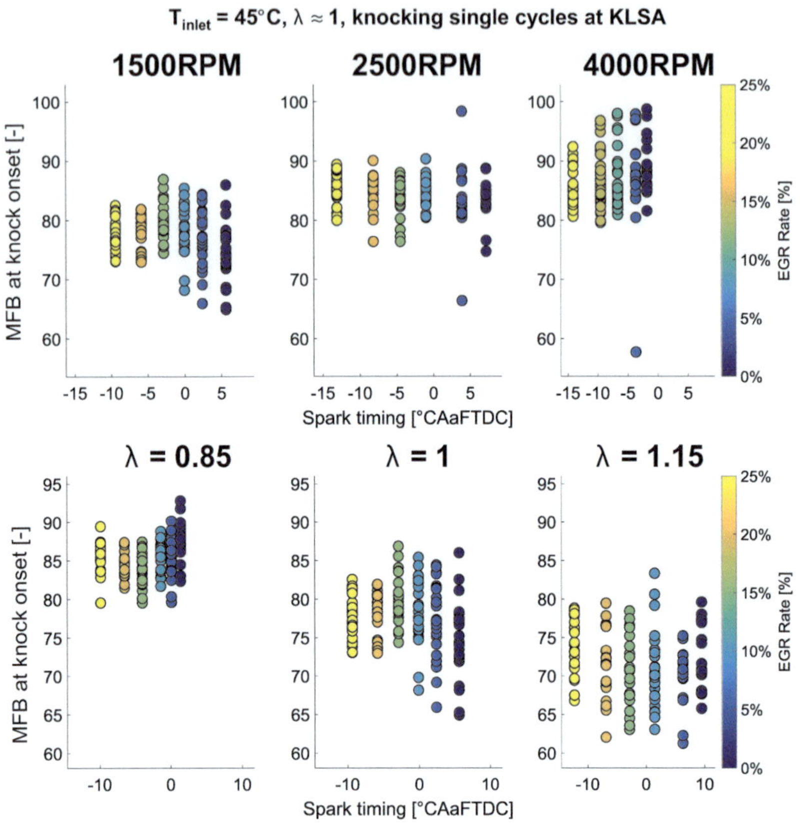

Fig. 20. Mass fraction burnt at measured knock onset of knocking single cycles at various operating conditions.

5.3 In-Cylinder Reaction Kinetic Simulations

In order to better understand the in-cylinder chemical processes resulting in auto-ignition and to investigate the performance of the commonly used simplified chemistry approach – the knock integral – a simulation model representing the unburnt zone of the Entrainment model [Gri06] (or a small spot in it, as the model is scalable) was developed and implemented in Matlab/Cantera. The simulations were performed with the detailed mechanism and the proposed surrogate composition from Sect. 2.

Simulation model

The developed simulation model is based on an adiabatic reactor containing the surrogate-air-exhaust gas mass, Fig. 21. In addition, a moving wall was installed to compress and expand the mixture and hence reproduce the piston movement. Furthermore, the wall has heat transport properties that were used to recreate the wall heat losses calculated by the corresponding model as part of the performed PTA, Sect. 5.2.

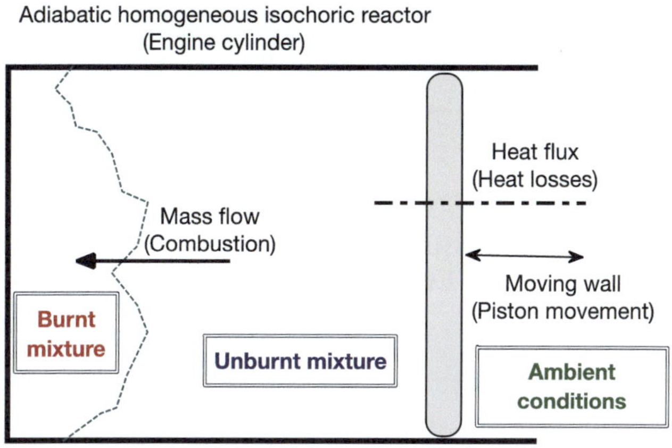

Fig. 21. Model for reaction kinetic simulations at in-cylinder conditions.

As a mass flow from the unburnt into the burnt zone is used in the Entrainment model to represent the combustion [Gri06], the same technique was chosen for this simulation model. Blow-by was considered by evaluating the mass flow rate through an orifice resulting from a pressure gradient. The exhaust gas composition can be calculated from the surrogate component fractions and the AFR by assuming post-catalyst extraction and perfect catalytic conversion. Thus, the initial conditions needed at simulation start are the mixture composition (surrogate, AFR and EGR) as well as the initial volume (piston position), pressure and temperature. The simulations yield if the mixture auto-ignites and when as well as the production and consumption of species causing changes in temperature and pressure. The model was validated by comparing simulated temperature profiles with PTA results in the compression stroke, where no heat release takes place. The curves showed very good agreement.

Investigation findings

By simulating thousands of measured single cycles at various boundary conditions, it was found out that the local auto-ignition in the unburnt mixture resulting in knock can occur in two stages. This phenomenon has already been observed in various studies in the context of gasoline Homogeneous Charge Compression Ignition (gHCCI) [Tan03], but has not been extensively investigated in the context of engine knock. The temperature profiles in Fig. 22 illustrate the two-stage ignition behavior. Obviously, the amount of heat released during the low-temperature ignition depends on the boundary conditions, e.g. engine speed and EGR rate as shown in this case. Generally, two-stage ignition behavior characterizes iso-octane and n-heptane [Tan03]. As these two hydrocarbons, together with toluene and ethanol, compose the surrogate fuel, they also cause the observed low-temperature heat release. The phenomenon is related to the negative temperature coefficient (NTC) zone clearly visible in Fig. 23, where the rate and extent of the hydrocarbon oxidation reactions, which at first increase rapidly with temperature, begin to fall with further rise in the temperature, causing an increase in the ignition delay times in this temperature region. The resulting temperature increase

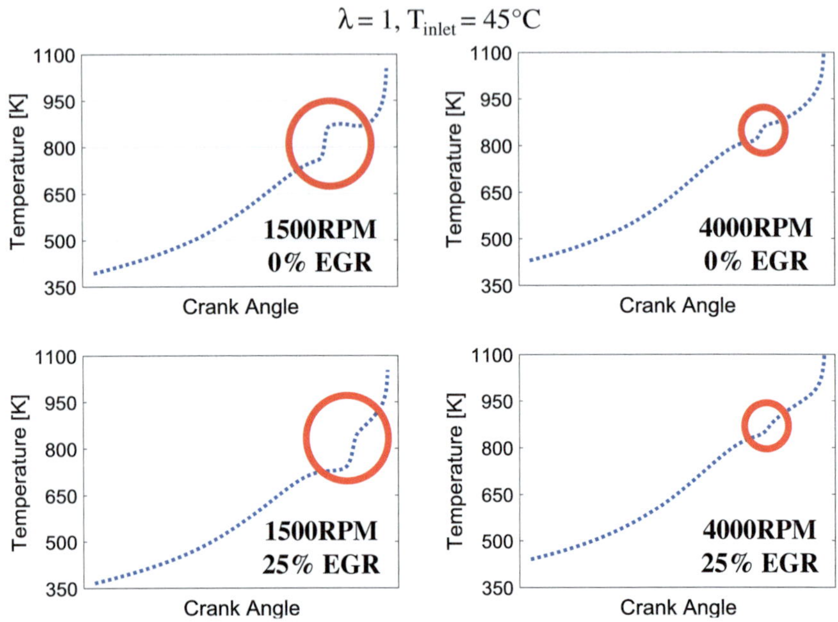

Fig. 22. Simulated temperature profiles of different working cycles with auto-ignition in two stages depending on the boundary conditions.

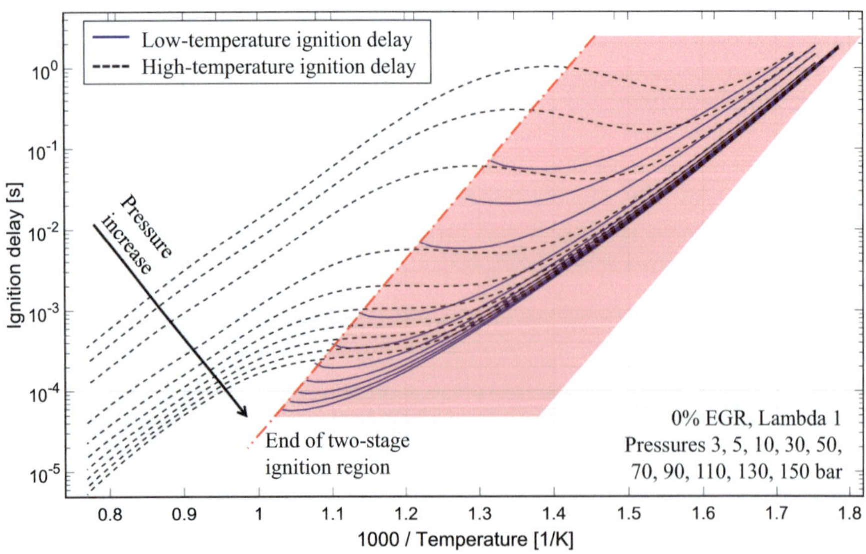

Fig. 23. Two-stage ignition region and ignition delay times of the low- and high-temperature ignition regimes (surrogate from Sect. 2).

influences the following chemical reactions and thus the ignition delay of the mixture significantly [Tan03].

The boundary condition region where two-stage ignition of the surrogate-air mixture occurs is illustrated in Fig. 23. The existence of an ignition delay time of the low-temperature ignition primarily depends on temperature and pressure and defines the two-stage ignition region of a given surrogate composition. Furthermore, Fig. 23 clearly shows that, at high temperatures, the two-stage ignition phenomenon does not occur at all. This observation is relevant for both knock and gHCCI operation with high internal EGR rates, as these lead to high unburnt temperatures.

Knock integral auto-ignition prediction performance

Because of the significant influence of the low-temperature heat release on the mixture's auto-ignition behavior, the question arises, if the commonly used knock integral can consider this effect and thus predict the auto-ignition accurately in case the mixture ignites in two-stages. Figure 24 shows a comparison between the crank angles of auto-ignition simulated with the detailed mechanism with the model presented in Sect. 5.3 and those predicted by a knock integral, Eq. (6), at various boundary conditions. As in Sect. 5.2, the ignition delay times used as integral inputs were obtained with the three-domain modeling approach presented in Sect. 5.4.

Obviously, the occurrence of two-stage ignition results in partially huge errors and hence poor performance of the auto-ignition prediction. For some cycles, no auto-ignition was predicted at all, although the detailed mechanism auto-ignited. Hence, the knock integral cannot account for the low-temperature ignition and its influence on the mixture's ignition delay. These results clearly show that the theoretical approach used for modeling knock in 0D/1D SI engine simulations is not capable of reproducing the in-cylinder behavior of the detailed reaction kinetics mechanism (and thus the "real chemistry"). Thus, it cannot be used for the reliable knock onset prediction in SI engines running on fuels with two-stage auto-ignition behavior, e.g. gasoline. The general knock modeling approach rather has to be changed or improved, in order to consider the occurrence of two-stage ignition.

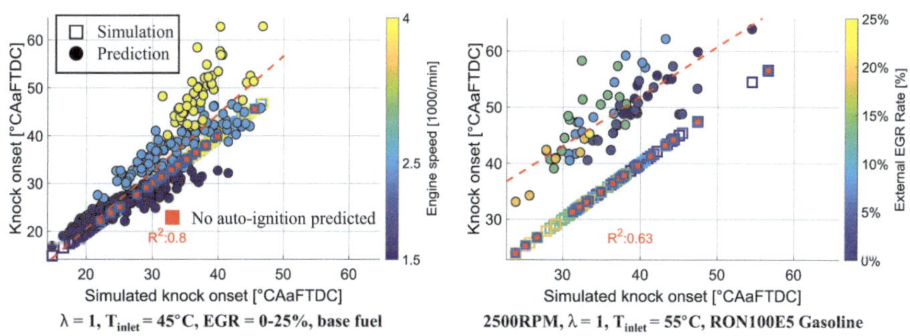

Fig. 24. Times of auto-ignition simulated with the detailed mechanism and predicted by a knock integral at various boundary conditions. For some working cycles, the knock integral predicted no auto-ignition at all.

5.4 Two-Stage Ignition Modeling

An approach for modeling knock that accounts for a low-temperature ignition possibly taking place requires the development of suitable models for the three main parameters characterizing the two-stage ignition phenomenon. These are the ignition delay times of both ignition stages as well as the temperature increase resulting from the low-temperature ignition. First, the quantification of these three parameters is necessary. To this end, following definitions are introduced (Fig. 25):

- The location of the low-temperature ignition τ_{low} is defined as the point $max(T_{grad})$, where the temperature gradient reaches its maximum before the auto-ignition of the mixture.
- The high-temperature or the auto-ignition delay τ_{high} is detected based on a threshold for the temperature gradient; the value was set to 25 K/microsecond.
- The temperature increase T_{incr} resulting from the low-temperature ignition is defined as temperature at simulation start subtracted from the temperature at the point $min(T_{grad})$, where the temperature gradient reaches its minimum between the two ignition stages.

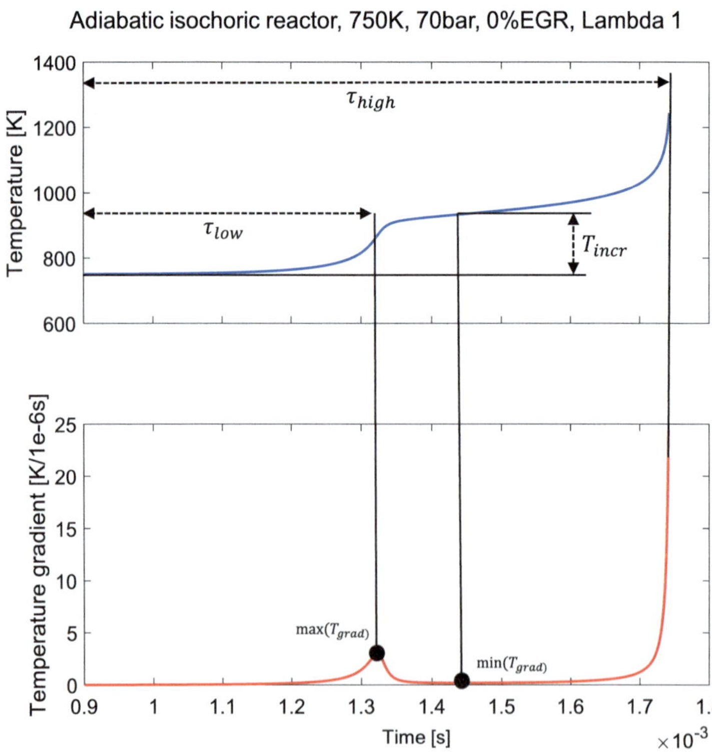

Fig. 25. Definition of low- and high-temperature ignition as well as the temperature increase resulting from the first ignition stage in an adiabatic isochoric reactor at 750 K, 70 bar and $\lambda = 1$, no exhaust gas mass.

The values of the parameters characterizing the two-stage ignition phenomenon were obtained in Cantera/Matlab simulations of an isochoric adiabatic homogeneous reactor with the detailed mechanism already brought up in Sect. 2. An algorithm based on the production of OH radicals followed by their consumption and consequentially temperature increase before the mixture's auto-ignition detected the occurrence of two-stage ignition. The simulated boundary conditions (temperature, pressure, AFR, EGR, ethanol content) were varied according to in-cylinder values typical for SI engines today with the main goal being to cover a sensible value range as wide as possible. Obviously, the exhaust gas composition has to be changed in accordance with the AFR, Sect. 5.3. Moreover, the fractions of the three base surrogate components were varied too, in order to capture the influence of the fuel characteristics on the ignition delay.

The simulation data were used to develop new models for the calculation of the three main parameters characterizing the two-stage ignition phenomenon, with the two main goals being the achievement of high accuracy of the calculated values and short computational times. Clearly, the NTC zone has to be modeled as precisely as possible because of its significant influence on the ignition delay times, Sect. 5.3. The developed models can afterwards be installed in a new knock modeling approach that accounts for the possible occurrence of two-stage ignition and its influence on the mixture's ignition delay. All estimated model parameters should only be recalibrated if values of the parameters characterizing the two-stage ignition phenomenon obtained with a reaction kinetics mechanism different than the one used in this work have to be considered. Further details on the models and their development can be found in [Fan17]; for the sake of completeness, here the model equations are included. The high-temperature ignition delay can be expressed as shown in Eq. (7). This correlation is particularly suitable for fuels with pronounced NTC behavior. Each of the three timescales $\tau_{i,high}$ can be evaluated using a single Arrhenius-type correlation, Eq. (8). Here, the influences of various changes in the boundary conditions are assembled in just two parameters $A_{i,high}$ and $B_{i,high}$. Thus, changes in the boundary conditions are reflected in different values of the activation energy of the chemical reactions leading to auto-ignition (as it is proportional to $B_{i,high}$) [Fan17]. This yields a simple, flexible and yet very powerful modeling approach.

$$\frac{1}{\tau_{high}} = \frac{1}{\tau_{1,\mathrm{high}} + \tau_{2,\mathrm{high}}} + \frac{1}{\tau_{3,\mathrm{high}}} \tag{7}$$

$$\tau_{i,high} = A_{i,high} \cdot e^{\left(\frac{B_{i,high}}{T}\right)}, A_{i,high}, B_{i,high} = f(p, \lambda, EGR, fuel) \tag{8}$$

For the low-temperature ignition delay τ_{low}, it is convenient to choose a modeling approach similar to the one for the auto-ignition delay times, Eqs. (9) and (10). Finally, for the temperature increase resulting from the first ignition stage, the fitted parameter $T_{incr,fit}$ is the sum of the temperature increase T_{incr} and the temperature T_{low}, at which the low-temperature heat release occurred, Eqs. (11) and (12).

$$\frac{1}{\tau_{low}} = \frac{1}{\tau_{1,low} + \tau_{2,low}} \tag{9}$$

$$\tau_{i,low} = A_{i,low} . e^{\left(\frac{B_{i,low}}{T}\right)}, A_{i,low}, B_{i,low} = f(p, \lambda, EGR, fuel) \tag{10}$$

$$T_{incr} = T_{incr,fit} \cdot 100 - T_{low} \tag{11}$$

$$T_{incr,fit} = C_1 \left(\frac{T_{low}}{100}\right)^4 + C_2 \left(\frac{T_{low}}{100}\right)^3 + C_3 \left(\frac{T_{low}}{100}\right)^2 + C_4 \left(\frac{T_{low}}{100}\right)^1 + C_5, \quad C_{1..5} = f(p, \lambda, EGR, \text{fuel component fracs}) \tag{12}$$

Additionally, an automatic parameter estimation and optimization of the estimated values was implemented in Matlab to minimize the normalized root mean square deviation of the empirically calculated ignition delay times. Thus, very high model accuracy could be achieved, as shown in Fig. 26 for the high temperature ignition delay. All models account for effects of temperature, pressure, AFR, EGR and the fuel properties (represented by the corresponding surrogate component fractions) [Fan17].

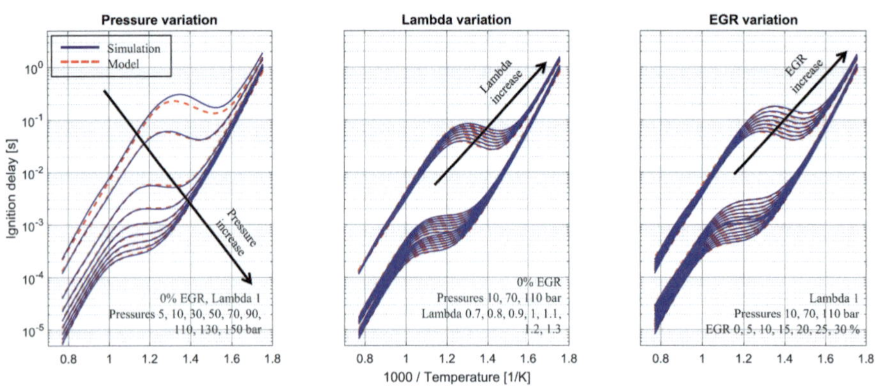

Fig. 26. Simulated and modeled high-temperature ignition delay times at various boundary conditions (surrogate from Sect. 2.2).

5.5 Two-Stage Knock Modeling Approach

The new auto-ignition approach must have the ability to predict the occurrence of two-stage ignition and to consider the significant influence of the low-temperature heat release on the auto-ignition behavior of the mixture, resulting in a reliable prediction of the time of auto-ignition of air-fuel mixtures for fuels with different characteristics at various boundary conditions.

General formulation

The occurrence of each of the ignition events can be predicted by a single knock integral similar to the correlation in Eq. (4), thus yielding a two-stage knock modeling approach, Eq. (13) [Fan17].

$$1 = \int_{0}^{t_1} \frac{dt}{\tau_{low}(BC)} \xrightarrow[\text{Progress of aggregate reaction}]{\text{Temperature increase } T_{incr} \text{, Pressure increase } p_{incr}} 1 = \int_{t_1}^{t_2} \frac{dt}{\tau_{high}(BC, T_{incr}(BC), p_{incr}(T_{incr}))} \quad (13)$$

The inputs of the two coupled integrals are the values of the ignition delay for the corresponding ignition stage as a function of the current boundary conditions (denoted by BC in Eq. (13)), calculated by the respective model presented in Sect. 5.4. The estimated time of low-temperature ignition t_1 is both upper limit of the first and bottom limit of the second integral. Thus, the integral representing the high-temperature ignition starts as soon as the first integral has reached 1 and the low-temperature ignition has occurred. The resulting temperature increase T_{incr}, as well as the corresponding pressure increase p_{incr}, then have to be added to the respective unburnt mixture values at each integration step of the second integral, Fig. 27. Thus, the influence of the low-temperature ignition on the ignition delay of the mixture is considered. Auto-ignition is defined as the point where the second integral reaches 1, a value that is independent of the boundary conditions representing the constant pre-reaction state of the mixture that has to be reached and exceed for auto-ignition to occur, Sect. 5.1. The integration is performed in the time domain with a small, constant step size of 10^{-7} s.

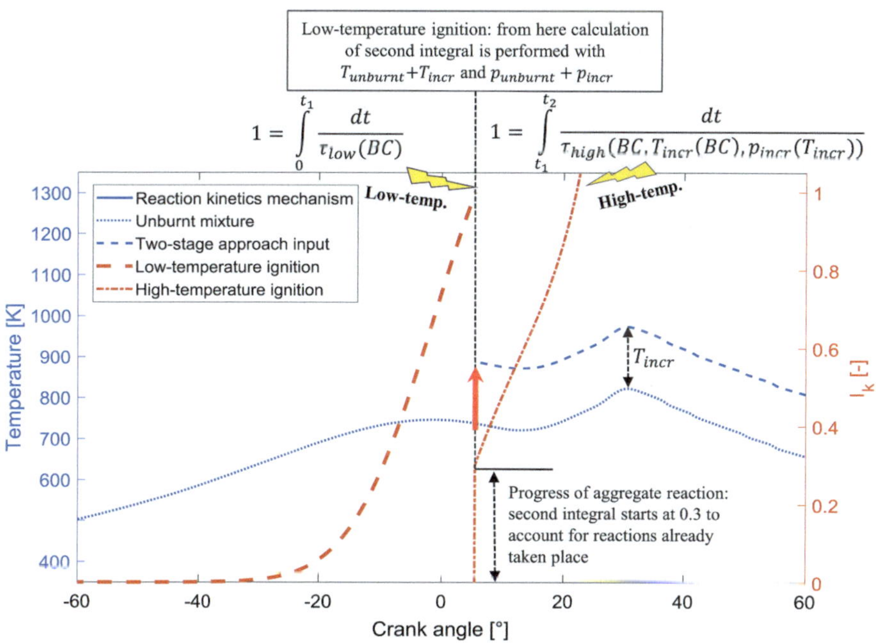

Fig. 27. Auto-ignition prediction for one working cycle considering the occurrence of low-temperature ignition with the newly developed two-stage approach.

Additionally, the progress of the aggregate auto-ignition reaction after the low-temperature ignition has to be considered, as at the starting point of the high-temperature reactions (and the corresponding integral in Eq. (13)), the pre-reaction state is not 0 [Her14]. This is because chemical reactions have already taken place and many intermediate species have been produced, which influences the following reactions significantly. This can also be observed in Fig. 27: the mechanism temperature almost does not change after the first ignition stage, although the cylinder volume is increasing (expansion stroke), thus suggesting that chemical reactions are already taking place and some heat is being released before the auto-ignition. The progress of the aggregate auto-ignition reaction after the first ignition stage was taken into consideration by setting the starting value of the second integral to 0.3, a value independent of the boundary conditions [Fan17].

Thus, the two-stage knock modeling approach involves the calculation of two integrals for the low- and high-temperature ignition respectively, where the seconds one starts at 0.3 and the temperature and pressure changes caused by the low-temperature ignition are added to the corresponding high-temperature integral inputs. Obviously, for the cycle shown in Fig. 27, the time of auto-ignition could be predicted very accurately, as the locations where the detailed mechanism auto-ignites and the high-temperature ignition integral reaches 1 coincide.

Prediction of two-stage ignition occurrence

The low-temperature ignition integration process in Eq. (13) can only take place as long as a low-temperature ignition delay exists for the respective boundary conditions during integration. This requirement is fulfilled inside the two-stage ignition region shown in Fig. 23. Consequently, if a point outside of this region is reached during the low-temperature ignition integration process (that is still running as the integral has not reached 1), a corresponding low-temperature ignition delay will not be available. Hence, in this case the low-temperature ignition region has been passed through without two-stage ignition occurring. On these terms, the high-temperature integration process in Eq. (13) is directly started at the beginning of the model calculation ($t = 0$) and neither a low-temperature ignition integral nor a temperature increase are considered. Thus, if no two-stage ignition occurs, the common single stage knock integral approach is used. Of course, in this case the ignition delay times are further calculated with the corresponding model proposed in Sect. 5.4.

Approach validation

The new two-stage auto-ignition prediction approach has been extensively validated at various boundary conditions [Fan17]. Validation results comparing the times of auto-ignition (in °CA) simulated with the detailed reaction mechanism at in-cylinder conditions with those estimated with the newly developed two-stage approach can be found in Fig. 28.

Generally, the predicted times of auto-ignition match the corresponding points calculated with the detailed mechanism very good within all boundary condition variations performed. Additionally, the prediction accuracy was validated for 20 different fuels with RONs between 91 and 102 as well as ethanol contents between 0 and 20 vol.%. It is important to remark that neither the critical pre-reaction state (equaling 1), nor the pre-reaction state after the low-temperature ignition (equaling 0.3) were

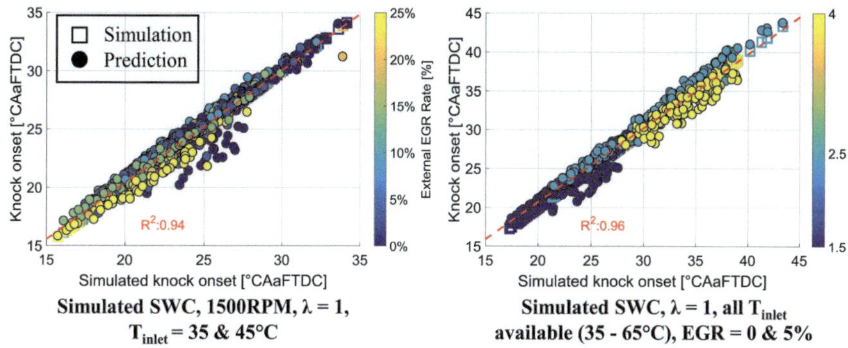

Fig. 28. Times of auto-ignition simulated with the detailed mechanism and predicted by the newly developed two-stage approach at various boundary conditions.

changed within the performed variations of boundary conditions. Hence, the accurate prediction of the time of mixture auto-ignition is ensured, even if the two-stage approach is fully decoupled from the detailed chemistry, and no recalibration of the approach constants is required. Consequently, temperature curves from PTA calculations or 0D/1D combustion simulations can be used as knock model inputs, without any recalibration of the auto-ignition model constants. Furthermore, extensive investigations were performed regarding the temperature inhomogenieties ("hot spots") in the unburnt zone estimated iteratively for measured single cycles and obtained from 3D-CFD stratification analysis, Sect. 4. The results showed that, if the exact values of all knock model inputs are known at each integration step (and not only the mean ones for the unburnt zone), the reliable prediction of knock occurrence based on single and averaged cycles with the two-stage approach is possible without any limitations. Hence, the newly developed modeling approach is generally suitable for the knock boundary prediction in 0D/1D engine simulations. The comparison of the auto-ignition prediction performance of the commonly used single integral (Eq. (6)) and the newly

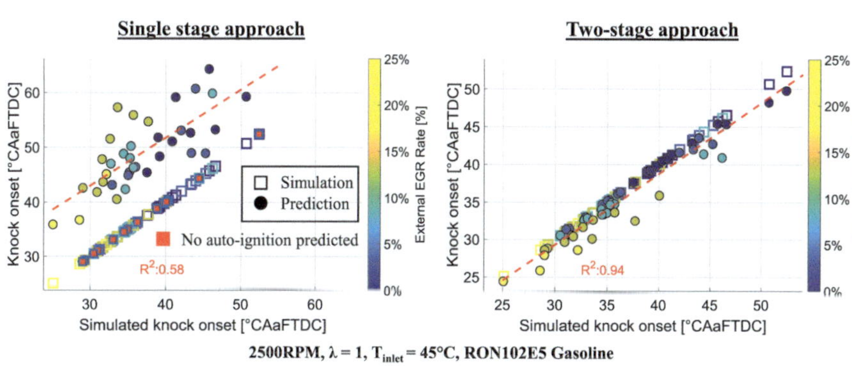

Fig. 29. Auto-ignition prediction with the commonly used single stage and the newly developed two-stage approaches, RON102E5 gasoline. For some working cycles, the single integral predicted no auto-ignition at all.

developed two-stage (Eq. (13)) approaches shown in Fig. 29 makes obvious the huge gain in accuracy achieved in this work. However, a local auto-ignition does not necessarily result in knock [Kön90]. Hence, a criterion for occurrence of knock as a result from the auto-ignition is still needed.

5.6 Knock Occurrence Criterion

As already discussed in Sect. 5.2, a constant end-of-integration point, as assumed in the commonly used knock models, leads to huge prediction errors, because the MFB-point of knock onset changes significantly with the boundary conditions, Fig. 20. Hence, a cycle-individual "latest MFB-point where knock can occur" has to be determined as a function of the boundary conditions. At this point, using an empirical fit will cause problems as soon as the simulated engine configuration or the used simulation models used are changed. Therefore, a knock occurrence criterion based on the unburnt mass fraction in the thermal boundary layer at the time of auto-ignition was developed. This boundary layer develops on the cylinder head, liner and piston over time as a function of a handful of parameters [Lyf84]. As soon as the propagating flame reaches the cylinder walls, the boundary layer at the corresponding locations of contact demolishes. Hence, estimating the unburnt mass fraction in the boundary layer requires spatial information about the flame propagation over time. To this end, a hemispherical flame propagation starting at the spark plug is assumed [Gri06], Fig. 30.

The boundary layer temperature T_{bl} is chosen as the mean value of the unburnt T_u and the wall temperatures [Lyf84]. The layer thickness δ_t at a certain location and a specified point of time can be estimated with Eq. (14) [Lyf84]. The elapsed time t for locations below the FTDC piston position begins at the moment they are uncovered by the piston. For all locations in the clearance volume, FTDC is defined as the starting time.

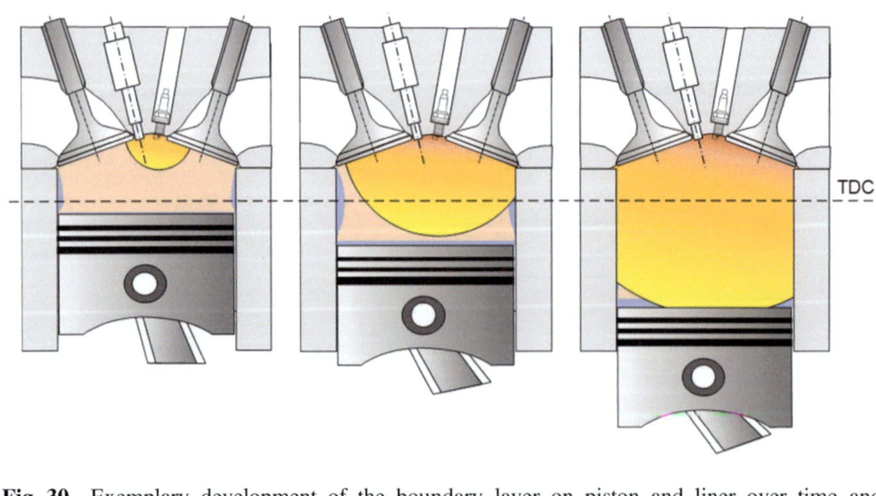

Fig. 30. Exemplary development of the boundary layer on piston and liner over time and counteraction with the flame.

$$\delta_t = 0.6 \cdot Re^{0.2} \cdot \sqrt{\alpha t} \tag{14}$$

$$x_{ub,bl} = \left(\frac{T_{bl}}{T_{ub}} \cdot \left(\frac{1}{V_{ub,bl}} - 1 \right) + 1 \right)^{-1} \tag{15}$$

Equation (14) further involves the calculation of some thermodynamic properties of the gas such as the gas viscosity, thermal conductivity and diffusivity α as well as the Prandtl number. The values of these parameters were evaluated at the boundary layer temperature with the correlations proposed in [Man80]. The local gas velocity needed for the calculation of the Reynolds number *Re* was evaluated by assuming a one-dimensional axial gas flow with velocity varying linearly from zero at the cylinder head to the piston velocity up at the piston crown [Lyf84]. Consequently, the boundary layer volume $V_{u,bl}$ and thus the fraction of unburnt mass in the boundary layer at a specified point of time $x_{u,bl}$ can be calculated with Eq. (15). The development of the thermal boundary layer is estimated for every working cycle. In this way, the current operating conditions are considered, as they influence the boundary layer development significantly. Hence, the criterion proposed here yields a working cycle individual latest MFB-point where knock can occur. Furthermore, it considers a handful of cylinder parameters, such as the TDC clearance, bore, stroke, spark plug position and the piston diameter. Thus, the knock model also accounts for differences in the engine geometry. A threshold for the unburnt mass fraction at auto ignition is the only calibration parameter of the knock model and it does not change with the operating conditions. It is assumed that, if the unburnt mass fraction in the boundary layer at auto-ignition is higher than the pre-defined threshold calibrated at the measured knock boundary, no knock can occur. This is because a significant amount of the unburnt mixture has a temperature that equals the one of the boundary layer and hence is much lower than the mean unburnt temperature.

5.7 Short Discussion on the Knock Model Inputs

Because of the mixture and temperature inhomogeneities in the unburnt mass, the question arises, if and how these should be considered. The temperature is by far the most critical knock model input (exponential influence on the ignition delay times of both ignition stages), so small temperature differences can cause significant changes in the auto-ignition behavior predicted by Eq. (13). To this end, extensive investigations were performed to evaluate if temperature and mixture inhomogeneities have to be accounted for in order to reliably predict the knock boundary with the knock onset criterion presented in Sect. 5.6.

Generally, the higher the unburnt temperature level is, the earlier the predicted auto-ignition will occur, and therefore the smaller the unburnt mass fraction in the boundary layer will be at auto-ignition. Thus, inaccuracies in other simulation models or the values of their calibration parameters (e.g. cylinder wall temperatures or wall heat transfer) are reflected in the knock model calibration parameter and can be compensated for by adjusting its value in the course of the model calibration. Consequently, the performed investigations showed that the knock occurrence proposed in

Sect. 5.6 yields accurate prediction results, even if the calculations are performed with the mean unburnt temperature and the global AFR, as will be demonstrated in Sect. 5.8. This can sometimes result in perhaps unrealistic thresholds for the percentage of unburnt mass in the boundary layer at KLSA of about 10–15%, depending on the boundary conditions and the chosen simulation models. However, in this way a submodel for the mixture and temperature inhomogeneities is not necessarily needed for the reliable prediction of the knock boundary – a fact that eliminates the need for any empirical submodels, as currently no quasi-dimensional inhomogeneities approaches exist. Hence, unlike all commonly used knock models that involve the evaluation of empirical fits containing engine-specific effects, the new model can be applied to different engines without any limitations.

Additional investigations have proven that a similar prediction quality is achievable with different wall heat transfer approaches. Currently, the simulation of single cycles with a respective model does not improve the knock boundary prediction accuracy, as it requires information about the local hot spot temperatures, and leads to a significant increase in computational time. Consequently, the proneness to knock in this case is also assessed based on the respective averaged cycle.

Overall, the flawless model use with different engine configurations is ensured, as no empirical fits derived from measurement data are included in the new knock model. Because of the significant influence of other simulation models on the knock model calibration parameter however, its absolute value is not necessarily meaningful and should not be interpreted.

5.8 Knock Model Validation

The performance of the newly developed knock model has been evaluated extensively at different boundary conditions on three different engines. The validation process involved the comparison of the measured MFB50-points at the knock boundary with the ones predicted by the knock model in the course of 0D/1D engine simulations. For this purpose, all measured operating points were simulated with 0D/1D and three pressure analysis models of the investigated engines. All simulation models (e.g. combustion, turbulence etc.) were calibrated so that they reproduce the behavior of each engine as accurately as possible. Thus, the simulations yield temperature and pressure curves as well as heat release rates similar to those estimated from the test bench data. The single knock model calibration parameter was estimated at one operating point for each of the investigated engines respectively and subsequently kept constant during the validation process.

The main validation goal was to ensure the model responds correctly to changes in the engine configuration and operating conditions represented by the direction of shift of the MFB50-points at the knock boundary. A further objective was to achieve an error in the predicted MFB50-point at the knock boundary of 2°CA or less with the constant threshold for the unburnt mass fraction in the cool boundary layer, calibrated at the measured knock boundary. Furthermore, because the knock model does not include any empirical submodels that could contain engine-specific effects, it is expected that the quality of knock boundary prediction does not vary significantly between different engines, provided that the calibration quality of the rest of the simulation models is equal.

Engine 1

To begin with, the knock model performance was evaluated for the single-cylinder engine setup presented in Sect. 3.1, Fig. 31. The operating points include variations of engine speed, EGR rate, inlet temperature and AFR (full load enrichment replacement by EGR). Overall, the prediction performance of the new knock model is very good and the effects of all boundary conditions on the KLSA are predicted correctly. When assessing the prediction performance of the new model, it should be considered that each measured point at the knock boundary has a different knock frequency, as the boundary was defined as a knock frequency window, Sect. 3.1. Nevertheless, the calculated MFB50-points at the knock boundary are within the defined limits of 2°CA. This high prediction quality enables the investigation of different combustion systems, engine configurations and operating conditions.

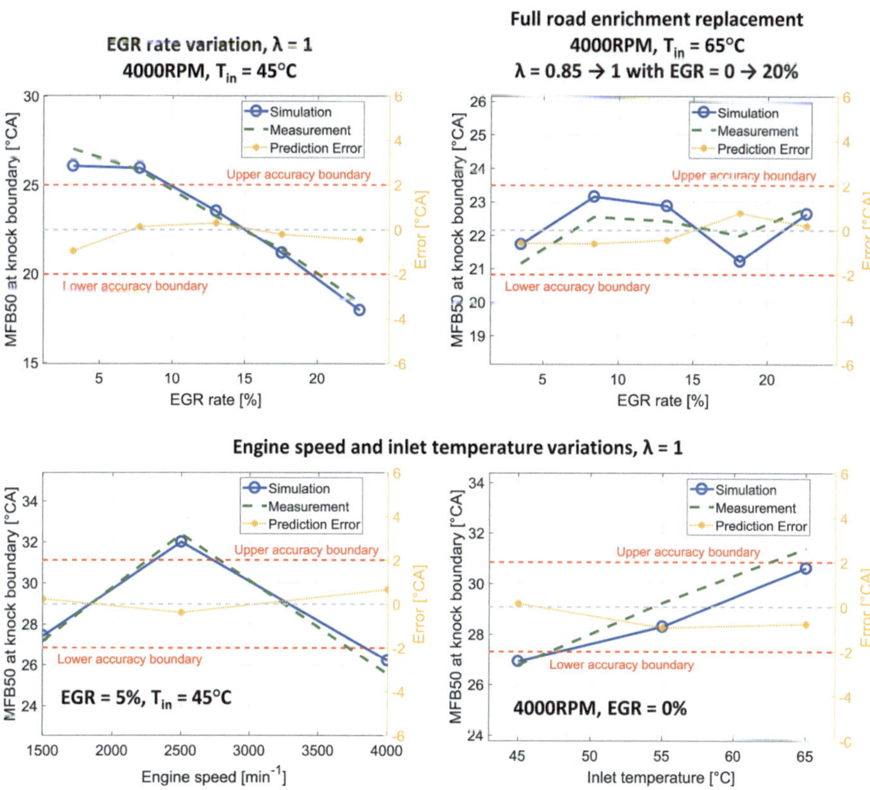

Fig. 31. Knock model validation, engine 1.

Engine 2

The second engine is a single-cylinder configuration that differs significantly from engine 1, as it has a different cylinder head with a tumble runner design and piston with a higher compression ratio of 11.84, Sect. 3.1. The corresponding model validation results for a load increase can be found in Fig. 32. Again, the prediction quality is very high, despite the significant change in the unburnt temperature level with engine load. Hence, the model is capable of reliably predicting the influence of different loads (and thus compression ratios, as this parameter affects the knock model inputs in a similar way) on the knock behavior.

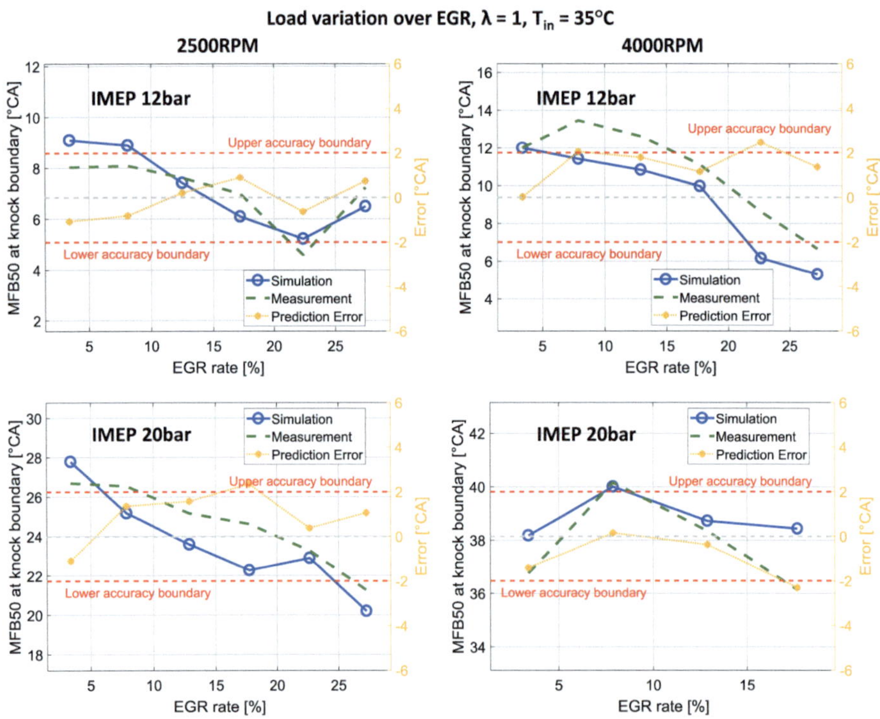

Fig. 32. Knock model validation, engine 2.

Engine 3

The third engine the knock model performance was evaluated on is a three-cylinder configuration described in [Hop15]. Because of the EGR cooler's limited cooling power, at higher EGR rates, the inlet temperature could not be maintained constant, Fig. 33. This on the other hand led to a decrease in the effective mean pressure at the knock boundary. Hence, this measurement series contains significant simultaneous changes in a handful of the operating conditions at a very high engine speed of 5500RPM. Despite the complex interrelations these result in, the performance of the knock model is very good, with both the direction of the shift and the absolute values of the MFB50-points at the knock boundary being predicted accurately by the new model, Fig. 33.

Overall, the model validation has shown that the model considers changes in the engine configuration and operating conditions correctly by accurately predicting the direction of the shift of the MFB50-points and their absolute values at the knock boundary. Moreover, the model has been successfully validated on three different engines with no significant variations in the knock boundary prediction quality. It has been demonstrated that an excellent knock prediction performance can be achieved with the mean values of the unburnt mixture's parameters being used as model inputs, even if the operating conditions change significantly. This is possible thanks to the lack of any empirical submodels, the very accurate two-stage auto-ignition prediction and the cycle-individual knock occurrence criterion that takes cylinder geometry, flame propagation and the operating conditions into account.

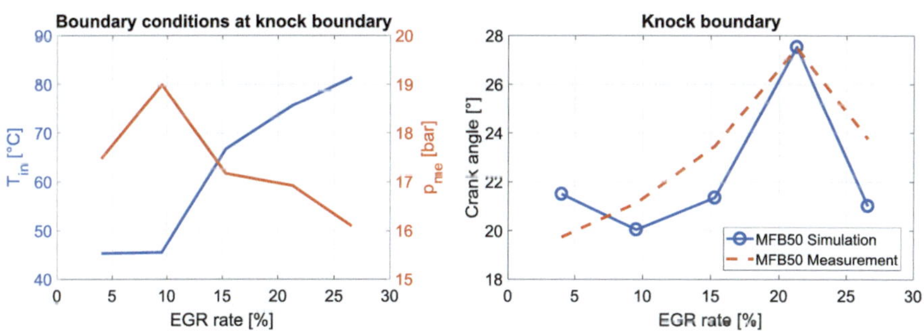

Fig. 33. Knock model validation, engine 3.

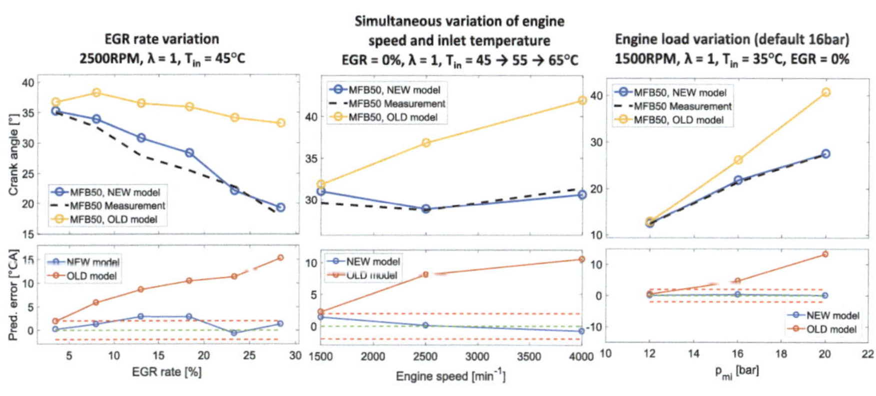

Fig. 34. Knock boundary prediction with a commonly used and the newly developed knock models.

Finally, Fig. 34 shows a comparison between the knock boundaries predicted with the newly developed knock model and an approach commonly used today [Sch10], demonstrating the overall huge gain in prediction accuracy achieved in this work.

6 Conclusion and Outlook

With the help of fundamental investigations the auto-ignition delay times of a gasoline fuel under the presence of EGR were measured. Based on recent research a surrogate mixture has been defined, which was successfully validated in a CFR engine for RON and MON. Subsequently a kinetic mechanism for gasoline surrogate fuels was developed and validated with the measurement data. For application in simulation models this mechanism has been reduced while maintaining a high accuracy. With the help of this mechanism problems involving real gasoline fuels especially in the field of EGR diluted combustion systems can be analyzed in more detail.

Single cylinder experiments were performed to investigate the influence of EGR on combustion in a detailed way. EGR was found to successfully inhibit knock allowing an up to 9° CA advanced center of combustion and an about 4% increased indicated efficiency at an EGR rate of 25%. For the number of knocking cycles, an exponential increase was observed in the area of KLSA. With spark timings relative to the KLSA similar knocking behavior with and without EGR was observed. The influence of nitric oxide on engine knock was studied as well. Trace amounts of NO (100 ppm) were found to lead to stronger knock limitation. With either decreased or increased NO concentration the knock tendency was reduced. With these findings experiments with catalyzed exhaust gas were carried out. Experimental results revealed that post catalyst extraction is favorable when the cylinder NO concentration can be reduced well below 100 ppm allowing a 2° CA advanced center of combustion at 25% EGR.

The kinetic mechanism was used in a new approach to predict the occurrence of knock in 3D-CFD simulations. For the investigated operating points the knock limit of an EGR combustion system could be detected with sufficient accuracy. Detailed analysis of the simulation results showed that two-stage ignition is playing a major role during knock occurrence for EGR combustion systems. A pool of formaldehyde and hydrogen peroxide is formed during first stage ignition which is then consumed during the second stage of ignition. Subsequently a pressure wave is triggered.

By developing a model reproducing the unburnt mixture at in-cylinder conditions and simulating thousands of working cycles with a detailed reaction kinetics mechanism, it was confirmed that two-stage ignition can occur in the unburnt mixture in conventional SI engines, resulting in knock. Subsequently, a two-stage knock integral approach was proposed for the prediction of the time of auto-ignition. Furthermore, a criterion for knock occurring as a result from the auto-ignition based on the unburnt mass fraction in the cool thermal boundary layer was developed. Finally, the excellent prediction performance of the new knock model was demonstrated on three different engines. The newly developed knock model for the 0D/1D SI engine simulation contains no empirical submodels and has just one calibration parameter that does not change with the boundary conditions. Therefore, the new model can be applied to different engines without any limitations, after a simple recalibration.

Acknowledgements. The presented work has been performed as part of the research tasks within the project "Knock with EGR at full load" defined and financed by the Research Association for Combustion Engines (FVV) e.V. (FVV Project 6301). The authors would like to thank the working group that accompanied the research work and all the companies involved for their support as well as the FVV for granting the financing.

Abbreviations

α_{h50}	Center of combustion
λ	Relative air-to-fuel ratio
AFR	Air-to-Fuel Ratio
aTDC	After top dead center
bTDC	Before top dead center
c_p	Specific heat capacity at constant pressure
CA	Crank angle
CFD	Computational fluid dynamics
CH_2O	Formaldehyde
CO	Carbon monoxide
CO_2	Carbon dioxide
D	Diameter/Bore
DI	Direct injection
DOHC	Double overhead camshaft
EGR	Exhaust gas recirculation
gHCCI	Gasoline Homogeneous Charge Compression Ignition
HC	Hydrocarbons
HOOH	Hydrogen peroxide
KLSA	Knock Limited Spark Advance
KO	Knock Onset
KPP	Knock Peak-to-Peak
MFB	Mass Fraction Burned
MON	Motor octane number
n	Engine speed
NO_X	Nitrogen oxides
NTC	Negative Temperature Coefficient
O_2	Oxygen
p_{mi}	Indicated mean effective pressure
PN	Particulate number
PTA	Pressure Trace Analysis
RANS	Reynolds-averaged Navier-Stokes equations
RCM	Rapid compression machine
RGF	Residual gas mass fraction
RON	Research octane number
s	Stroke
ST	Shocktube
TDC_{firing}	Firing Top Dead Center
x_{EGR}	EGR rate

References

[Alg07] Alger, T., Gingrich, J., Mangold, B.: The effect of hydrogen enrichment on EGR tolerance in spark ignited engines, SAE Technical Paper 2007-01-0475 (2007)

[And10] Andersen, V.F., Anderson, J.E., Wallington, T.J., Mueller, S.A., Nielsen, O.J.: Distillation curves for alcohol-gasoline blends. Energy Fuels **24**, 2683–2691 (2010)

[Bap05] Babajimopoulos, A., Assanis, D.N., Flowers, D.L., Aceves, S.M., Hessel, R.P.: A fully integrated computational fluid dynamics and multi-zone model with detailed chemical kinetics for the simulation of premixed charge compression ignition engines. Int. J. Engine Res. **6**, 497–512 (2005)

[Ben76] Benson, S.W.: Thermochemical Kinetics, 2nd edn. Wiley, New York (1976)

[Bur15] Burke, S.M., Simmie, J.M., Curran, H.J.: Critical evaluation of thermochemical properties of C1–C4 species: updated group-contributions to estimate thermochemical properties. J. Phys. Chem. Ref. Data **44**, 013101 (2015)

[Cai16] Cai, L., Pitsch, H.: Optimized chemical mechanism for combustion of gasoline surrogate fuels. Combust. Flame **162**, 1623–1637 (2015)

[Cai16_2] Cai, L., Pitsch, H., Mohamed, S.Y., Raman, V., Bugler, J., Curran, H., Sarathy, S. M.: Optimized reaction mechanism rate rules for ignition of normal alkanes. Combust. Flame **173**, 468–482 (2016)

[Cur02] Curran, H., Gaffuri, P., Pitz, W.J., Westbrook, C.K.: A comprehensive modeling study of iso-octane oxidation. Combust. Flame **129**, 253–280 (2002)

[Deb06] Dubreuil, A., Foucher, F., Mounaïm-Rousselle, C.: Effect of EGR chemical components and intake temperature on HCCI combustion development, SAE Technical Paper 2006-32-0044 (2006)

[Fan17] Fandakov, A., Grill, M., Bargende, M., Kulzer, A.C.: Two-stage ignition occurrence in the end gas and modeling its influence on engine knock, SAE Technical Paper 2017-24-0001 (2017)

[Fra91] Franzke, D.: Beitrag zur Ermittlung eines Klopfkriteriums der ottomotorischen Verbrennung und zur Vorausberechnung der Klopfgrenze, Ph.D. thesis, Technical University of Munich (1991)

[Goo15] Goodwin, D.G.; Moffat, H.K., Speth, R.L.: Cantera: an object-oriented software toolkit for chemical kinetics, thermodynamics, and transport processes, Version 2.2.1, Zenodo (2016)

[Gri06] Grill, M., Billinger, T., Bargende, M.: Quasi-dimensional modeling of spark ignition engine combustion with variable valve train, SAE Technical Paper 2006-01-1107 (2006)

[Her14] Hernández, J., Lapuerta, M., Sanz-Argent, J.: Autoignition prediction capability of the Livengood–Wu correlation applied to fuels of commercial interest. Int. J. Engine Res. **15**(7), 817–829 (2014)

[Hop15] Hoppe, P., Lehrheuer, B., Pischinger, S.: Downsizing with Biofuels II, Final Report, FVV Spring Conference, Research Association for Combustion Engines (FVV) e. V., Bad Neuenahr (2015)

[Höp12] Hoepke, B., Jannsen, S., Kasseris, E., Cheng, W.: EGR effects on boosted SI engine operation and knock integral correlation. SAE Int. J. Eng. **5**(2), 547–559 (2012)

[Jak11] Jakob, M., Pischinger, S., Adomeit, P., Brunn, A., Ewald, J.: Effect of intake port design on the flow field stability of a gasoline DI engine, SAE Technical Paper 2011-01-1284 (2011)

[Kön90] König, G., Maly, R., Bradley, D., Lau, A., et al.: Role of exothermic centres on knock initiation and knock damage, SAE Technical Paper 902136 (1990)

[Lee12] Lee, C.S., Vranckx, S., Heufer, K.A., Khomik, S.V., Uygun, Y., Olivier, H., Fernandez, R.X.: On the chemical kinetics of ethanol oxidation: shock tube, rapid compression machine and detailed modeling study. Z. Phys. Chem. **226**, 1–28 (2012)

[Liv55] Livengood, J.C., Wu, P.C.: Correlation of autoignition phenomena in internal combustion engines and rapid compression machines. In: Symposium International Combustion (1955)

[Lyf84] Lyford-Pike, E.J., Heywood, J.B.: Thermal boundary layer thickness in the cylinder of a spark-ignition engine. Int. J. Heat Mass Transf. **27**(10), 1873–1878 (1984)

[Man80] Mansouri, S.H., Heywood, J.B.: Correlations for the viscosity and Prandtl number of hydrocarbon-air combustion products. Combust. Sci. Technol. **23**, 251–256 (1980)

[Mor05] Morley, C.: Gaseq. http://www.gaseq.co.uk/

[Mor10] Morgan, N., Smallbone, A., Bhave, A., Kraft, M., Cracknell, R., Kalghatgi, G.: Mapping surrogate gasoline compositions into RON/MON space. Combust. Flame **157**, 1122–1131 (2010)

[Par15] Parsons, D., Akehurst, S., Brace, C.: The potential of catalysed exhaust gas recirculation to improve high-load operation in spark ignition engine. Int. J. Eng. Res. **16**, 592–605 (2015)

[Pep07] Pepiot-Desjardins, P., Pitsch, H.: An efficient error-propagation-based reduction method for large chemical kinetic mechanisms. Combust. Flame **154**, 67–81 (2008)

[Sch10] Schmid, A., Grill, M., Berner, H.-J., Bargende, M.: A new approach for SI-Knock prediction. In: 3. IAV International Conference on Knocking in Gasoline Engines, Berlin (2010)

[She14] Shock and Detonation Toolbox. http://shepherd.caltech.edu/EDL

[Sun14] Sung, C.-J., Curran, H.J.: Using rapid compression machines for chemical kinetics studies. Progress Energy Combust. Sci. **44**, 1–18 (2014)

[Tan03] Tanaka, S., Ayala, F., Keck, J.C., Heywood, J.B.: Two-stage ignition in HCCI combustion and HCCI control by fuels and additives. Combust. Flame **132**, 219–239 (2003)

[Zha16] Zhang, K., Banyon, C., Bugler, J., Curran, H.J., Rodriguez, A., Herbinet, O., Battin-Leclerc, F., B'Chir, C., Heufer, K.A.: An updated experimental and kinetic modeling study of n-heptane oxidation. Combust. Flame **172**, 116–135 (2016)

Optical Measurement Technique

Visualization of Fuel Wall Wetting, Oil Dilution by Fuel, and Oil Transport Mechanisms in an Optically Accessible Engine by LIF Imaging

Stefan Wigger[1], Torben Müller[1(✉)], Hans-Jürgen Füßer[1], and Sebastian Kaiser[2]

[1] Daimler AG, Ulm, Germany
torben.mueller@daimler.com
[2] University of Duisburg-Essen, Duisburg, Germany

Abstract. Fuel impingement on the cylinder wall causes problems in both lubrication and combustion due to interaction of liquid fuel and engine oil. To analyze this issue with temporal and spatial resolution, a laser-induced fluorescence (LIF) system for simultaneous kHz-rate imaging of fuel and oil films on the cylinder wall is presented. Experiments were performed in a research engine with optical access along the entire stroke of the cylinder in fired and motored engine operation. Besides fuel wall wetting, fuel transport across the piston ring pack and the impact on the lubricant conditions in the piston group can be seen in the images.

1 Introduction

Modern direct-injection spark-ignition engines can exhibit fuel wall wetting, influencing mixture formation and combustion. Piston fuel films may cause pool fires and generate soot [1, 2]. Also, fuel impingement on the cylinder wall causes fuel-oil interactions, and a modification of the tribologic properties of the lubricant, such as wear resistance and viscosity [3]. As a consequence, adverse mixture formation due to fuel rich zones, as well as lower oil viscosity cause emissions of hydrocarbons [4, 5]. Also, mixtures of fuel and oil, particulate substances originating from oil deposits and oil droplets in the combustion chamber, are suspected to cause abnormal combustion, such as low-speed pre-ignition [6–8]. Therefore, a deeper understanding of the phenomena occurring in the piston group of DISI engines is needed.

In previous approaches, the oil dilution with fuel [9, 10] and the oil emission in the exhaust gas [5] have been investigated by mass spectrometry. This technique is non-invasive, thus it can be applied in production engines. However, the system response versus the variation of the input variables, like the injection timing or the engine load, is delayed. Furthermore, the origin of the examined phenomena is difficult to specify due to the indirect character of this technique.

To provide data instantaneously and at high resolution in time and space, laser-induced fluorescence (LIF) has been used in optically accessible engines. Non-fluorescing surrogate fuels were doped with fluorescing "tracers" such as ketones or aromatics with well-known photophysical properties [11]. Hereby temperature, concentrations and fuel/air ratios could be measured [12–14]. Tracer LIF has also been used to

M. Günther and M. Sens (eds.), *Knocking in Gasoline Engines*,
https://doi.org/10.1007/978-3-319-69760-4_10

investigate liquid fuel films in internal combustion engines [15]. Beside the fuel film on the piston crown [1, 2, 16], the fuel wall wetting on the cylinder liner was visualized with LIF [17, 18].

In addition to the imaging of liquid fuel films, the imaging of oil in optical engines with LIF is also well-known. For this purpose, a fluorescence dye was added to the engine oil. Weimar presents a measurement technique with fiber optics to observe oil on the cylinder wall [19]. To provide a larger field of view, an LIF system was developed to image the oil in the piston group of an optically accessible engine with a camera [20]. Wigger improved this camera-based approach towards a quantitative measurement of the oil film thickness in a model experiment [21] and applied LIF in an optically accessible engine [22]. Using a high-speed camera and laser, Zanghi [23] extended the technique record crank-angle resolved data series throughout a single cycle.

This paper deals with the combination of a tracer and a dye LIF to simultaneously investigate fuel and oil. Therefore, tracer-doped surrogate fuel and oil are stimulated by two laser beams with different wavelengths. The fluorescence light from the two fluids is spectrally separated and detected by two cameras. The experiments were carried out in an optically accessible engine in motored and fired engine operation. Beside basic studies of oil transport mechanisms and fuel wall-wetting, the ability to visualize the interaction of fuel and oil in the piston group is demonstrated.

2 Optically Accessible Single Cylinder Research Engine

All investigations were performed in a purpose-built designed research engine. The single-cylinder engine is equipped with two honed-in sapphire windows (height: 93 mm, width: 18 mm), which provide optical access to the cylinder on the thrust and antithrust side. Figure 1 shows the upper part of the engine with one of these windows. This enables the observation of the piston including the piston ring area and the piston skirt as well as the cylinder surface. Lubrication in the piston group as well as fuel wall-wetting can be observed.

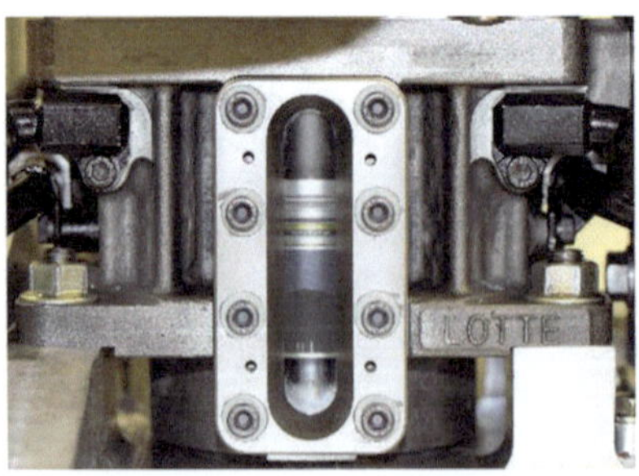

Fig. 1. Sapphire window on the thrust side.

First and second order mass balance allow engine speeds up to 6000 min^{-1}. The dimensions of the power cylinder were chosen according the 2.0 l version of the Daimler M274 gasoline engine and are listed in Table 1. The maximum indicated mean pressure

(IMEP) possible in the optical engine depends on the window size. The window design shown in Fig. 1 was tested up to 14 bar IMEP, while with slightly smaller windows the engine was tested up to 25 bar IMEP.

Table 1. Engine parameters.

Compression ratio	9.8
Displacement [cm^3]	498
Bore [mm]	83
Stroke [mm]	92
Max. speed [min^{-1}]	6000
Max IMEP [bar]	14 (25)
Fuel injection	piezo DI

3 LIF Imaging

3.1 Laser-Induced Fluorescence of a Liquid Film

The Lambert-Beer law describes the absorption of light by a transparent medium. Incident radiation with the intensity I_0 and the wavelength λ is attenuated as a function of the thickness d of the absorbing layer, the concentration c of the absorber, and the molar absorption coefficient ε. The intensity I_0 of the absorbed light is given by:

$$I_a = I_0 \cdot (1 - e^{\varepsilon cd}) \tag{1}$$

A fluorescent absorber (fluorescent dye or tracer) converts a portion of the absorbed light into fluorescence with intensity I_f. The ratio between the emitted and the absorbed photons is the quantum yield Φ. Due to rotational-vibrational relaxation, most of the fluorescence light is of lower photon energy and thus red-shifted to longer wavelengths with respect to the excitation wavelength. This red-shift can be used for separating fluorescence and scattered laser radiation by optical filters. The collection efficiency η of the detection system is determined by the optical path, i.e., filters, lenses, detector, etc.

$$I_f = \Phi \cdot \eta \cdot I_a = \Phi \cdot \eta \cdot I_0(1 - e^{\varepsilon cd}) \tag{2}$$

Since I_f depends on the concentration and the thickness of the absorbing medium, the film thickness can be derived from the LIF intensity. The quantification of the film thickness in fired engine operation is challenging, as laser-induced fluorescence may be influenced by temperature and oxygen concentration. Here, we will restrict the analysis to supply qualitative information of the phenomena in the piston group.

3.2 Optics, Fuel, and Oil

Figure 2 schematically shows the LIF imaging system and in Fig. 3 the principal spectral characteristics of LIF excitation and detection are plotted. A diode-pumped Nd:YAG-laser emits bursts of frequency-doubled (2ω = 532 nm) and quadrupled (4ω = 266 nm)

10-ns pulses (1). The maximum pulse energy is 8 mJ at 532 nm and 2.2 mJ at 266 nm at a repetition rate of 10 kHz. Both laser beams are guided towards the cylinder by a "laser guiding arm" with several mirrors (2). The transverse beam profiles are homogenized and expanded in (3). Both beams pass the sapphire window of the optical engine and excite the oil and fuel in the cylinder (4). The fluorescence of both fluids is spectrally separated by the long-pass filter (5). Light with a wavelength shorter than 509 nm (fuel tracer) is reflected, longer wavelengths (oil dye) pass the filter. The oil detection channel consists of a second long-pass filter (6) at 547 nm blocking scattered laser light. The oil dye fluorescence is imaged by a high-speed CMOS camera with a lens (105 mm focal length, f/2.8) (7).

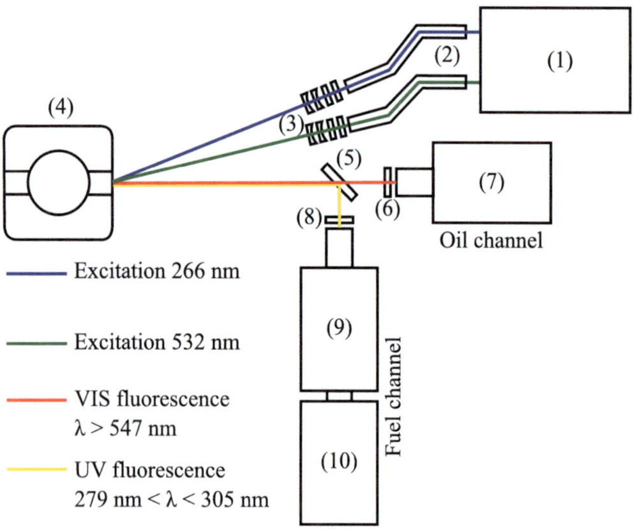

Fig. 2. Optical path of the LIF imaging system (numbered components are described in the text).

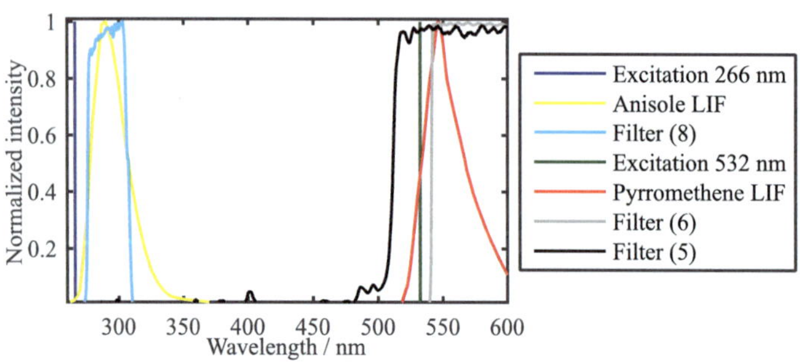

Fig. 3. Overview of the spectral features in the LIF system. Emission spectrum of pyrromethene 567 from [24], emission spectrum of anisole from [25], and transmission spectra of the filters from [26].

Fuel-tracer fluorescence in the range of 278-305 nm passes the bandpass filter (8) and is imaged by a UV lens (100 mm, f/2.8) onto a second camera equipped with a high-speed intensified relay optic (9 + 10). In the following, we will refer to the two images from the combined LIF system as "fuel channel" and "oil channel". The projected pixel size of the oil channel and of the fuel channel are about 85 μm/pixel.

Since commercial pump fuel contains a mixture of numerous species with different properties in terms of evaporation and photophysics, non-fluorescing surrogate fuels mixed with one (or more) tracer(s) are often used for fuel-tracer LIF. Here, a non-fluorescing, multi-component fuel with suitable properties such as research octane number (RON), evaporation characteristics, and density was used. Anisole (methoxybenzene) [13] and was used as a tracer.

To suppress oil fluorescence in the fuel channel, a customized, low-additive engine oil was used instead of first-fill oil. Pyrromethene 567, a laser dye with suitable optical properties, was mixed into the oil in trace amounts.

Optical Design of the LIF Imaging System

In Fig. 3 the light spectrum is shown in the range of 260–600 nm. The diagram includes the excitation wavelength of the two laser beams (266 nm and 532 nm), the transmission characteristics of the filters and the peak normalized fluorescence spectra of the fuel tracer and of the oil dye. The first separation of UV and VIS fluorescence is implemented by filter (5). Whereas the light with a wavelength shorter than 509 nm is reflected, the radiation with a longer wavelength passes the filter. The fluorescence of the dye used in this application (pyrromethene 567) shows a large overlap with the transmission spectrum of the long-pass filter (filter (6) in Fig. 2) after excitation with 532 nm. In addition, the fluorescence spectrum of anisole, excited by 266 nm, does not interfere with the oil channel.

Note that Fig. 3 only includes the spectra of fuel *tracer* and oil *dye*. Additionally, fluorescence and absorption from and by the fuel and oil *themselves* may occur. Whereas the surrogate fuel without a tracer did neither influence the fuel nor the oil detection, the surrogate oil did. Particularly in the fuel detection, the surrogate oil caused an interfering signal upon excitation at 266 nm, as will be shown in Sect 3.3.

3.3 Example Images

Figure 4 shows two background-corrected LIF images taken simultaneously in the fuel and in the oil channel on the anti-thrust side in motored engine operation. At a repetition rate of 10 kHz, one image was acquired per one or two degrees crank angle, depending on the engine speed. A separate linear color scale is adapted for each channel and figure. Lower film thicknesses are shown in blue and greater thicknesses in red. Note that the color scales are not calibrated against each other, thus a given color may represent completely different film thicknesses in each channel.

In the right image in Fig. 4, a photography of the optical access and the piston on the inside of the sapphire window mid-stroke shows the characteristic areas of the piston. This nomenclature is used in the following for spatial orientation.

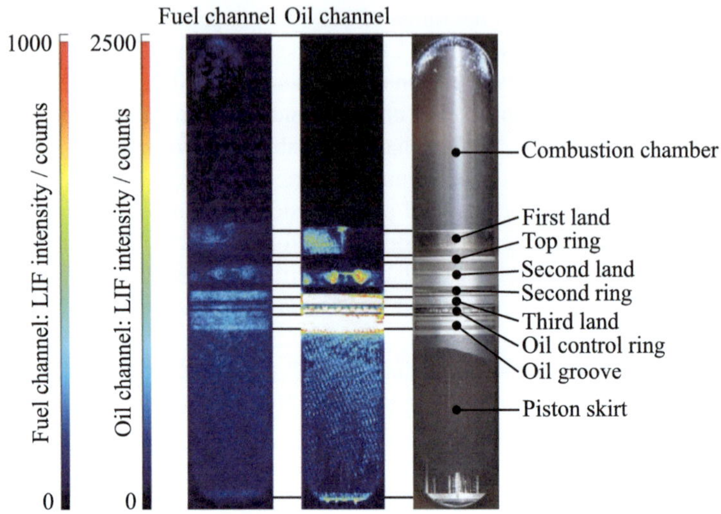

Fig. 4. Characteristic parts of the piston and exemplary false-color LIF images of the fuel and oil channels.

In the oil channel, oil was detected on the first land. The cylinder wall above the piston as well as the contact zones of the first and second ring show only weak signal, indicating minimal oil film thickness. Above and under the oil control ring, the signal is high and detector saturation occurred. The pattern on the piston skirt originated from the laser illumination and was not a physical pattern on the piston skirt (compare the piston skirt in the photography).

In the fuel channel, edges were imaged less sharp due to the intensifier, but the areas of the piston were still visible. Although the engine was motored without injection, fluorescence light could be detected in the fuel channel. The spatial distribution of the detected signal correlated with that of the oil channel. This indicates oil fluorescing in the spectral range of the fuel channel after 266 nm excitation. However, with the injection activates, much more signal was detected in the fuel channel, such that this cross-talk from oil fluorescence is negligible.

4 Measurements and Results

Fuel wall wetting and oil transport mechanisms depend on multiple parameters, such as the piston, piston rings, charge motion, and temperature, and thus on the performance of the entire system. Here, selected phenomena are investigated at different engine speeds and loads to show principal mechanisms in the piston group.

4.1 Motored vs. Fired Engine Operation

Oil Channel

In Fig. 5 images in the oil channel from the compression and the expansion stroke are shown. The engine speed is 2200 min^{-1} and the temperatures of oil and water were 50 °C. In this experiment, the top ring was pinned to hold the ring gap in the field of view. The piston position is given by crank angle in degree before (bTDC) and after top dead center (aTDC). Whereas in the first and second row motored engine operation with different intake pressures is shown, the third row shows fired engine operation.

In all images it was evident that only a small amount of oil remained on the cylinder wall after the piston rings had passed by and scraped off the oil. Also, a large amount of oil was in the oil groove and the third land. After upward movement of the piston, oil droplets originating from turbulences in the crankcase and from the piston-cooling jet were detected below the piston.

Figure 5(a) shows motored conditions with a closed throttle. A large oil volume was detected on the first and the second piston land independent from the position of the piston. In the expansion stroke, oil droplets were transported to the combustion chamber. This can be explained by means of pressure differences and an imperfect sealing of the piston rings. Since a closed throttle led to a low intake pressure and a higher pressure in the crankcase than in the combustion chamber, oil is transported upwards on the surface of the piston. Compression led to an increase in the pressure in the combustion chamber, but also in the second land of the piston. This pressure decreased in the expansion stroke, but slower in the second land than in the combustion chamber. This resulted in an upward gas force (reverse blowby), whereby oil was transported through the ring gap into the combustion chamber.

In Fig. 5(b) the amount of oil adhering to the first and to the second land decreased and oil droplets were not detected in the combustion chamber. In this case, the throttle was partly opened and the intake pressure and therefore the compression pressure increased. This led to stronger blowby gases pushing oil downwards on the surface of the piston and towards the crankcase. Since reverse blowby still occurred in the expansion stroke, the absence of oil on the first and on the second land prevented the transport of oil droplets to the combustion chamber.

In Fig. 5(c) the fired condition is shown, recognizable by the luminous flame above the piston in the expansion stroke. The LIF intensity on the piston lands is comparable to (b) and oil droplets were not detected during expansion. The lightly lower signal intensity on the second land is caused by two phenomena: First, the combustion pressure resulted in a higher blowby gas flow and therefore in a higher downward force pushing oil towards the crankcase. Second, the LIF intensity of the fluorescence dye decreased with a higher temperature. In this case the temperature of the oil increased by the combustion heat transferred to the piston and the oil adhering to the piston lands.

Fuel Channel

The images shown in Fig. 6 were taken in the compression stroke at 100 °CA bTDC, while the engine was running at 1200 min^{-1} in motored (left) and fired (right) operation (the water and the oil temperature were set to 40 °C). As discussed in Sect. 3.3, the oil fluorescence interfered with the fuel channel and caused a signal even when the engine

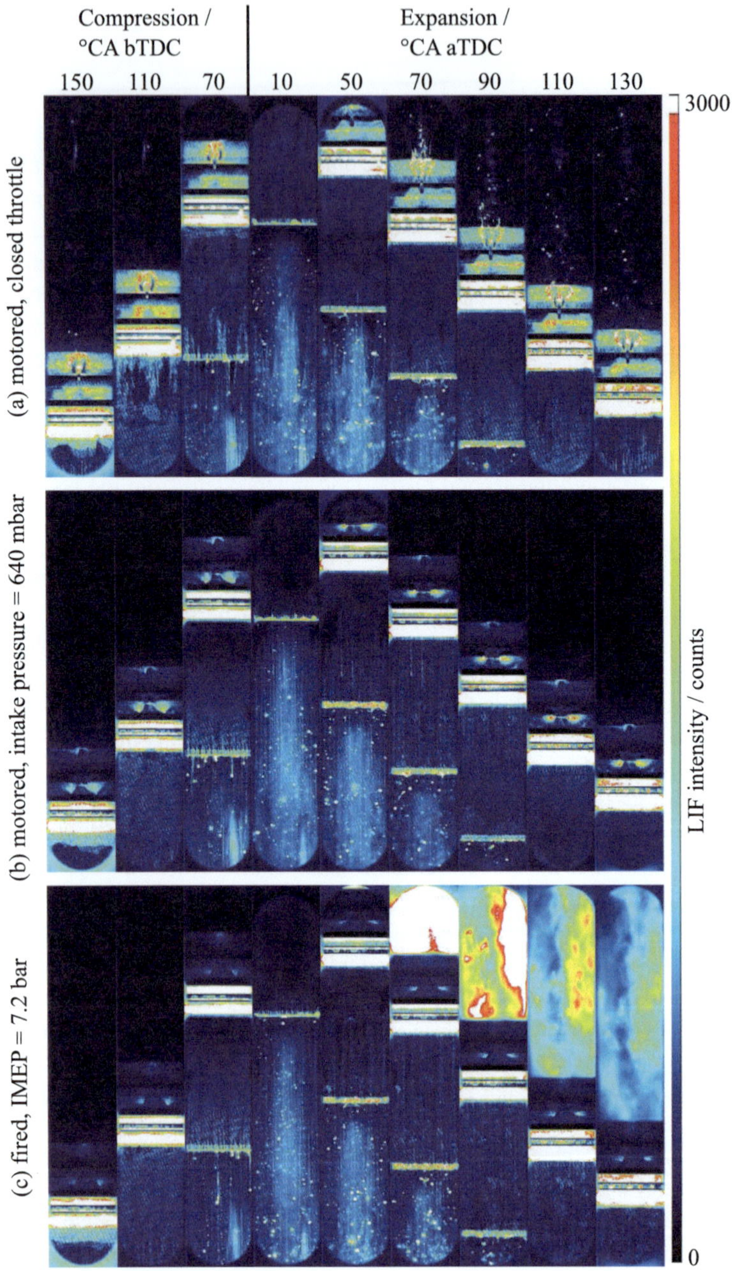

Fig. 5. LIF images in the oil channel: Intake and compression stroke for motored and fired engine operation.

was motored, and in absence of fuel (see left in Fig. 6). But there was also an increase in the detected signal intensity due to oil dilution by fuel, when the engine was fired and fuel mixed in with the oil on the piston lands (right image). In the right image, the detected fluorescence intensity in a single pixel is thus the depth-integrated fluorescence of fuel and of oil.

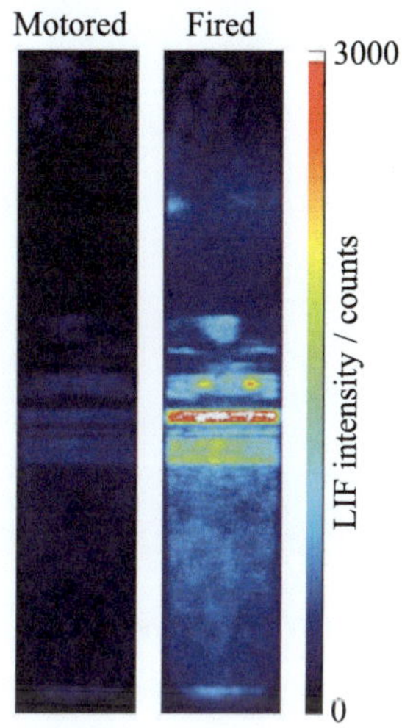

Fig. 6. Exemplary LIF images in the fuel channel: motored vs. fired engine operation.

4.2 Visualization of Fuel Wall Wetting and Fuel Transport into the Piston Lands

The images in Fig. 7 show the intake stroke while the engine was fired at 1500 min^{-1}, IMEP = 5 bar, and $\lambda = 1$ (water and oil temperature were 40 °C). The fuel was injected in a single injection at 210 °CA bTDC. The hollow-cone fuel spray reached the field of view at 205 °CA bTDC and fuel impinging on the cylinder wall resulted in a specific pattern in form of three spots at 185 °CA bTDC.

In the compression stroke shown in Fig. 8, the three fuel spots on the cylinder liner visualized in Fig. 7 could still be seen indicating incomplete evaporation. Whereas the amount of fuel on the wall remain equal in the range of 86–81 °CA bTDC, the detected LIF intensity increased. This was caused by fluorescence reflected on the aluminum surface of the piston first land in the background of the fuel spots. When the top ring reached the fuel accumulations at 76–71 °CA bTDC, the peak LIF intensity on the upper side of the top ring increased. This can be explained by the fact that fuel was scraped off the cylinder wall by the ring and accumulated to a higher thickness. Afterwards, fuel was transported from the cylinder wall to the piston lands ("bridging"). Further upward movement of the piston (71–66 °CA bTDC) led to fuel draining off through the ring gap into the second land.

In addition, fuel on the cylinder wall reached the second land, when the ring gap of the top ring passed by and the fuel was scraped off by the second ring. This can be seen at the increase of LIF intensity on the upper side of the second ring at 66–61 °CA bTDC. Once the fuel reaches the second land of the piston, fuel mixes with oil and is transported to the third land and to the oil groove.

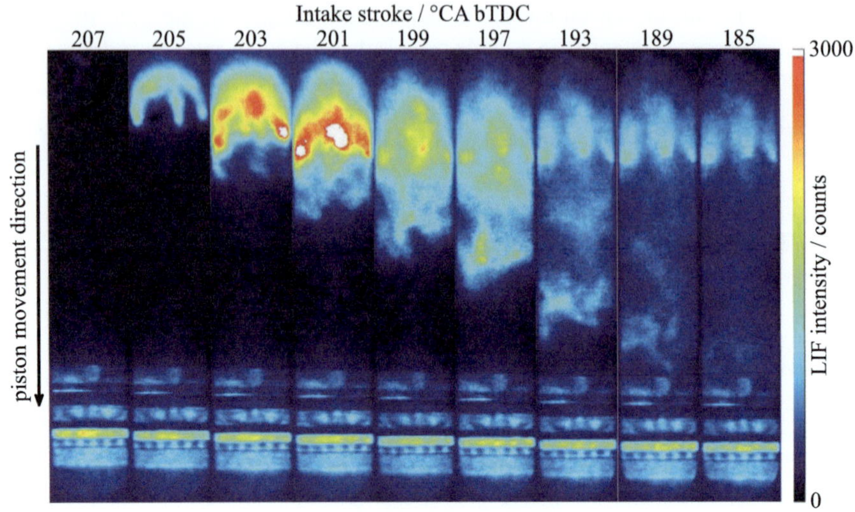

Fig. 7. LIF images in the fuel channel: Visualization of the fuel injection and fuel wall wetting during the intake stroke.

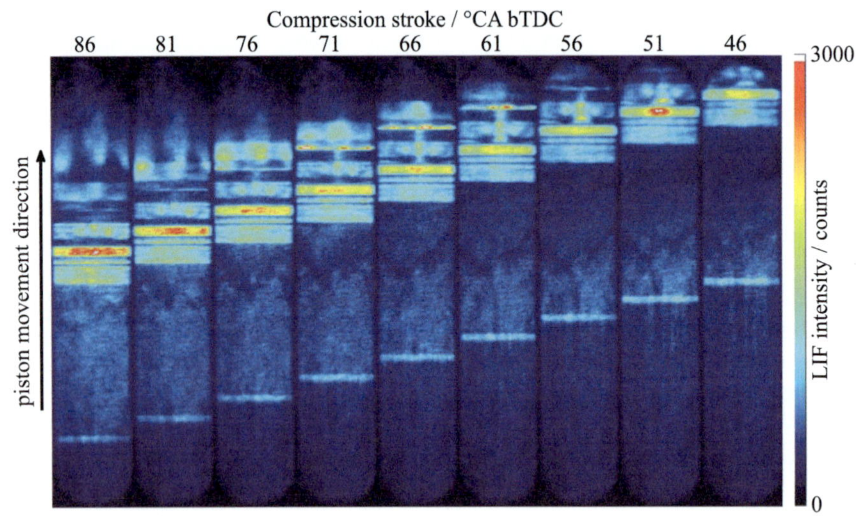

Fig. 8. LIF images in the fuel channel: "Bridging" of fuel from the cylinder wall on the piston lands due to ring scraping.

4.3 Variation of Injection Timing

To investigate the consequences of different injection strategies, the compression stroke of the base injection timing (first row in Fig. 9) was compared with an optimized timing (second row in Fig. 9). The engine load was 7.2 bar IMEP at 2200 min^{-1} (water and oil temperatures were 50 °C). Before imaging, the operating point was held for 10 s. The fuel, which hit the cylinder wall after base injection timing, remained there during the compression, as can be seen in the first row. Under these operating conditions the fuel did not spread into three single spots (compare Fig. 7), but into a single spot spread over a larger area. When the top ring reached the fuel spot at 86 °CA bTDC, fuel was scraped off and bridged from the cylinder wall onto the piston lands. During the further upward movement of the piston, the fuel accumulation drained off from the upper side of the top ring to the second land.

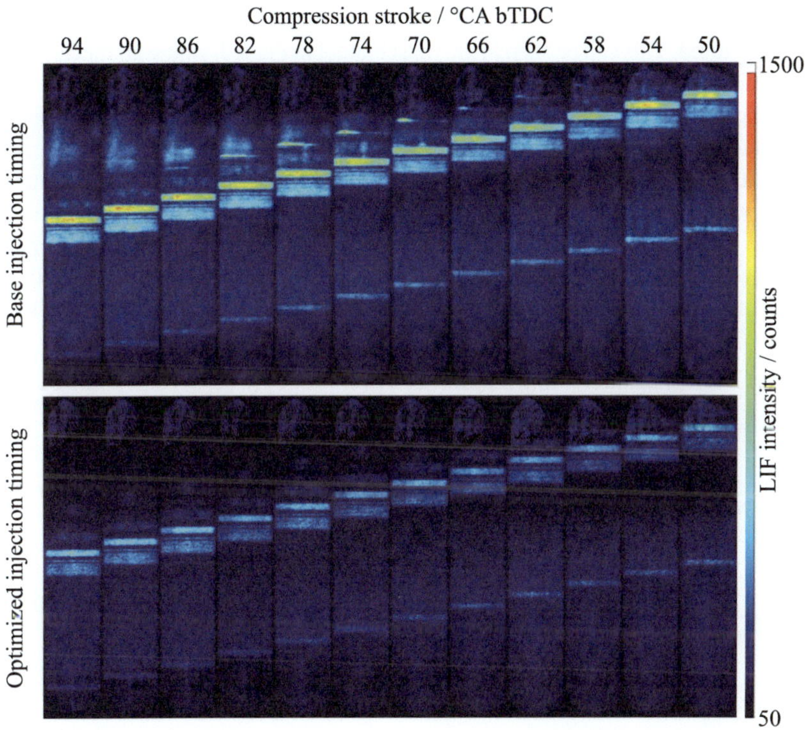

Fig. 9. LIF images in the fuel channel: Oil dilution with fuel for base and optimized injection timing.

With optimized injection timing, the fuel wall-wetting could be reduced and there was no fuel detected on the cylinder wall in the field of view during compression. Lower signal intensity in the area of the third land and in the oil groove indicates lower oil dilution by fuel. As discussed in Sect. 3.3, the detected fluorescence intensity was depth-integrated, and includes oil and fuel fluorescence. However, fuel wall wetting, and oil dilution by fuel, could be reduced by optimizing the injection timing.

4.4 Interaction of Fuel and Oil

In Fig. 10 fuel and oil channel are compared to show oil transport mechanisms caused by fuel wall wetting. The engine speed was 1500 min^{-1} with 500 mbar intake pressure in motored operation, while the IMEP was 5 bar with $\lambda = 0.9$ when the engine was fired (water and oil temperatures were 40 °C). To investigate long-term effects, one image was taken in each cycle at 100 °CA bTDC (stroboscopic measurement) simultaneously in the oil and in the fuel channel. The top ring was pinned on the opposite side in this experiment, thus there was no ring gap in the field of view. In the first step, the engine was motored with a partly opened throttle. We can see oil on the first and on the second land (see cycle 1 in the oil channel). This oil distribution remained equal for the first 100 cycles. From cycle 101 on, the engine was fired for 200 cycles, using the optimized injection timing. By using the optimized injection timing a reduced fuel wall wetting could be ensured. This can be seen in the fuel channel in Fig. 10, where no fuel is detected on the cylinder wall above the piston. However, the LIF intensity on the third land and in the oil groove increased during the fired engine operation. This means that fuel still accumulated in the piston group despite the optimized injection timing. Whereas fuel accumulation was detected on the piston lands when the engine was fired, the impact of fuel wall wetting on the oil distribution was minimal compared to motored engine operation (see oil channel in cycles 101–301). The latter fact changed when the injection timing was switched to the base timing (cycles 302–497). As can be seen in the fuel channel in the area above the piston, wall wetting occurred. Also, the LIF intensity

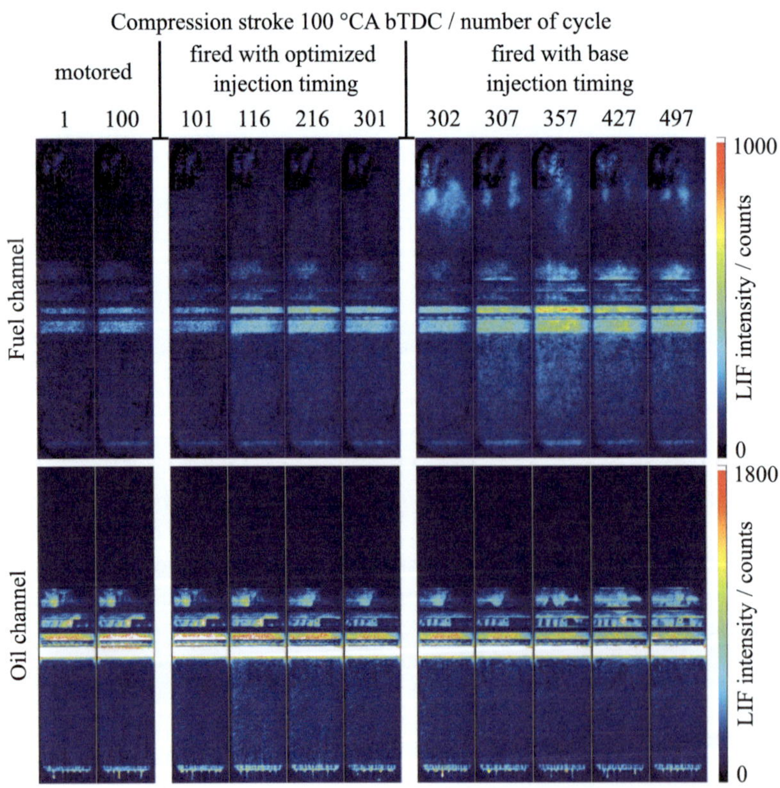

Fig. 10. LIF images in fuel and oil channels: Impact of fuel wall wetting on oil transport on the piston lands.

further increased in the area of the third land and in the oil groove due to a higher oil dilution by fuel.

Regarding the oil channel, it is mentioned that the LIF intensity increased in the area of the first land due to an accumulation of oil. This was caused by oil being flushed out of the ring groove, which resulted from the oil dilution by fuel, and thus from the decrease in viscosity.

5 Conclusions

In this work an optical experiment was presented to simultaneously visualize fuel and oil with laser-induced fluorescence (LIF) at high resolution in time and space. LIF imaging was performed in motored and fired engine operation in an optically accessible single-cylinder engine. It was shown that oil was transported through the ring pack into the combustion chamber in motored engine operation. In a second step, fuel wall wetting and fuel transport mechanisms into the piston lands were analyzed. It was also shown that optimized injection timing could reduce fuel wall-films and oil dilution with fuel. Also, long-term impacts of fuel wall wetting on oil transport mechanisms could be monitored. Here, the base injection timing caused an increase over the optimized timing in the oil transport towards the combustion chamber due to fuel-oil interaction.

This dual-channel imaging system offers new insights in the principal mechanisms that cause oil emissions, oil dilution with fuel, or unusual combustion events such as LSPI, under real engine operating conditions. While the current work focused on relatively low loads, experiments at up to 25 bar IMEP have recently been performed and will be reported on in future publications.

References

1. Stevens, E., Steeper, R.: Piston wetting in an optical DISI engine: fuel films, pool fires, and soot generation. SAE Technical Paper, 2001-01-1203 (2001)
2. Drake, M.C., et al.: Piston fuel films as a source of smoke and hydrocarbon emissions from a wall-controlled spark-ignited direct-injection engine. SAE Technical Paper, 2003-01-0547 (2003)
3. Schwarze, H., et al.: Lubricant degradation and wear behaviour in a spark-ignition engine. MTZ worldw. **69**(10), 60–67 (2008)
4. Landsberg, G.B., Heywood, J.B., Cheng, W.K.: Contribution of liquid fuel to hydrocarbon emissions in spark ignition engines. SAE Technical Paper, 2001-01-3587 (2001)
5. Krause, S.: Massenspektrometrisches Verfahren zur Charakterisierung der Ölverdampfung im Brennraum von Ottomotoren. Institut für Messtechnik, Technische Universität Hamburg-Harburg (2009)
6. Kassai, M., et al.: Research on the effect of lubricant oil and fuel properties on LSPI occurrence in boosted S. I. engines. SAE Technical Paper, 2016-01-2292 (2016)
7. Dingle, S.F., et al.: Lubricant induced pre-ignition in an optical SI engine. SAE Technical Paper, 2014-01-1222 (2014)
8. Okada, Y., et al.: Study of low-speed pre-ignition in boosted spark ignition engine. SAE Int. J. Engines, **7**(2) (2014)

9. Behn, A., et al.: Fuel transport across the piston ring pack: measurement system development and experiments for online fuel transport and oil dilution measurements. SAE Technical Paper, 2015-24-2535 (2015)
10. Artmann, C., et al.: A new measurement technique for online oil dilution measurement. SAE Int. J. Fuels Lubr. **6**(3), 594–604 (2013)
11. Schulz, C., Sick, V.: Tracer-LIF diagnostics: quantitative measurement of fuel concentration, temperature and fuel/air ratio in practical combustion systems. Prog. Energy and Combust. Sci. **31**(1), 75–121 (2005)
12. Luong, M., et al.: Toluene laser-induced fluorescence for in-cylinder temperature imaging in internal combustion engines. Appl. Phys. B **91**(3), 669–675 (2008)
13. Faust, S.: Characterisation of organic fuel tracers for laser-based quantitative diagnostics of fuel concentration, temperature, and equivalence ratio in practical combustion processes. Universität Duisburg-Essen (2013)
14. Koban, W., Schorr, J., Schulz, C.: Oxygen-distribution imaging with a novel two-tracer laser-induced fluorescence technique. Appl. Phys. B **74**(1), 111–114 (2002)
15. Schulz, F., et al.: Systematic LIF fuel wall film investigation. Fuel **172**, 284–292 (2016)
16. Park, S., Ghandhi, J.B.: Fuel film temperature and thickness measurements on the piston crown of a direct-injection spark-ignition engine. SAE Technical Paper, 2005-01-0649 (2005)
17. Mittal, M., et al.: A study of fuel impingement analysis on in-cylinder surfaces in a direct-injection spark-ignition engine with gasoline and ethanol-gasoline blended fuels. SAE Technical Paper, 2010-01-2153 (2010)
18. Shibata, G., et al.: Impingement and adhesion on cylinder liners with post diesel fuel injections. SAE Technical Paper, 2016-01-2193 (2016)
19. Weimar, H.-J.: Entwicklung eines laser-optischen Messsystems zur kurbelwinkelaufgelösten Bestimmung der Ölfilmdicke zwischen Kolbenring und Zylinderwand in einem Ottomotor. Institut für Kolbenmaschinen, Universität Karlsruhe (2002)
20. Thirouard, B.: Characterization and Modeling of the Fundamental Aspects of Oil Transport in the Piston Ring Pack of Internal Combustion Engines. Massachusetts Institute of Technology, Massachusetts (2001)
21. Wigger, S., et al.: Quantitative two-dimensional measurement of oil-film thickness by laser-induced fluorescence in a piston-ring model experiment. Appl. Opt. **55**(2), 269–279 (2016)
22. Wigger, S.: Charakterisierung von Öl- und Kraftstoffschichten in der Kolbengruppe mittels laserinduzierter Fluoreszenz. In: Fakultät für Ingenieurwissenschaften, Abteilung Maschinenbau und Verfahrenstechnik. Universität Duisburg-Essen (2014)
23. Zanghi, E., Tian, T.: Development of a High Speed Laser Induced Fluorescence (HSLIF) system in a single cylinder engine for oil transport studies. SAE Technical Paper, 2016-01-1203 (2016)
24. Assor, Y., Burshtein, Z., Rosenwaks, S.: Spectroscopy and laser characteristics of copper-vapor-laser pumped Pyrromethene-556 and Pyrromethene-567 dye solutions. Appl. Opt. **37**(21), 4914–4920 (1998)
25. Faust, S., Dreier, T., Schulz, C.: Photo-physical properties of anisole: temperature, pressure, and bath gas composition dependence of fluorescence spectra and lifetimes. Appl. Phys. B **112**(2), 203–213 (2013)
26. Semrock (2017). https://www.semrock.com/product-families.aspx. Accessed Oct 2017

Irregular Combustion Events in RDE Test Situations Diagnostics – Analysis – Improvements

Ernst Winklhofer(✉), Alois Hirsch, and Paul Kapus

AVL List GmbH, Graz, Austria
ernst.winklhofer@avl.com

Abstract. The paper describes a method to identify and analyze pre-ignition (PI) events appearing in engine and vehicle RDE drive cycle tests or similar transient and stationary test schedules. A specific requirement in such long duration drive cycles is the large amount of data to be recorded for each combustion cycle and the data processing during and after a test procedure. Identification of PI events is accomplished with measurement of ignition delay time, i.e. spark discharge timing and the instant when flame growth is detected. PI events show up with negative ignition delay. How this simple concept is implemented in a combustion measurement procedure with many thousand cycles is explained in some detail. Classification of PI events and root cause analysis is then supported with evaluation of the optical signals of the tests. Specific evaluation procedures to accomplish such task have been described in earlier publications.

1 Introduction

Temperature, engine speed and load and, especially, dynamic drive sequences from engine start up to high load transients have become part of intensive development and calibration efforts in view of RDE test situations. Such testing includes situations with irregular combustion events: knock, pre-ignition, misfire.

On each of such events, there is ample understanding derived from engine development. The challenge in engine or vehicle calibration, however, arises from the RDE test requirements. The constraints imposed by regulations as well as users' drivability expectations create the need for combustion analysis techniques applicable to the dynamics and the duration of these user defined drive situations in test cycles of up to and above 2 h duration.

A test and analysis procedure to identify, evaluate and systematically improve irregular combustion events in RDE situations exploits the information gained with cylinder pressure and flame sensor measurements. The methodology includes

- Test preparations to access each cylinder with pressure sensors and apply fiber optic spark plug sensors for flame radiation measurement
- Accessing ECU signals for comparison of actuator set values with actual combustion signals
- Accessing ECU and combustion signals for test statistics and protocol

M. Günther and M. Sens (eds.), *Knocking in Gasoline Engines*,
https://doi.org/10.1007/978-3-319-69760-4_11

- Evaluation of signals from general overview down to individual crank angle resolved combustion cycles with signal classification for normal, misfire and irregular combustion cycles
- Evaluation of root causes for misfire and pre-ignition events
- Comparison of test variants and improvements

The paper presents the test methodology applied to vehicle testing on a chassis dynamometer ("chassis dyno"). Examples are given including misfire and pre-ignition events under conditions of specific RDE drive sequences. Emphasis is given to test sequences typical for the dynamic change of engine conditions at low as well as high engine temperatures.

The benefit of such cycle- and cylinder specific analysis is seen in the identification of required improvements. Such identification starts with test overview statistics to first focus on most serious irregularities. The procedure then relates irregular combustion events to the actual conditions as identified by the cycle and crank angle resolved ECU and combustion signals. This identification then guides any adaptation of calibration parameters. Test repetitions then yield the data to understand the outcome of any modification introduced by variation of ECU parameters.

2 RDE Test Situations

RDE testing includes the entire engine map at various temperature conditions in stationary as well as transient drive situations. Prerequisite to RDE testing is the correct handling and, whenever possible, prohibiting the appearance of irregular combustion events such as knock, pre-ignition together with megaknock as well as misfire cycles.

- Handling of knock situations is a well established procedure.
- Avoiding the appearance of pre-ignition and megaknock cycles is first of all a task to be accomplished in the engine development phase. However, as some RDE situations are beyond the scope of predictive engine testing, vehicles are operated at the risk of PI events, and even at the absence of PI cycles, a response protocol has to be established in order to minimize the risk of potential engine damage.
- At engine start, especially in low temperature environment, misfire has been found to not only impair the start procedure and tailpipe emissions, but such misfire can also be linked to pre-ignition in subsequent cycles.

How then can we operate a vehicle, identify irregular combustion situations and define actions to safely guide combustion back to normal operation?

3 Sensors and Instrumentation

The measurement environment is defined by vehicle test procedures on chassis dyno test rigs, see Fig. 2. The engine is accessed with a combination of sensors as summarized in Table 1. The main data required for the intended analysis are the instant of spark discharge (SA – "spark advance") and the instant of noticeable flame kernel

Table 1. Signals and sensors to enable PI diagnostics in chassis dyno tests

Signal	Best source	alternative
SA	Current clamp	
Flame	Visiolution spark plug sensor	
Cyl. Pressure, IMEP	Cyl. sensor	Next cyl spark plug sensor
Deg CA, rpm	Shaft encoder	ECU
Useful signal		
Injection timing	*Current clamp*	*ECU*

growth (OS – "optical signal"). SA is determined from the ignition coil's feed current by means of a current clamp sensor, OS is evaluated from optical signals recorded with fiber optic spark plug sensors. Signal examples to identify OS are included in the graphics of Fig. 5. Variants of fiber optic spark plug sensors applicable for such tests is shown in Fig. 3

Instrumentation required to record such signals together with further crank angle resolved pressure and timing signals comprises of multichannel data recorders especially configured to handle large and continuous data streams at up to 100 kHz data rate per signal channel. Signal converter bandwidth and dynamic range of sensors and A/D converters must be matched to the dynamics of the signals of interest.

4 Flame Signals and Data Reduction

The speed – load map of Fig. 1 shows the continuous change of engine operation. Identifying irregular combustion events requires the continuous recording of an entire test cycle. In case of RDE tests, this includes engine start followed by up to 2 h of engine operation. Such tests produce data streams as is shown in Fig. 4 for a short cutout of 500 cycles. The data box covers a tip-in event starting at 3 bar IMEP with a ramp up to 24 bar IMEP. The data comprise of flame radiation signals recorded with an 80-channel spark plug sensor. Each combustion cycle is represented with the integral flame radiation record per sensor channel. Such integrals are gained from crank angle resolved radiation signals as shown in the data insert of Fig. 2.

The base flame radiation data are shown in the data box A including all flame radiation signals. Such flame data comprise of premixed and diffusion flame radiation and are routinely used to identify sources of soot formation. For this purpose, each cycle's flame data are subjected to a "PN filter" procedure, essentially subtracting premixed flame signals from the total signal [2]. The residual "diffusion flame signal" is integrated for each cycle, results are shown in Fig. 4B.

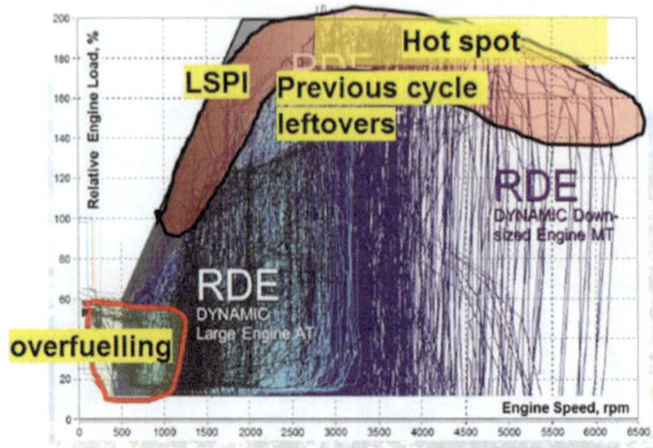

Fig. 1. Speed – Load maps in RDE tests. Irregular combustion risks arise in high load conditions (LSPI, hot spot, etc.) and in cold start

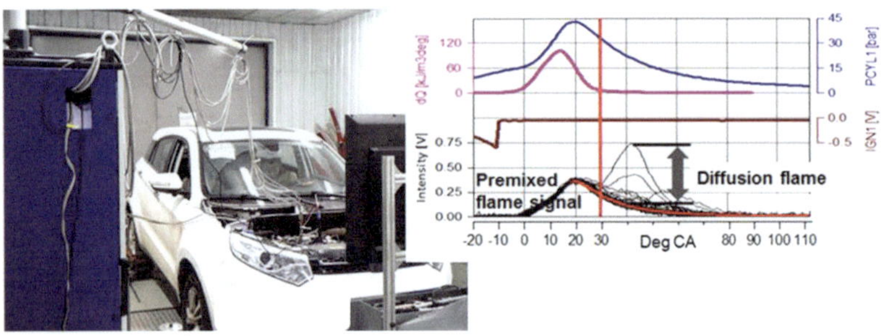

Fig. 2. The chassis dyno test environment: sensors access individual cylinders for flame, pressure and timing signals to record crank angle resolved flame and combustion signals.

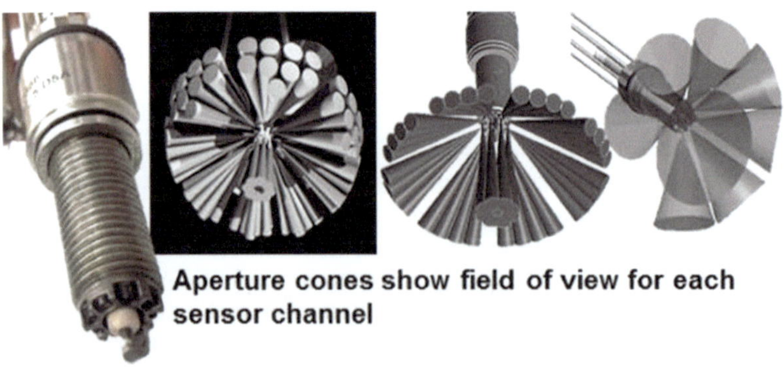

Fig. 3. Fiber optic spark plug sensors with up to 80 sensor channels

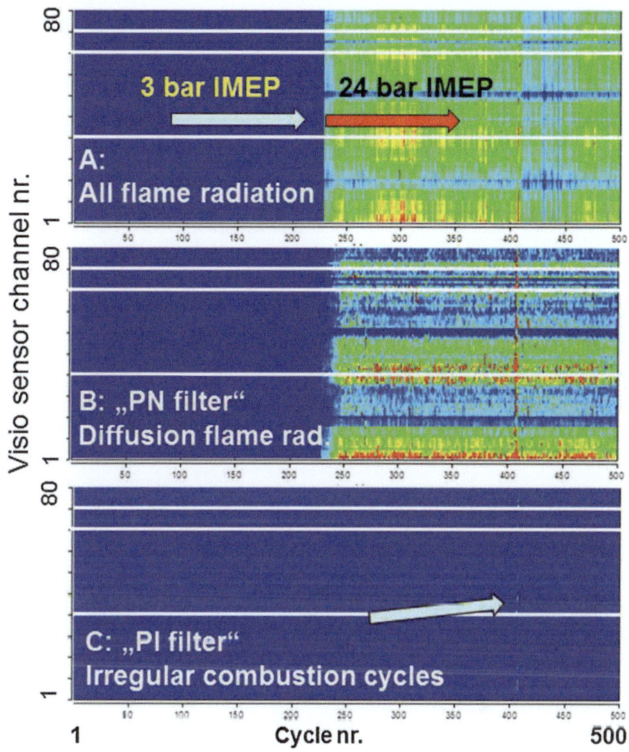

Fig. 4. flame signals in a tip- in event with IMEP rising from 3 to 24 bar. Cycle integral flame radiation signals showing all flame radiation (A), diffusion flames only ("PN filter" B) and identifying one cycle with a pre-ignition event ("PI filter" C)

The same set of "all flame radiation" data, Fig. 4A, also includes information on pre-ignition cycles. Identification of PI cycles is achieved with a procedure as is explained in Fig. 5:

- Record the instant of spark discharge: SA [deg CA]
- Use flame signals to identify the instant of flame kernel growth: OS [deg CA]
- Use ID = OS – SA to identify normal or PI ignition

Spark timing (SA) is evaluated from the current signal feeding the ignition coil. The instant of flame kernel growth (OS) is determined from the optical signals, examples are shown in Fig. 5 for a PI and a normal ignition event. Criterion for OS is the flame signals growth above its noise level. This is easily detected in any PI situation and it even is well suited to distinguish between normal ignition cycles and cycles with significantly short ignition delays, so called "borderline" PI cycles, see Fig. 7. Of course, as the spark plasma radiation appears in the optical signals, the "PI filter" procedure accounts for this radiation event.

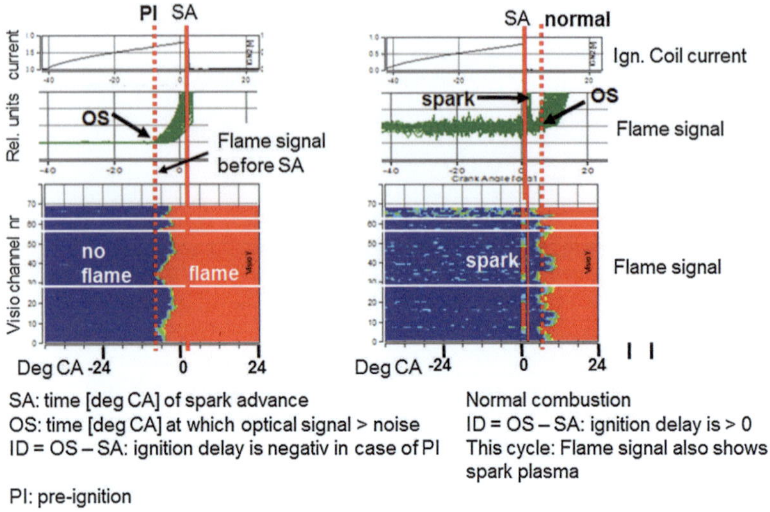

Fig. 5. "PI Filter" - data reduction for PI event detection. Ignition delay (ID) is key parameter, it is derived from flame signal detection (OS = optical signal) and spark timing (SA). Examples show a PI cycle and a normal cycle, respectively

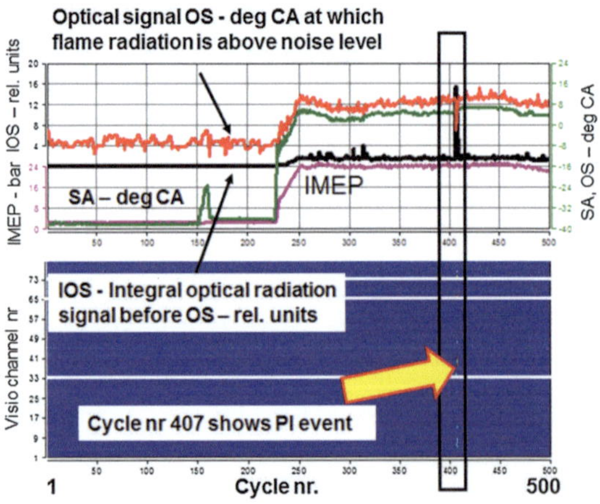

Fig. 6. "PI Filter" applied to a cycle series of engine nr. 1. Key data are spark advance (SA) and start of flame (OS – optical signal). IOS together with OS, SA helps to identify PI cycles

The examples in Fig. 5 summarize this PI filter procedure with the ignition coil current and the flame radiation signals at start of combustion. Flame signals are given in line graphs as well as in color codes in order to see the flame signals raising out of noise and the flame's local growth. Notice the spark signal in the normal combustion example.

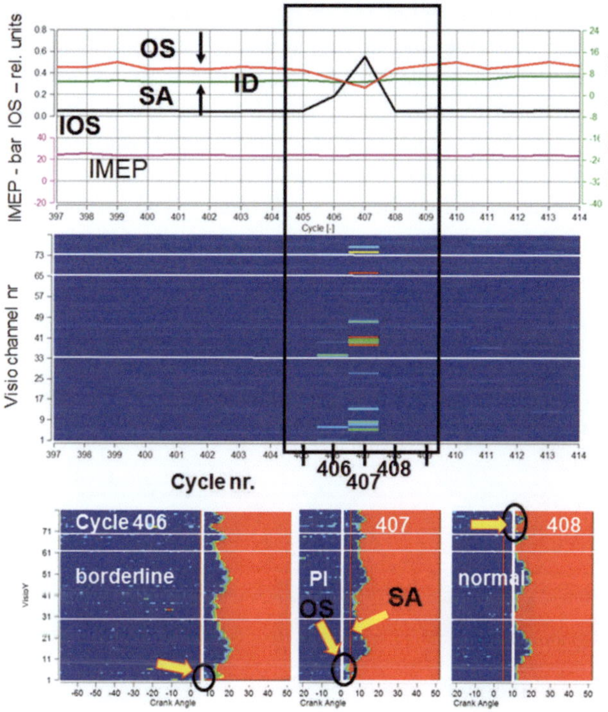

Fig. 7. Zoom into PI event of Fig. 6 with flame signals showing ignition and flame growth. Notice borderline cycle 406 with ID ~ 0 and PI cycle 407 with negative ID and large IOS signal.

5 Test Analysis Examples

Application of this PI filter to the flame radiation data in Fig. 4A yields the results of Fig. 6. The line graphics for SA and OS show the long ignition delay at low load followed by much shorter ID at high load. In cycle 407 the flame signal OS appears before spark discharge as an indicator for a PI event. A zoom into the data box around the PI cycle 407 is given in Fig. 7. The cross over between OS and SA data is well noticed for the PI cycle.

In addition to the OS and SA graphics, Figs. 6 and 7 show the graph "IOS", the flame signal integrated between TDC gas exchange and OS. This IOS signal, essentially integrates the background noise of all optical channels put together. IOS, however, is also an indicator of any glowing residuals surviving the exhaust stroke and thus presenting the risk of an erroneous ignition source. Such ignition by "glowing particles or deposits" is a well known cause for PI events [3].

The IOS graph in Fig. 7 thus shows the appearance of "glowing residuals" already in cycle 406 and furthermore in the PI cycle 407. Thus, the synchronicity of IOS and the OS-SA data in cycle 407 already points to the mechanism causing this PI event.

In comparison to this PI event in engine nr. 1, a group of cycles is shown in Fig. 8 with data for PI and near PI events in engine nr. 2. Data were again recorded with a multichannel spark plug sensor, the PI filter procedure was applied to yield OS and IOS signals for comparison with spark timing data SA. The group of cycles in Fig. 8 is taken at stationary high load operation, it shows one PI and two "borderline" PI cycles. Different to engine nr. 1, the IOS signal here never rises above noise level, flame growth in the PI cycle starts long before spark timing. The two borderline PI cycles start flame growth at spark ignition rather than about 6 deg CA after spark discharge which would be the usual ignition delay time at this operating point.

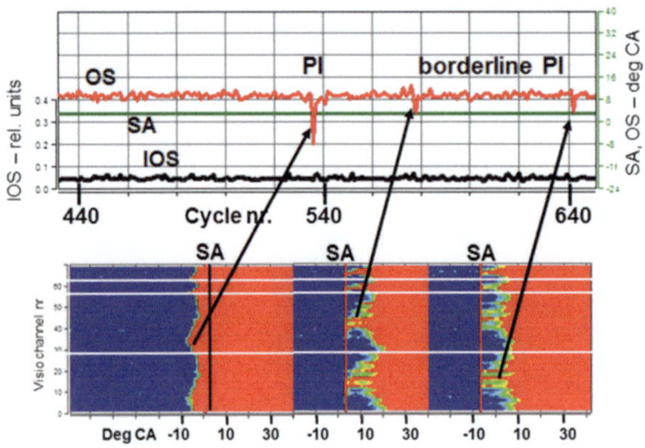

Fig. 8. PI event identification in engine nr. 2. IOS signal shows absence of any glowing particles (free deposits). Here, PI and borderline PI cycles appear sporadic without any signal traces in preceding or trailing cycles.

The conclusion on this short example for engine nr. 2, thus is as follows:

1. IOS: is at background noise level: no indication of any glowing residuals
2. PI event is typical for thermal sources: LSPI or hot spot
3. Borderline PI: engine operates near critical conditions

In summary, such sporadic events need to be collected for local ignition evaluation statistics in order to identify potential local areas of predominant self ignition. The procedure for such local ignition source evaluation has been described in Ref. [1].

Above examples have shown some short snapshots of combustion data out of large data sets collected in chassis dyno tests. A test evaluation, however, starts with the review of PI filter statistics, identifying PI cycles by their ID and IOS results. The basic content of such test statistics is shown in Table 2 for the above two engine's examples.

In any RDE or engine test, there is of course the supplementary information on engine operation modes, on temperatures and environmental or test bed conditions. Such data allow for the further classification of PI filter test results in order to identify patterns of PI occurrence.

Table 2. Cycle statistics derived from signal filters for ID and IOS. Together with indicating data (IMEP, rpm) such results enable PI root cause analysis in big data sequences.

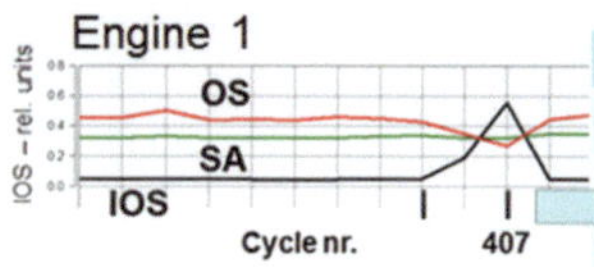

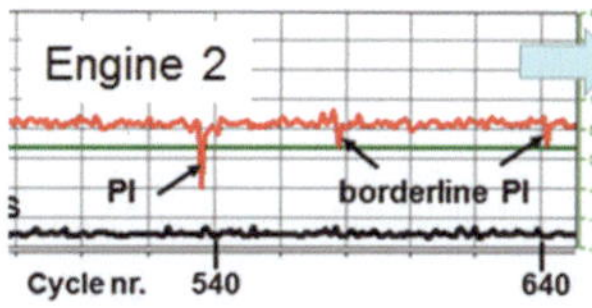

Cycle nr.	SA	OS	ID	IOS
Engine 1, Fig. 6				
406	6	6	0	0.2
407	6	3	-3	0.5
Engine 2, Fig. 7			PI: ID < 0	
537	3	-8	-11	0.06
578	3	3	0	0.06
642	3	3	0	0.05

IOS is significant

IOS at noise level

ID = 0: borderline PI

6 Irregular Combustion at Engine Start

Result charts as per Table 2 for entire test cycles may frequently point to irregular combustion cycles in an engine's start sequence. In an engine instrumented with a fiber optic spark plug and a pressure sensor in the same cylinder, such events yield signal sequences as given in Fig. 9. Pressure and rate of heat release signals show misfire cycles nr. 5,6,7 followed by normal combustion cycles. The flame signals, however, show that cycle nr. 6 has some flame activity, nr. 7 has combustion already in the early compression stroke, cycles 8 and 9 have normal flame radiation traces.

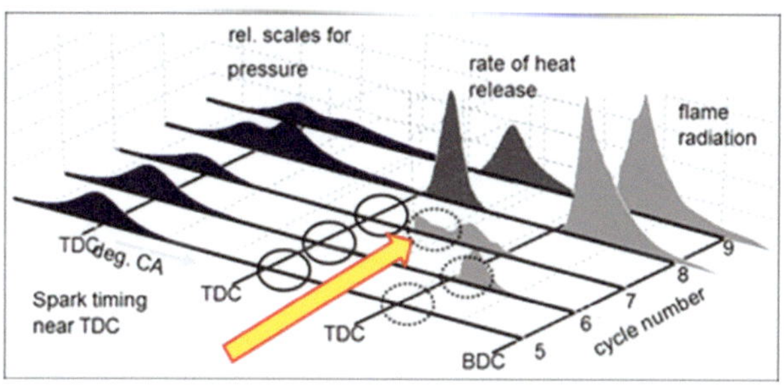

Fig. 9. Flame radiation signal shows irregular ignition/combustion at engine start

The mechanisms of such irregular cycles at engine start are clarified with tests in an optical engine, see Fig. 10. Here, a start cycle was initiated with considerable over-fuelling to support ignitability. However, with poor evaporation, flame propagation was impaired, resulting in very slow and ongoing combustion throughout the exhaust stroke. This combustion was fueled by the pool of liquid fuel on the piston surface and continued well into the intake/injection stroke of the subsequent cycle.

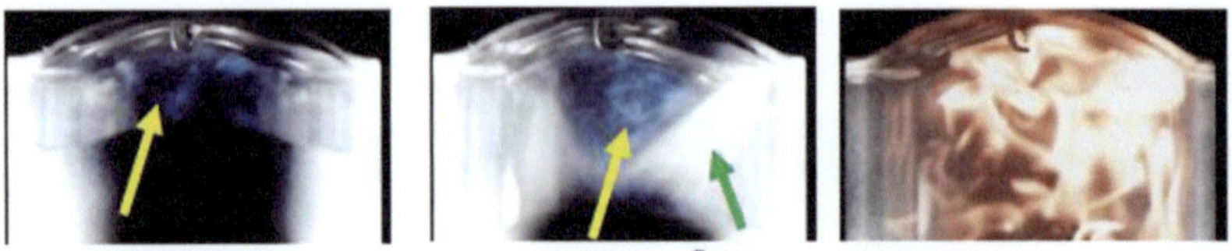

Fig. 10. Irregular ignition mechanism in cold engine: residual flames surviving the exhaust stroke ignite the charge in the next cycle.

The simultaneous presence of the residual flame, fresh air and fuel spray then resulted in combustion during the ongoing intake and compression stroke.

7 Improvements of a Combustion System

The purpose of all measurements and analyses presented in above examples is to derive measures to improve an engine's combustion system. In chassis dyno tests, the system variability is usually limited to actuator operation as controlled by ECU parameters. Test of parameter variation is done in design of experiment (DOE) procedures. However, as such "black box" optimization does not address the root causes for PI events, it is time consuming and may miss opportunities for systematic problem solving.

Above examples with the IOS parameter have shown that a simple evaluation of flame radiation prior to ignition is used to identify PI induced by deposits related effects. In the absence of the IOS "pointer signal", PI cycles are furthermore evaluated with respect to the local origin of the ignition source. Here the benefit of multichannel spark plug sensors becomes evident as the flame data allow reliable location of the ignition event. Procedures for signal analysis have been published recently [1].

8 Summary

A measurement procedure has been presented to enable PI detection and analysis in RDE and similar long term tests on a chassis dyno or an engine test bed. The measurement system comprises conventional combustion sensors together with fiber optic spark plug sensors applicable in standard SI engines. Optical signal evaluation accesses the same flame signals as are otherwise used to identify sources of soot formation [2]. PI detection is accomplished with a signal filter procedure evaluating radiation signals up to and including flame kernel growth.

9 Nomenclature

SA: spark advance = time instant of spark discharge
OS: time instant of flame front detection, see Fig. 5.
ID = OS – SA: ignition delay
IOS: integral of optical signals from TDC gas exchange until OS
PI: pre-ignition
ECU: engine control unit

References

1. Grüneberger, P., Hirsch, A., Philipp, H., Winklhofer, E.: Flame analysis supports combustion system development in normal SI engine test operations. In: 13 Internationaler Kongress "Motorische Verbrennung", Ludwigsburg, 16/17 März 2017
2. Miklautschitsch, M., et al.: Experimental analysis of soot formation in BMW TwinPower turbo engines by simultaneous use of high-speed endoscopy and fiber-optic instrumented spark plug. In: 12th International Symposium on Combustion Diagnostics, Baden, Germany, 10–11 May 2016
3. Hirsch, A., et al.: Irregular ignition events in TC GDI engines: phenomenology, analysis and engine development. In: PTNSS Congress, Poland (2010)

Quantitative Optical Measurement Techniques for Mixture Formation and Combustion Process Analysis

Thomas Berg(✉), Stefan Seefeldt, and Olaf Thiele

LaVision GmbH, Göttingen, Germany
tberg@lavision.de

Abstract. Optical imaging and sensor techniques provide detailed analysis of mixture formation, ignition and combustion processes. The evaluation of image sequences and sensor data obtains quantified measured values to characterize engine conditions.

In-cylinder infrared absorption measurement techniques are applied to record Lambda values, residual gas content (water and CO_2 concentration) and gas temperature with high time resolution. Especially the temporal analysis of the Lambda value evolution within engine cycles is a criterion for the quality of mixture formation and ignitability. This reveals quantified indices for the combustion stability.

Advanced processing of a series of highly time resolved images from an engine's cycle generates quantified data about spray pattern formation, flame propagation and soot temperature. Images of the early flame evolution are analyzed using a dynamic thresholding algorithm and reveal the flame propagation.

1 Introduction

Engine indication is typically referred to as cylinder internal pressure measurement. From the crank angle resolved pressure evolution engine and combustion relevant quantities are derived. The approach presented here is similar to that, but based on optical measurements. Instead of the pressure, a series of high-speed images and highly time resolved infrared absorption data are recorded. From these data again several quantitative measures are derived, like spray geometry, flame propagation speed, soot temperature, cylinder-internal Lambda value, temperature evolution, water concentration and EGR rate.

2 Highly Time Resolved Sensor Measurement of Transient Phenomena

Future legislation includes Real Driving Emissions (RDE) standards. To optimize engines under real driving conditions the evaluation of highly transient engine conditions is necessary. These conditions require techniques which can measure hundreds of consecutive cycles with crank angle resolution to analyze short time phenomena.

M. Günther and M. Sens (eds.), *Knocking in Gasoline Engines*,
https://doi.org/10.1007/978-3-319-69760-4_12

LaVision's ICOS (Internal Combustion Optical Sensor) systems provide simultaneous high time resolution optical indication of local air/fuel-ratio, gas temperature, water concentration and the EGR-rate. The systems are based on infrared absorption spectroscopy on gaseous molecules to measure the above quantities directly in the cylinder. Different types of probes are available for this. A modified spark plug probe can be used up to mid-load conditions to measure locally at the spark plug (Fig. 1). Only the original spark plug needs to be replaced and no modifications of the engine are necessary [1]. For all engine conditions, especially higher loads and knocking conditions, line-of-sight probes with small engine modifications are used. The line-of-sight probe will provide cylinder averaged information. These systems are specifically designed to be used in engine test bed environments for research as well as near production engines as shown in Fig. 2.

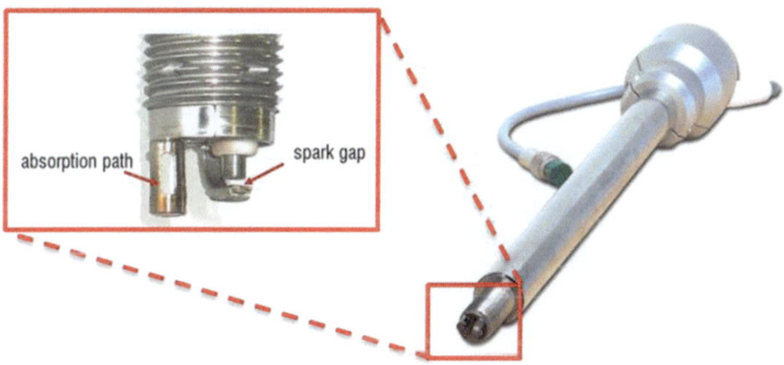

Fig. 1. ICOS system and M12 spark plug probe.

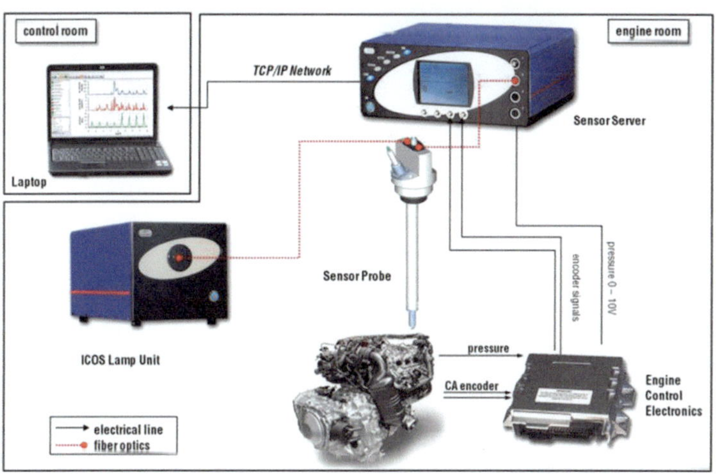

Fig. 2. Layout of an ICOS system with spark plug probe in engine test bed environment.

In the following a tip-in phase is shown as an example of a common transient engine operation during real driving conditions. The measurements of a load change after a gear change are carried out on a near production line engine at a transient test rig with the ICOS system and M12 spark plug probe [2].

2.1 Lambda Value and Mixture Formation

For analysis of transient phenomena high time resolved measurements of the Lambda value (Air/fuel equivalence ratio) development of consecutive cycles is shown in Fig. 3. Here Lambda = 1 at the time of ignition is only reached in cycle four and following. The mixture formation is visualized and shows a temporal cycle-to-cycle change (indicated by the white arrow), which results from ECU controlled valve timing changes.

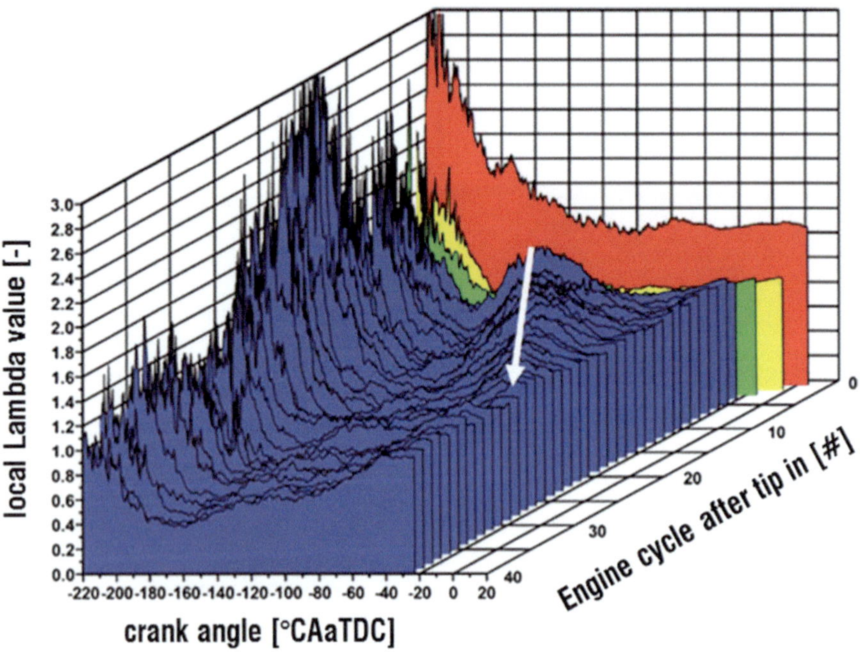

Fig. 3. Crank angle resolved development of the Lambda value at the spark plug position for the first cycles of an engine tip-in.

This shows not only the Lambda value at the time of ignition. The statistical cycle-to-cycle fluctuations indicate the engine operating stability. In addition the homogeneity of the fuel mixture is determined from the crank angle based evolution (see chapter 2.3).

2.2 EGR Rate and Valve Timing

The CO_2 concentration was measured simultaneously to the Lambda value using the same ICOS measurement system. From the CO_2 concentration the EGR-rate for each cycle can be determined. Since the measurement takes place in cylinder the EGR-rate incorporates all exhaust gas components: internal and external EGR. The effect of the valve timing changes during the tip-in procedure can be seen in the EGR rate, shown in Fig. 4.

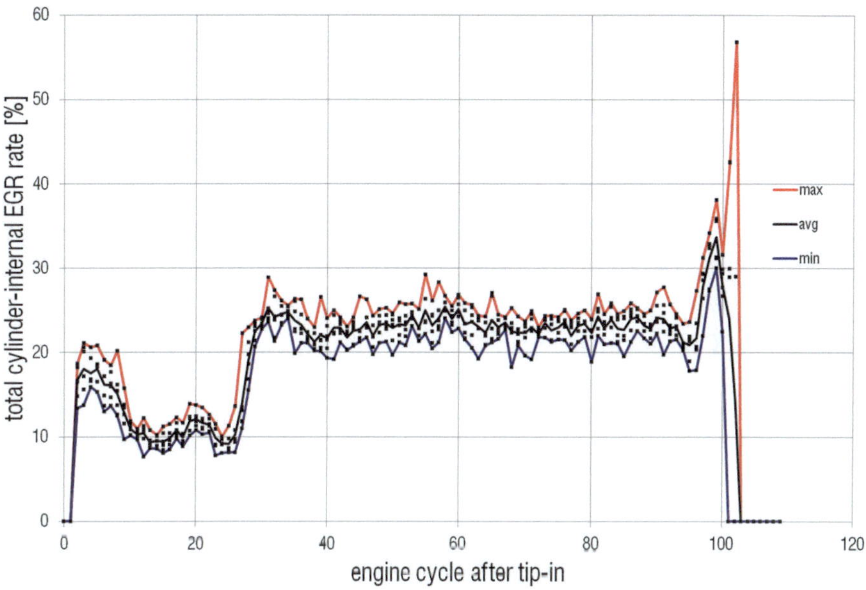

Fig. 4. In-cylinder measured EGR-rate for consecutive cycles covering the entire tip in procedure. EGR rates for 10 repeated tip-ins are plotted. Average, maximum and minimum values are highlighted by the 3 curves.

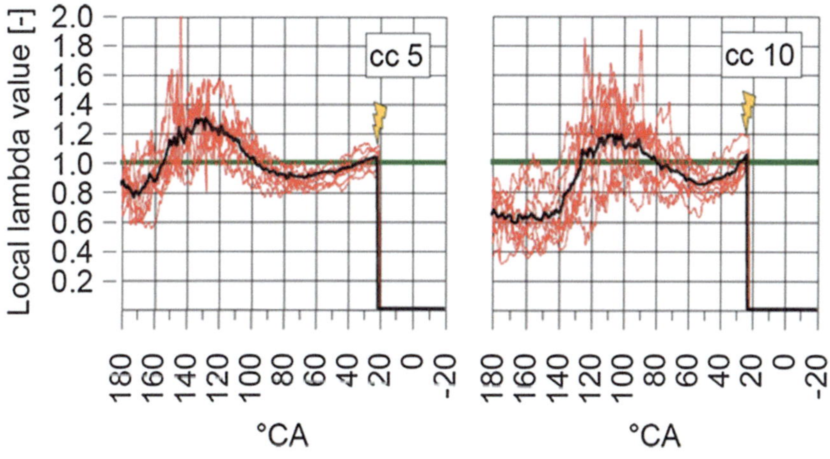

Fig. 5. Variations in the Lambda values during compression for combustion cycles #5 and #10 during 10 repeated tip-ins. Black curve shows average over 10 cycles and red curves represent individual cycles.

2.3 Operation Stability

Operation stability is an important indicator to analyze engine optimization. Particularly when coming close to operating limits, such as knock limits, it is essential to ensure that the operation is stable enough to avoid knocking. Operation stability can be analyzed by analyzing the variation of the crank angle resolved Lambda evolution over many cycles. As an example Lambda value variation over 10 consecutive tip-in procedures is shown in Fig. 5. In combustion cycle 5 (cc5) of the tip-in the spread in the Lambda curves up to ignition is reduced. In comparison to that the variation in the Lambda curves in cycle 10 (cc10) is stronger over the entire compression. A more detailed analysis of these variations is given in [2]. In [3] this type of crank angle resolved data is used to evaluate the flame speed.

3 Charge Temperature and Water Concentration Evolution

With modern gasoline engines closing in on knock limits, methods for increasing knock resistance need to be investigated such as charge dilution or water injection. The ability to measure temperature and water concentration will help to achieve the optimization process of these methods, such as charge cooling through water injection.

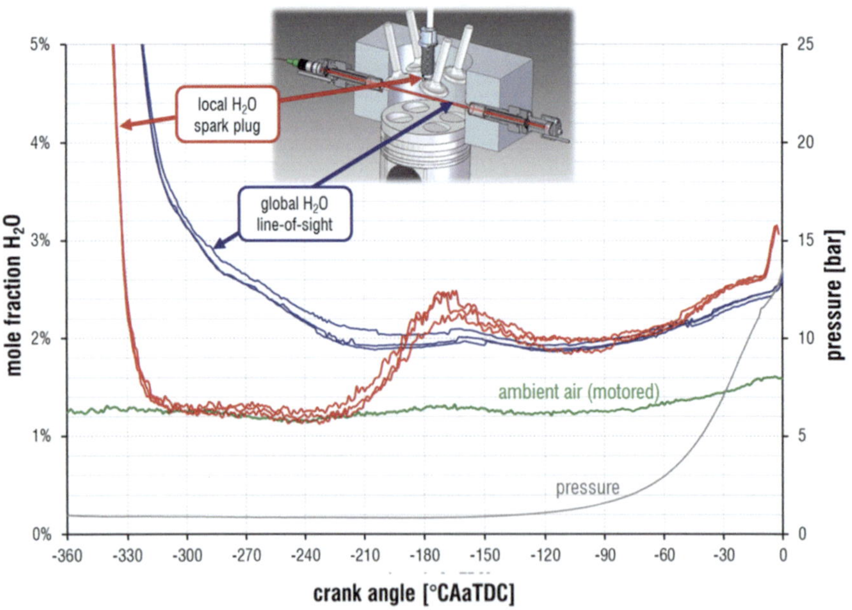

Fig. 6. Water concentration detection at the spark plug and averaged across the cylinder. Both locations show different temporal behavior.

With the ICOS-Temperature system the gas temperature as well as the water concentration and the temporal mixture formation of water are analyzed. In Fig. 6 the water concentration during first half of a DISI engine cycle is shown for two different locations in the cylinder: locally at the spark plug and cylinder averaged using a cross cylinder line-of-sight probe. Local and global differences and phenomena such as temporal evolution of mixing of fresh air at the spark plug are made visible.

Figure 7 shows the cylinder averaged temperature evolution for motored and fired engine operation measured with the line-of-sight probe. The intake temperature typically measured in the intake manifold is approximately close to the load change temperature under motoring conditions. Differences between intake temperature and in-cylinder gas temperature during the load change are always present but rarely known.

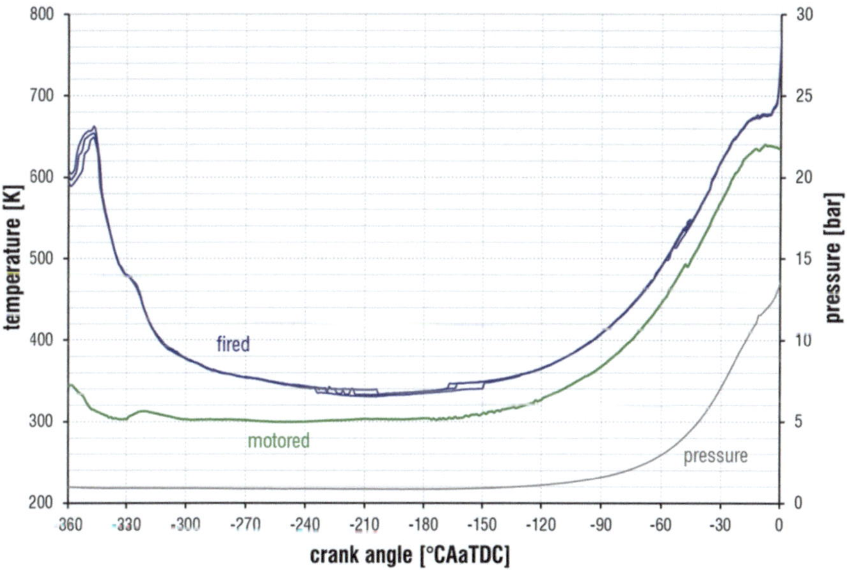

Fig. 7. Cylinder averaged temperatures for motored and fired conditions.

4 Quantitative High-Speed Endoscopic In-Cylinder Imaging

For the optimization of near-production engines endoscopic high-speed imaging can be applied to visualize in-cylinder phenomena. Key-hole imaging using endoscopes is a minimally invasive technique to monitor real-time in-cylinder processes such as fuel spray injection, ignition, combustion, and soot formation. The access to the combustion chamber is maintained by compact size pressure sealed window inserts. Figure 8 shows the typical layout of LaVision's EngineMaster inspex endoscopic imaging system.

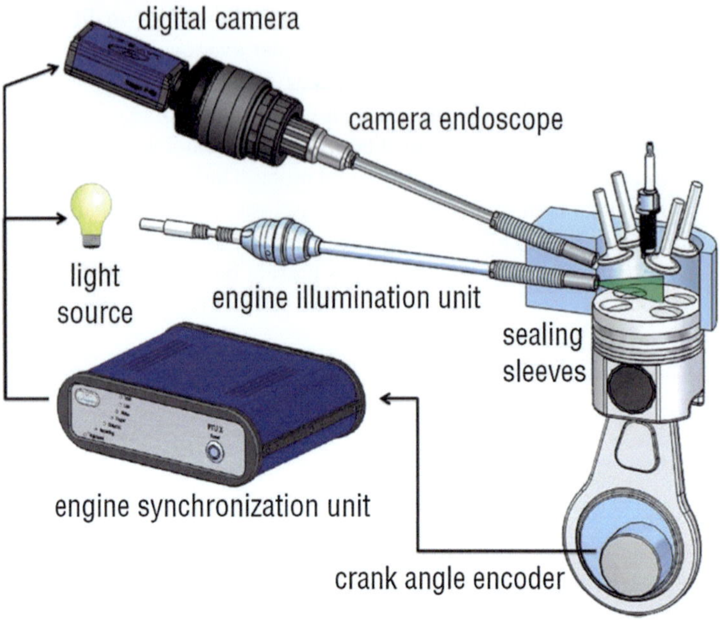

Fig. 8. Layout of an in-cylinder endoscopic imaging system.

Crank angle synchronized high-speed color cameras primarily record a crank angle based movie of the processes in the combustion chamber. Additional image processing routines generate quantitative temporally resolved data from these images. This leads to a comprehensive picture of the in-cylinder processes, which are beyond plain movie generation.

4.1 In-Cylinder Spray Characterization

During the injection, direct visualization of the spray evolution helps understanding the correct mixture formation and to avoid wall wetting. The in-cylinder spray breakup significantly depends on the internal pressure and nozzle temperature, which is hard to predict. Flash boiling drastically changes the spray pattern and propagation length. Figure 9 shows an example of different spray geometries at different in-cylinder pressures. From image processing the spray cone angle and shape can be quantified as it propagates over crank angle (Fig. 10).

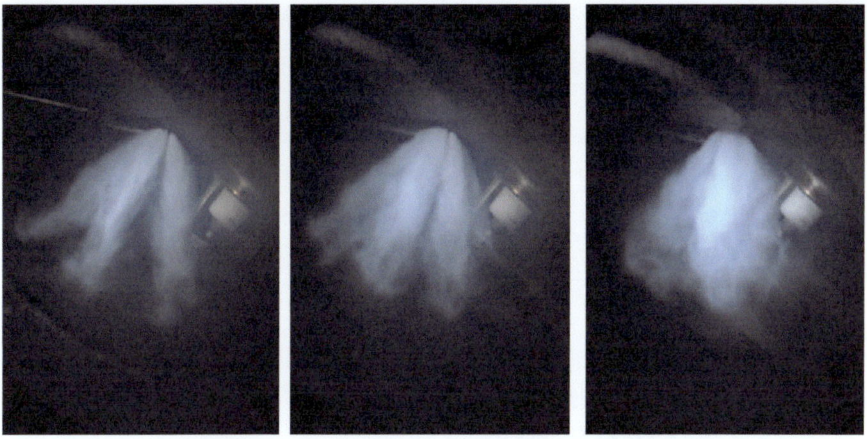

Fig. 9. Spray pattern at varying in-cylinder pressure.

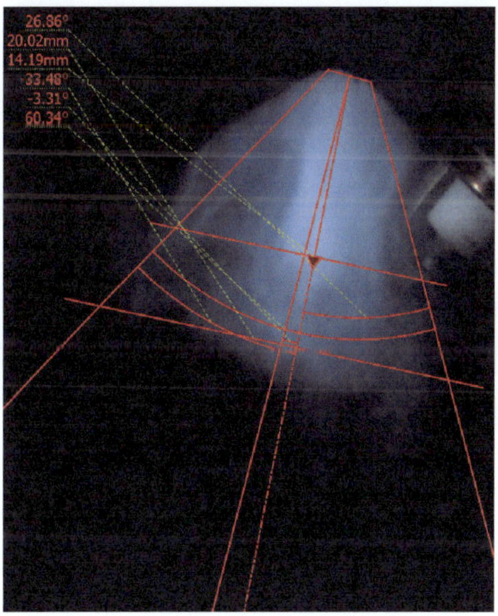

Fig. 10. Quantitative spray cone analysis.

Wall and piston wetting is the major source of unburnt hydrocarbon (HC) and particulate matter (PM) emission. Even though pressurized spray chambers are an important tool for the development of a fuel injector and the correct injection timing, the in-cylinder conditions and flow field can only be monitored directly inside the operating engine. The example in Fig. 11 shows the fuel spray impinging on the piston surface. During the combustion phase, the sooty diffusion flames are clearly visible as a major source of PM emission.

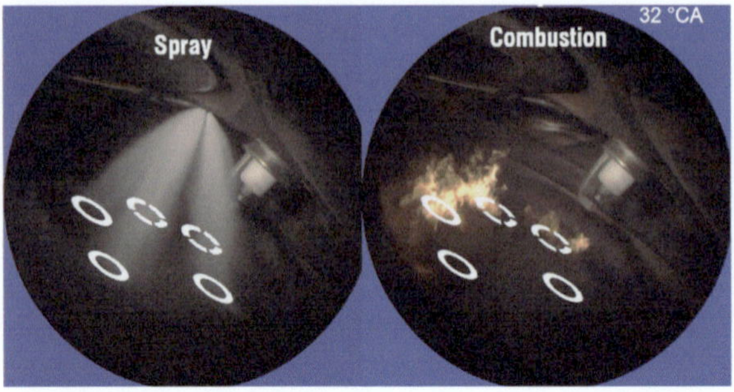

Fig. 11. Fuel spray piston impinging followed by sooty flames in the same cycle.

4.2 Ignition and Flame Propagation

The ignition process itself is highly dynamic and underlies strong cyclic variations. By image processing of the flame growth, the flame propagation speed can be deter-mined directly from a series of high-speed endoscopic images. A dynamic threshold scheme for image binarization automatically adapts to the varying intensity during flame evolution to determine the flame boundary (see Fig. 12). The area covered by the flame is then converted to an equivalent flame radius as plotted in Fig. 13.

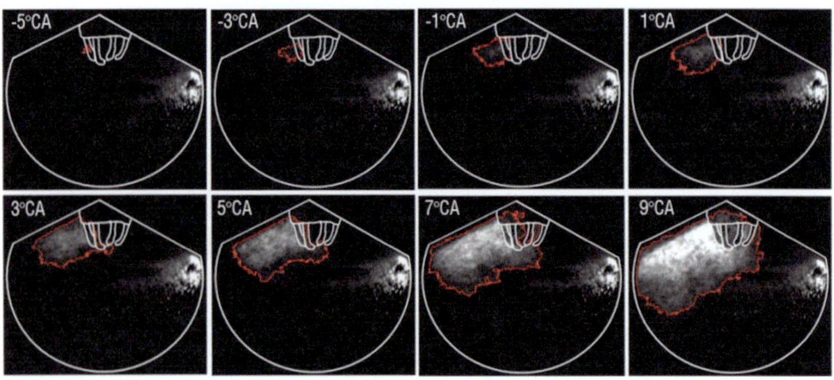

Fig. 12. Endoscopic high-speed cinematography. The red line indicates the projected flame boundary.

An in depth statistical analysis of the flame speed, i.e. the growth rate of the equivalent flame radius (see) and flame timing was analysed in [4] from a sequence of consecutive engine cycles.

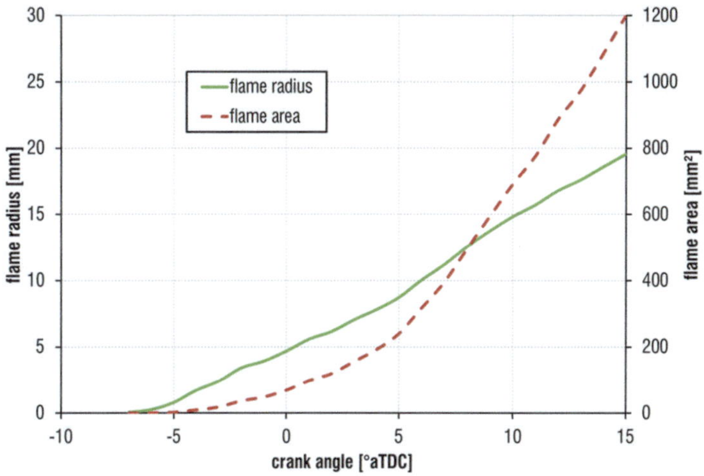

Fig. 13. Equivalent flame radius (solid line) and projected flame area (dashed line).

4.3 Combustion and Soot Formation

During the combustion premixed and diffusion flames are visible and can be distinguished from each other, using the color information. Calibrating the flame color and luminosity at a known Tungsten reference light source allows calculating flame temperature and soot concentration (KL-value) from these images. The example in Fig. 14 shows the temperature distribution of a sooty flame.

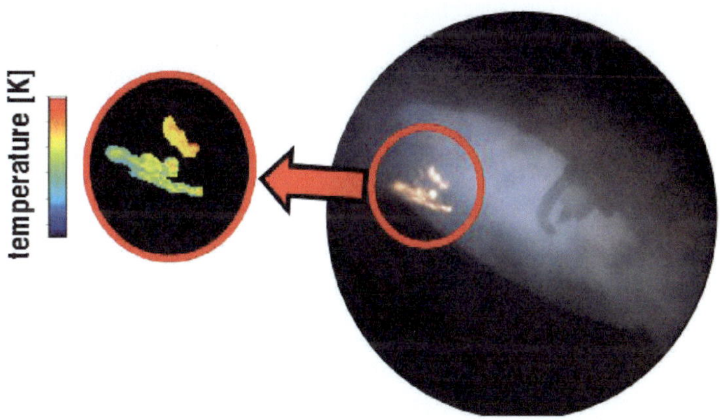

Fig. 14. Soot temperature derived from color images.

4.4 Cold Start Analysis

Finally, recording a sequence from the very beginning of an engine start reveals a clear understanding of the processes taking part for the majority of PM emissions during cold start conditions. The images in Fig. 15 show a pool fire in the first cycle, then sooty diffusion flames near the walls in the second cycle, turning to normal low emission combustion in the following cycles.

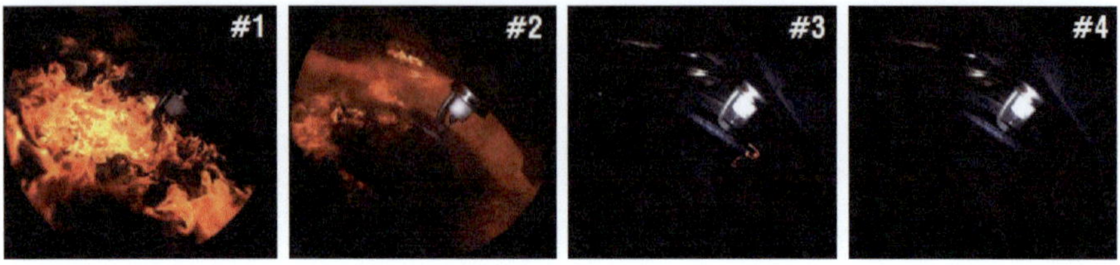

Fig. 15. First four cycles of a cold start with visible pool fire from piston and wall wetting. All images recorded at 25° CA after TDC.

5 Conclusion

ICOS optical sensors and minimally invasive endoscopic imaging delivers crank angle resolved quantified data about mixture formation, residual gases, temperature evolution, spray pattern, ignition and flame propagation, and soot temperature. This highly time resolved in-cylinder optical indication provides information to understand short term phenomena like emission sources or reasons for knock events. To demonstrate the potential of these measurement techniques a number of engine applications are presented in this paper.

While pressure indication mainly shows the knocking or combustion phenomena themselves, the presented optical techniques give an indication about the actual causes which lead to these events.

References

1. Grosch, A., Beushausen, V., Thiele, O., Grzeszik, R.: Crank angle resolved determination of fuel concentration and air/fuel ratio in a SI-internal combustion engine using a modified optical spark plug. SAE Technical Paper 2007-01-0644 (2007). https://doi.org/10.4271/2007-01-0644
2. Disch, C., Pfeil, J., Kubach, H., Koch, T., et al.: Experimental investigations of a DISI engine in transient operation with regard to particle and gaseous engine-out emissions. SAE Int. J. Engines **9**(1), 262–278 (2016). https://doi.org/10.4271/2015-01-1990
3. Hanabusa, H., Kondo, T., Hashimoto, K., Furutani, M.: Study on cyclic variations of laminar flame speed in homogenous lean charge spark ignition combustion. SAE Technical Paper 2016-01-2173 (2016). https://doi.org/10.4271/2016-01-2173
4. Shawal, S., Goschutz, M., Schild, M., Kaiser, S., et al.: High-speed imaging of early flame growth in spark-ignited engines using different imaging systems via endoscopic and full optical access. SAE Int. J. Engines **9**(2), 704–718 (2016). https://doi.org/10.4271/2016-01-0644

Detection and Analysis Methods for Irregular Combustion in SI Engines

Ulrich Spicher(✉)

APL Automobil-Prüftechnik-Landau GmbH, Landau, Germany
ulrich_spicher@web.de

Abstract. Since the invention of spark ignition engines, knocking has been the limiting factor in the development of new engines with increased power density and improved efficiency. The mechanisms that lead to knocking combustion have been investigated in numerous research studies that have produced, to some extent, controversial theories. Even though it is very certain that knocking involves a detonation caused by self-ignition processes in the end-gas region, there is a great interest in growing the understanding of knocking phenomena and the associated multitude of combustion irregularities. This article gives an overview of the contemporary state of knowledge of knocking and other irregular combustion phenomena. This article discusses the kind of detection and analysis methods to investigate irregular combustion in SI engines. Additionally, knowledge about irregular combustion is still not totally understood and the areas in which further research and development are necessary are discussed. For the purpose of describing and evaluating knocking and other combustion irregularities, not only experimental investigations will be employed, but also theoretical considerations.

1 Introduction

Knocking combustion is the best known irregular combustion phenomena in spark ignition engines. This kind of combustion leads to undesirable noise harshness and can cause fatal mechanical and thermal damage of engine structures. To avoid these irregular combustion phenomena, which are represented by rapidly accelerating combustion in more or less large areas of unburned mixture in front of the propagating flame, it is necessary to understand the phenomena involved with irregular combustion and to optimize the combustion in a way that irregular combustion does not occur. Sir Douglas Clerk first observed irregular combustion in the year 1882. Due to the knocking noise made by the engine, he called this combustion process "knocking combustion". Since the beginning of the development of internal combustion engines, many engineers have worked to understand the phenomena involved with and causes of knocking. As early as 1923, Sir Henry Ricardo [1] explained that knocking combustion is produced by auto-ignition within the unburned mixture during the late stage of combustion (end-gas region), Fig. 1 left. These pockets are formed by the turbulent flame front in close proximity to the combustion chamber walls in the last phase of flame front propagation. Some of these pockets have short ignition delays due to non-uniformities in temperature and

M. Günther and M. Sens (eds.), *Knocking in Gasoline Engines*,
https://doi.org/10.1007/978-3-319-69760-4_13

concentration, the so called "hot-spots", and are able to auto-ignite. Their location is strongly dependent on both the local temperature-time history and the mixture inhomogeneity in the end-gas.

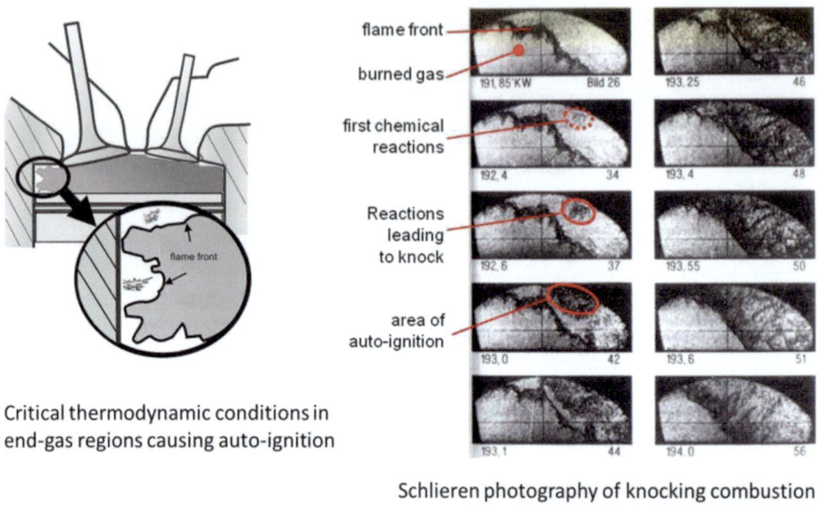

Fig. 1. Auto-ignition in end-gas pockets (left); knocking combustion detected with Schlieren photography (right)

In 1936, Withrow und Rassweiler [2] carried out measurements with high-speed cinematography of knocking combustion, which allowed observation of the entire combustion chamber of a single-cylinder engine. For that purpose, the engine was equipped with a quartz window in the cylinder head, which provided a view of the entire cylinder volume from the top of the engine. With these investigations, it was detected for the first time that knocking combustion can be initiated by auto-ignition, which supports Ricardo's statement in 1923. Further investigations by Withrow have shown that the location of auto-ignition can be different in the end-gas region. The occurrence of knocking starts both at different times and different locations within the end-gas region.

Extensive investigations of knocking combustion with high-speed cinematography were performed by Miller et al. from 1938 on at the National Advisory Committee for Aeronautics (NACA) in Cleveland. The first investigations were performed with frame rates of up to 40,000 frames per second [3]. In later years from 1943 on, special drum cameras were used, which allowed frequencies of up to 200,000 frames per second [4]. Further investigations of knocking combustion were performed in the following years by Male [5] at NACA. During these investigations two high-speed cameras have been used. With one camera, which measured at 40,000 frames per second, the entire combustion process was observed, while with the other ultra-fast camera (500,000 frames per second) only a small time sequence of the combustion was recorded. With these measurements, Male succeeded in detecting detonation waves for the first time in addition to the combustion process.

In the preface of the 4th edition of the book “The High-Speed Internal Combustion Engine”, published in 1953 by Sir Henry Ricardo [6], he wrote that he dedicated a considerable amount of discussion to knocking combustion and its relation to both engine parameters and fuel conditions in the 1st edition of his book, published in 1923. The reason was that at that time the knowledge of knocking was very limited. In the meantime a lot of research work had been done, and the literature about knocking combustion had already obtained a confusing scale. Additionally, from the standpoint of chemical considerations, it became more and more confusing. From the viewpoint of an engineer, the problem of knocking combustion in internal combustion engines had been well understood. Therefore, Ricardo would have claimed that knocking would not be a problem for future engine applications due to improved fuels and suitable fuel additives that would prevent the occurrence of knocking.

Further investigations of knocking combustion in both research and development of internal combustion engines have been performed until the middle of the last decade of the last century. With the reduction of the amount of lead in gasoline fuel due to the introduction of the 3-way-catalyst, knocking combustion again became a subject of increasing interest. Due to the research work performed until the beginning of the 1980 s, different conceptual models for the prediction of knocking combustion have been developed. Two main theories have emerged, which differ in that knock onset is located either directly inside the flame front (detonation theory) [7] or away from the flame front inside the unburned end-gas region (auto-ignition theory) [8, 9].

Kröger and Spicher [10] performed investigations of knocking combustion in which they proved the theory that knocking combustion is initiated by an auto-ignition in the unburned end-gas region during the later phase of combustion, shown in Fig. 1 (right). The figure represents knocking combustion in the end-gas region, detected by a high-speed Schlieren cinematography in a single-cylinder engine. The intensity of knocking during this combustion cycle has been defined as an intermediate level with a zero to peak amplitude of 1.1 MPa in pressure oscillations, created by knocking combustion.

The first picture (number 26) was taken at 191.85 CAD after bottom dead center (ABDC), respectively 11.85 CAD after top dead center (ATDC). At this time, only the flame front of the non-knocking combustion can be seen. In picture 34, the initial chemical reactions of auto-ignition are detected. From this auto-ignition location, a so-called reaction front quickly propagates through the end-gas region from pictures 37 to 44, which seems to push the non-knocking flame front back slightly (pictures 46 to 56). The normal flame front propagates at approximately 70 m/s, while the speed of the knocking reaction front increase from approximately 200 m/s at the beginning to nearly 950 m/s at 15 CAD ATDC. After that, the speed of the knocking reaction front decreases to 700 m/s and in the final phase to 200 m/s.

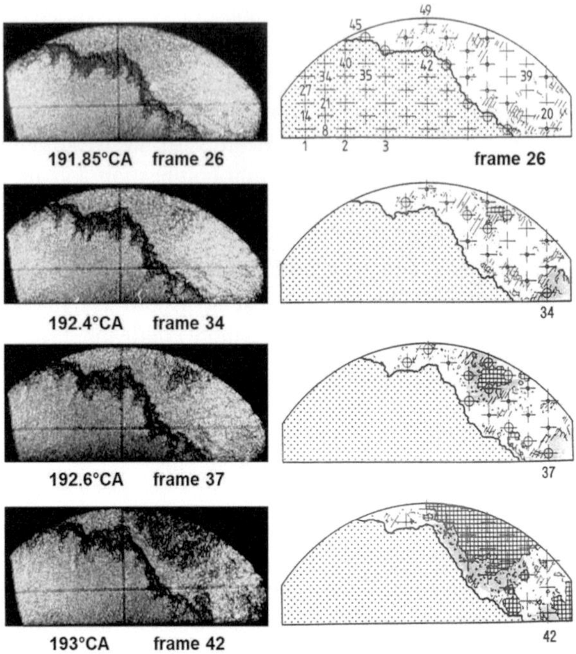

Fig. 2. Detection of auto-ignition location by Schlieren cinematography and optical fiber technique in the same combustion cycle

Simultaneously to the Schlieren measurements a special optical fiber measurement technique in conjunction with photomultipliers, which convert the optical signals of the different optical fibers into amplified electrical signals, was used (Fig. 2). With this specially developed optical fiber measurement device the combustion radiation in the ultra-violet (UV) wave length range was mainly detected. This wave length range characterizes mainly the radiation of combustion and auto-ignition regions. It can be seen from the results that both the Schlieren cinematography and the optical fiber technique detect the onset of auto-ignition at both the same location and the same time.

This optical fiber technique was used to investigate normal flame propagation as well as knocking combustion in several research engines and in a multitude of production engines. In a multitude of these general investigations of flame propagation additionally knocking combustion as well as the location of knock onset have been analyzed and detected with the aim of improving combustion chamber design as well as combustion characteristics to avoid knocking combustion in both 2-valve and 4-valve production engines. In nearly all production engines with external mixture preparation as well as with direct fuel injection, it could be detected that knocking normally occurs on the colder intake side of the combustion chamber [11]. As an example of the general behavior of the combustion process with knocking, Fig. 3 shows the flame propagation of the non-knocking combustion, starting at the spark plug, as well as the knock onset in the unburned end-gas region close to the intake valves of a 4-valve engine including the knocking reaction front with propagating pressure waves.

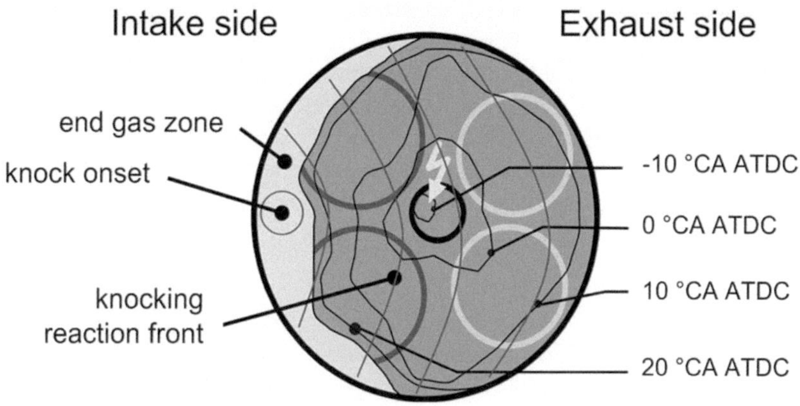

Fig. 3. Propagation of both normal flame front and knocking reaction front

2 Characterization of Knock Intensity

The occurrence of irregular combustion like knocking in SI engines depends significantly on the design of the combustion chamber (shape, compression ratio), the engine operating conditions (engine power, speed, air-fuel ratio, ignition timing) and the used fuel as well. It is well known that knocking combustion is considerably critical during fuel load operation with air-fuel ratios slightly below stoichiometric mixture conditions ($\lambda \approx 1$). Additionally, knocking combustion becomes more critical with increasing compression ratio and using fuels with reduced research octane numbers (RON).

The improvement of efficiency and fuel economy are one of the most important tasks in the development of internal combustion engines (ICE). The spark ignition engine with gasoline direct injection represents a measure to improve fuel consumption both at part load operating conditions as well as at full load operation. In addition, engine downsizing is a concept that can significantly reduce the fuel consumption of gasoline engines. This is achieved by reducing engine displacement in combination with boosting the engine. Due to the boosting and direct injection, the downsized engine is characterized by the same level of power output as the larger, naturally-aspirated engine. The high power density and thermal loading at full load operation can cause knocking combustion, which may extremely critical during certain engine operating conditions. However, the occurrence of knocking combustion is always limited in spark advance at high engine loads.

Knocking is normally characterized by its intensity. With higher knock intensities, knocking is more critical, because the engine may be damaged. In order to avoid both engine damage and knocking combustion without disadvantages in fuel consumption, many spark-ignited engines are equipped with a knock control system. In case of knocking, a sensor mounted at the cylinder block detects the mechanical vibrations and the ECU retards the spark timing until safe engine operation is established. However, with increasing spark advance, the knock tendency of different engines can vary.

Figure 4 shows the pressure traces of both a non-knocking combustion cycle and a knocking combustion cycle. The non-knocking combustion is represented by a smooth pressure trace without high frequency oscillations (red curve), while the pressure trace of knocking combustion shows characteristic pressure oscillations. The resulting structural vibrations are responsible for the characteristic knocking sound which gave its name to the phenomenon. The pressure wave interferes with the surface layer at the combustion chamber walls, which leads to a locally elevated heat transfer. In conjunction with the mechanic loading due to the high pressure gradients and vibrations, this can cause severe engine damage.

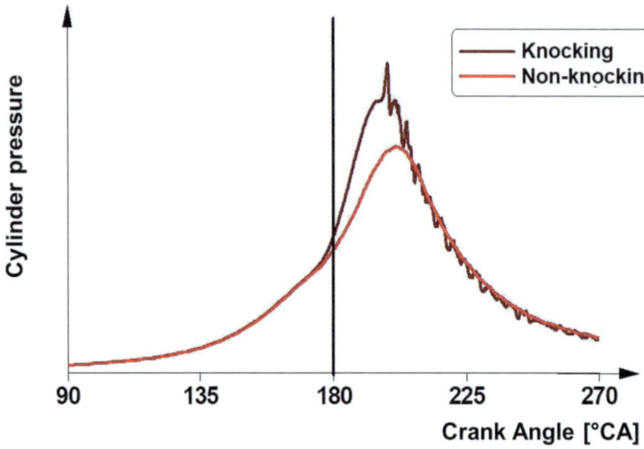

Fig. 4. Pressure traces of normal combustion and knocking combustion

To quantify the knocking phenomenon, it is helpful to introduce a variable which enables a characterization of the knock intensity. This can be done by processing the pressure trace in different ways. Figure 5 illustrates a pressure trace with high temporal resolution during the time interval in which knocking occurs. The knocking is indicated by the high frequency oscillations. This high frequency pressure trace, $p_i(t)$, can be smoothed resulting in a low frequency signal, $L_i(t)$, represented by the dashed line in the Figure. Using this resulting low frequency signal, it is possible to derive several variables to evaluate knock intensity KI. The maximum pressure peak $/\Delta P_{max}/$is equal to the difference between the maximum pressure of the high frequency signal and the corresponding value of the low frequency signal. This procedure of evaluation is mainly used to characterize knock intensity in development of SI engines [9].

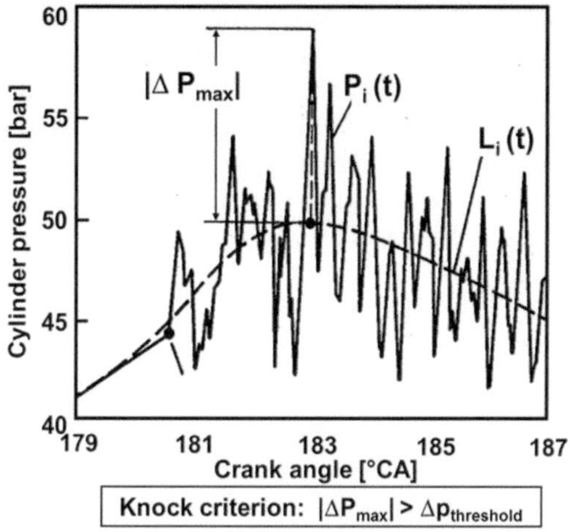

Fig. 5. Maximum pressure amplitude as one knock criterion KI

One other criterion to define knock intensity KI is to use the so called knock signal energy $E_{\Delta P}$ within a time window (t_1, t_2), represented in Fig. 6 [9]. This knock criterion can be calculated as follows:

$$KI = E_{\Delta P} = \int_{t1}^{t2} (p(t) - L(t))^2 \, dt$$

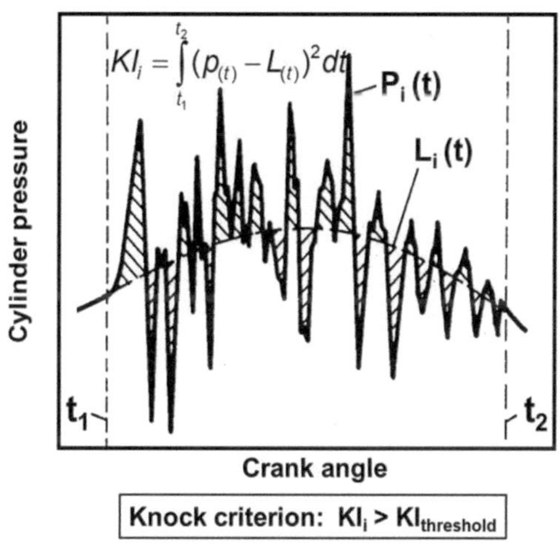

Fig. 6. Signal energy as one other knock criterion KI

This knock criterion is a little bit more sensitive with regard to the occurrence of knocking and is normally used for the definition of knock intensity of individual knocking combustion cycles in research activities.

Figure 7 shows knock intensity $KI = /\Delta P_{max}/$ for two spark-ignition engines with different knock limits and different increases in knock intensity versus spark advance at the same engine speed. While engine 1 shows a slight rise in knock intensity with spark advance, which can be easily managed by a knock control system, engine 2 shows a significant rise in knock intensity with spark advance, which is more difficult to control. Therefore, engine 2 will run into heavy knocking conditions with a small range of higher spark advance after the knock limit. This can generally be defined as strong knocking or mega-knocking behavior.

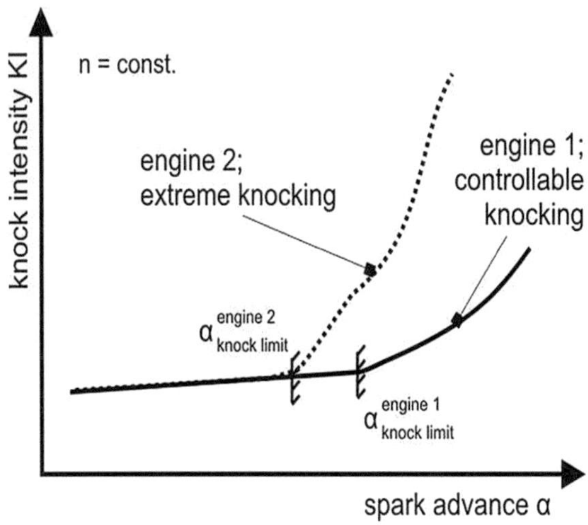

Fig. 7. Knock intensity versus spark advance for two spark ignition engines

Also the flame radiation (light emission) from the combustion process is suitable for a good extraction of knock information. Extensive studies have shown light emissions of knocking combustion and non-knocking combustion are clearly different from each other [9, 11, 12]. In addition, the flame propagation velocity is much higher when engine knock occurs. Therefore, the detection of knocking combustion with an optical fiber sensor is a very good way to develop a suitable knock control system for SI engines both during the research process and the development process. Figure 8 shows a definition of the knock intensity KI from an optical signal detected by an optical fiber installed in a sensor, which is located inside the end-gas region, in which the auto-ignition of knocking occurs [12]. The knock intensity KI is defined by the gradient in the signal rise (tan α). The velocity of the knocking flame can be calculated by the timing when the strong increase in the light signals occurs.

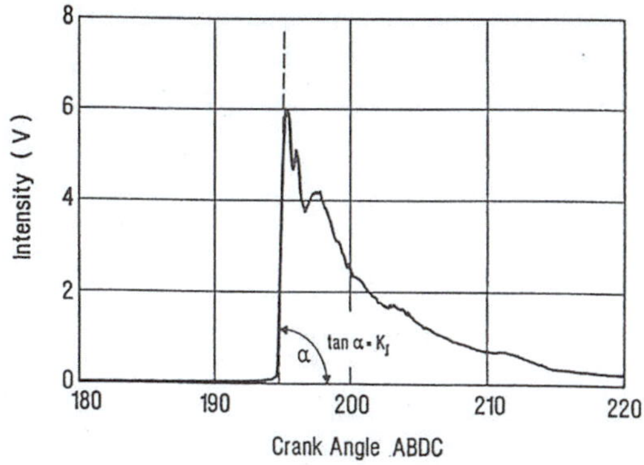

Fig. 8. Knocking detection by an optical fiber sensor in end-gas region

Figure 9 shows the optical signal detected by an optical fiber in a sensor, which is located in the area in which the non-knocking combustion already takes place before knocking combustion occurs. Due to the normal combustion during the early combustion phase the flame signal rises with a lower gradient compared to the signal rise shown in Fig. 8. The reason is that the speed of the normal flame propagation is significantly lower than the speed of the knocking flame front. When knocking occurs during the later combustion phase, a second steep rise in the optical signal from the optical fiber sensor can be recognized. The second steep gradient can be used to define the knock intensity KI, as indicated in Fig. 9.

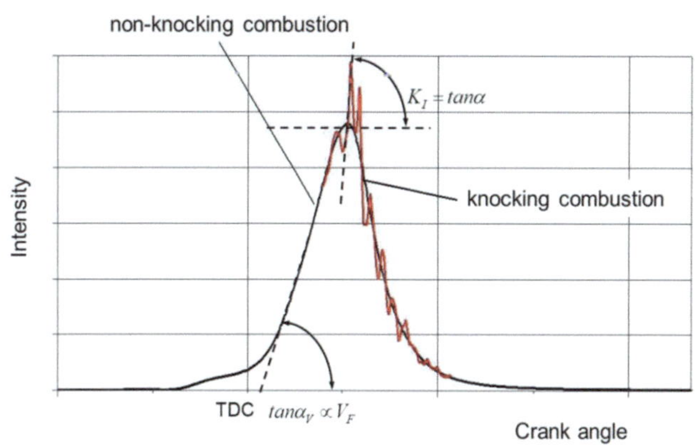

Fig. 9. Knocking detection with optical fiber sensor near the spark plug

To detect the location of knock onset, optical fiber spark plugs with multiple optical fibers in the body of the spark plug can help to obtain information about the start of knocking combustion or a sensor which contains at least 3 optical fibers with a triangle arrangement of location [12]. Figure 10 shows the pressure trace and the flame signal at the optical fiber in the spark plug which first detects the propagation of the knocking reaction zone for a combustion cycle with mega-knocking. Due to the moving direction of this reaction front, given from the order of the different optical signals and the occurrence of the pressure fluctuations due to knocking the location of knock onset can be determined [13].

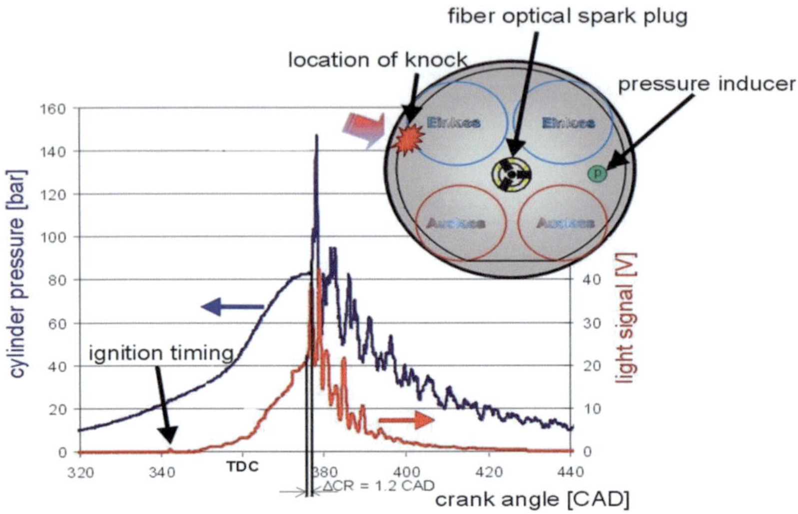

Fig. 10. Mega-knocking combustion for a gasoline direct injection engine

It can be seen from the pressure trace, respectively the maximum amplitude in pressure oscillation, that knock intensity is nearly 7 MPa. Knock onset occurs close to the cylinder liner wall at the left intake valve. The reason for that is the structure of the in-cylinder flow generated by the design of the intake ports. While the right intake port creates a swirl flow, the left intake port is designed for high volumetric efficiency. Due to these two different flow conditions, a weak flow movement with richer fuel-air ratios occurs at the location of knock onset. The difference in the steep increase between oscillations in flame signal and oscillations in the pressure trace is 1.2 CAD. From this range, the engine speed and the distance between the location of both the first optical fiber with an optical signal from the oscillating knocking reaction front and the pressure trace the speed of the knocking combustion front can be estimated to approximately 1,150 m/s, which seem to be the sonic speed or even supersonic speed. Knocking combustion with such high knock intensities is defined as "Extreme Knocking" or "Mega-Knocking", with can lead to a strong engine damages after some or even one of such irregular combustion cycles.

Calculations of the mass fraction burned function (X_B or MFB) have shown that the rate of burned mass can be reduced to about 60% at the crank angle when auto-ignition occurs in the end-gas region (knock onset) [13]. That means approximately 40% of unburned air-fuel mixture may be consumed by the occurrence of knocking combustion, shown in Fig. 11.

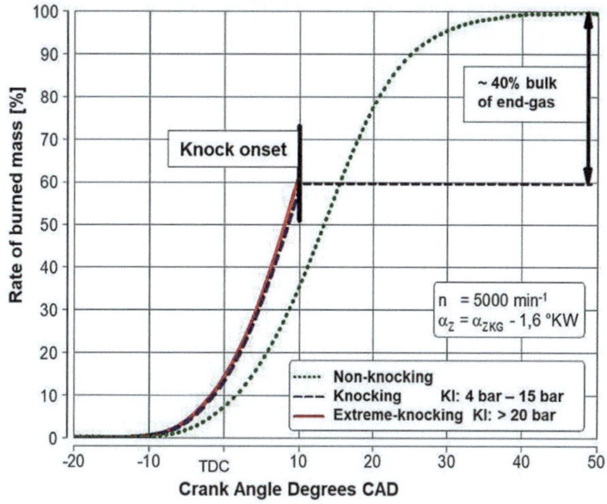

Fig. 11. MFB function (non-knocking, averaged knocking, extreme-knocking)

Figure 12 shows the temperature traces in the unburned mixture during non-knocking combustion and knocking combustion until knock onset, received from the corresponding thermodynamic calculations of mass fraction burned [13].

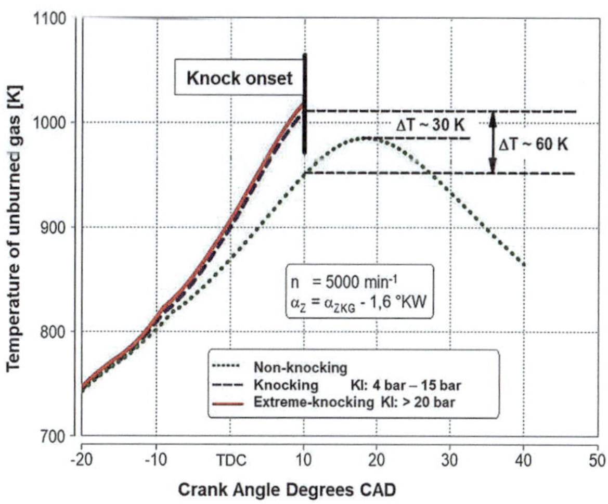

Fig. 12. Temperature behavior of unburned gas (non-knocking, averaged knocking, extreme-knocking)

It can be seen that the temperature of the unburned gas in the combustion chamber at the time when knocking combustion occurs by auto-ignition in the end-gas bulks can be 60 K higher at knock onset compared to the temperature of the unburned gas in the case of non-knocking combustion and 30 K higher than the maximum temperature of the non-knocking unburned gas.

3 Pre-ignition and Knocking

During the last years, with the appearance of highly supercharged downsizing engines, a new combustion phenomenon has been observed in these engines, especially in full load operation at low engine speed. This phenomenon, referred to as premature ignition or pre-ignition, shows some similarities to extreme knock as well as to surface ignition. As in the case of surface ignition, combustion obviously starts before the ignition timing. But the phenomenon acts neither self-preserving nor self-amplifying. Instead, pre-ignition occurs stochastically, just like extreme knock. All three phenomena of irregular combustion (knocking, mega-knocking, and pre-ignition) have in common that they are characterized by very high knock amplitudes. Figure 13 shows pressure traces for cycles with pre-ignition and regular combustion at the same operating point. Spark timing is adjusted to 3 CAD ATDC. It can be seen that pre-ignition has already started at nearly 5 CAD BTDC, which yields fast inflammation with strong rates of pressure rise. In the later phase of combustion, mega-knocking with maximum amplitude of more than 15 MPa in pressure oscillations occurs. It is obvious that such irregular combustion events can lead to engine damage.

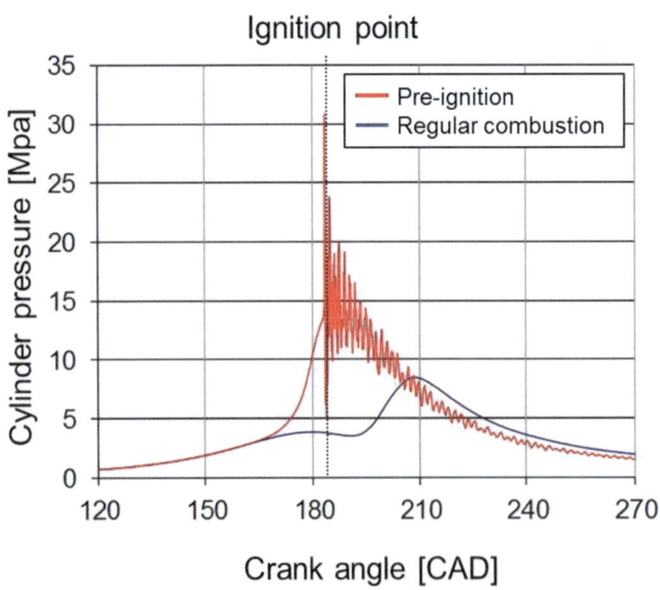

Fig. 13. Pressure traces of regular combustion and pre-ignition

Concerning the phenomenology of pre-ignition, two different aspects can be distinguished: on the one hand the characteristics of its occurrence, on the other hand its development. Both aspects shall be briefly discussed in the following section. Pre-ignition is a phenomenon occurring in a stochastic manner at operating points with low engine speed and high load. Until now, it is not possible to predict the occurrence of pre-ignition reliably, not even in a test engine equipped with pressure transducers in the cylinders and the ports.

Based on the data recorded with a high speed camera system, the progress of combustion in cycles with pre-ignition could be investigated. Furthermore, the distribution of pre-ignition origins could be evaluated. The result is illustrated in Fig. 14. The picture shows the view through the endoscope into the combustion chamber. Each dot in the picture represents the origin of one optically detected pre-ignition.

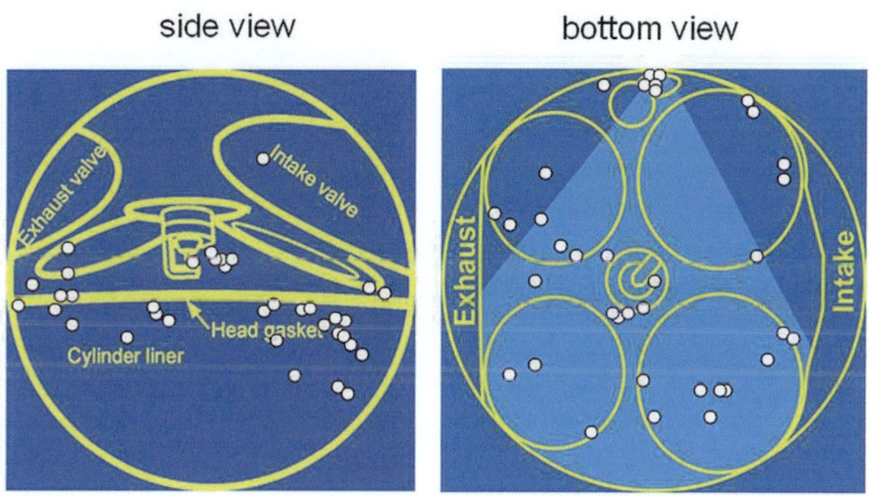

Fig. 14. Individual combustion cycles with several locations of pre-ignition

However, the characteristics of the occurrence of pre-ignition are remarkable. It sometimes occurs in the form of intermittent sequences consisting of multiple cycles with pre-ignition alternating with regular burning cycles [15]. It is clearly visible that there is no accumulation of ignition locations at any component of the combustion chamber. Instead, pre-ignition origins are spread over a wide range along the plane of the cylinder head gasket. In a different experiment, the high-speed camera system and an optical fiber spark plug were applied simultaneously to localize pre-ignitions. The additional use of an optical fiber spark plug makes it possible to determine the starting point of a pre-ignition also in the plane which is perpendicular to the optical fibers.

Figure 15 shows examples of pre-ignition locations during different combustion cycles. It can be recognized that in some cycles several pre-ignition locations appear in different locations approximately at the same time. Other cycles show movable pre-ignition kernels, while pre-ignitions outside of the viewing area of the combustion chamber could also be observed. Altogether, the optical investigations of pre-ignition have shown that the locations of auto-ignition can be designated by an enormous variability. That behavior can be considered as a strong variability of spark location. Due to that the flame propagation starts normally at the wrong location inside the combustion chamber and additionally to early. These two effects provoke the fulminating knocking combustion (Fig. 15).

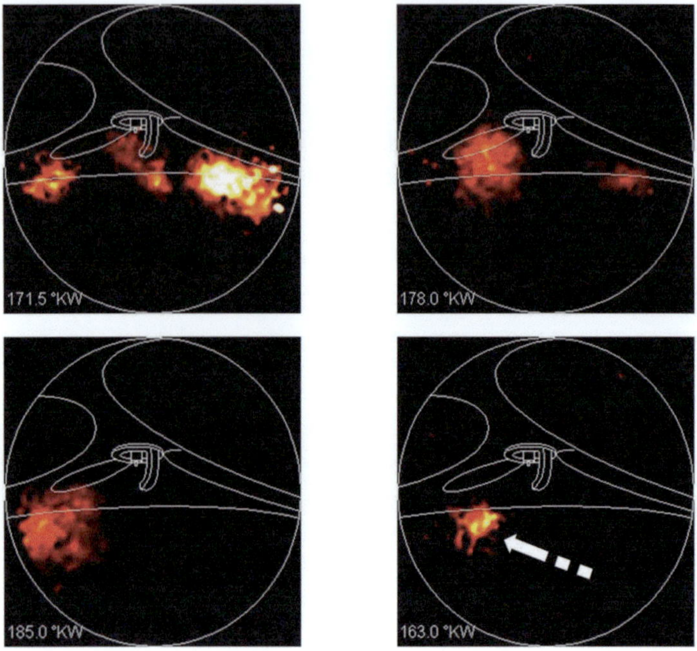

Fig. 15. Individual combustion cycles with different kinds of auto-ignition

Due to the enormous knock amplitudes, pre-ignition is considered one of the most important issues in the development of future downsizing engines. Three main questions arise while considering auto-ignition phenomena:

- When does auto-ignition occur?
- What are the effects of the auto-ignition?
- How can auto-ignition be avoided?

Each of these questions has to be answered and require intensive research work in this field. But as it is the newest and thus least explored of the described phenomena, there is still very little literature about it. To this end, a combination of experimental and numerical investigations was employed [17]. Numerical simulations were carried out to investigate certain aspects concerning auto-ignition, and these results were combined with the results of key experiments and general observations in the test bench to assess possible causes for pre-ignition. Due to their retarded spark timing, supercharged engines show auto-ignition characteristics which are substantially different from those of naturally-aspirated engines. As Fig. 16 shows, the ignition delay fraction has a local maximum before spark timing. For the load conditions achieved in actual engines, neither the mean thermodynamic state nor even the overlaid scatter band is close enough to the auto-ignition limit to be critical. But stochastic effects such as droplets can cause additional deviations from the scatter band, so that in single cycles, auto-ignition can occur already around the local maximum.

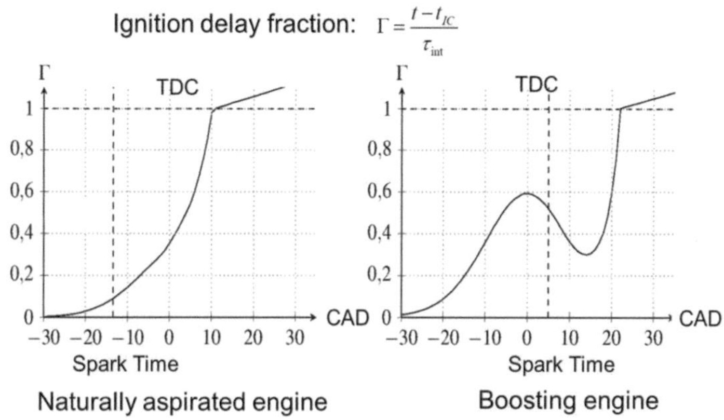

Fig. 16. Ignition delay fraction for NA- and Boosted engine

Depending on when the ignition delay fraction reaches the auto-ignition limit, phenomenon can be separated, Fig. 17.

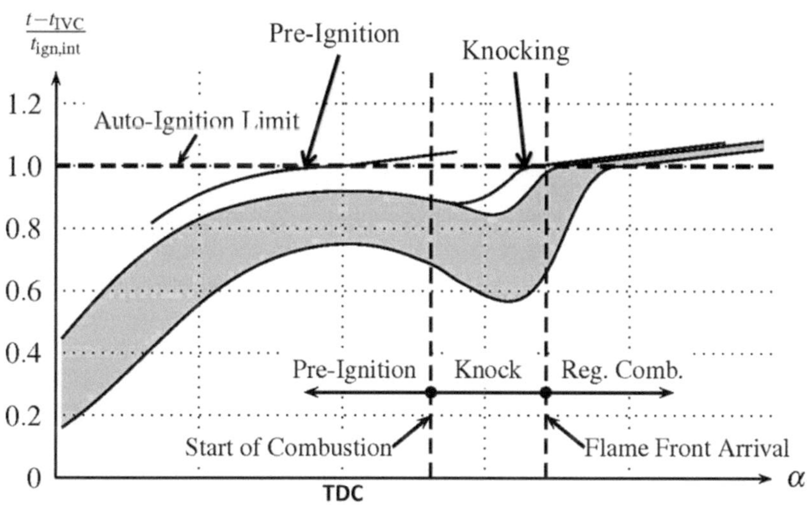

Fig. 17. Ignition delay fraction behavior during compression and combustion stroke

As already mentioned the domain to the right of the line of flame arrival marks regular combustion. At the left side of it, the domain of knock (and extreme knock) is attached, followed by the domain of pre-ignition and surface ignition, respectively, around top dead center. By definition, the border between pre-ignition and knock is given by the start of combustion [17].

Once auto-ignition occurs, its progress is strongly dependent on the thermodynamic state of the surrounding mixture. The most important factor is the homogeneity of reactivity, which is strongly correlated to temperature and air/fuel ratio. A further important criterion on the occurrence of pre-ignition is the time duration from the start of compression until spark time. This interval is dependent on engine speed, which means that this time interval increases with reduced engine speed. Due to that behavior the probability that pre-ignition occur increases with lower engine speed. Figure 18 shows the curves of ignition delay fraction for two different engine speeds. It can be seen from the results that with the engine speed of 1750 rpm the probability that pre-ignition takes place increases compared to the engine speed of 2500 rpm.

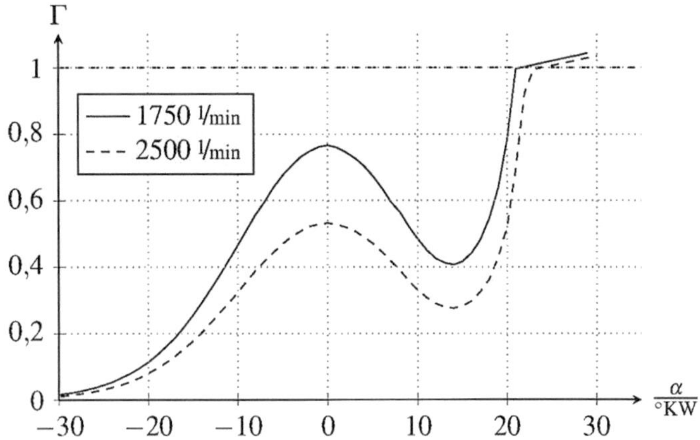

Fig. 18. Ignition delay fraction dependence on engine speed

Currently, the allowable load in downsized engines with low engine speeds is limited by pre-ignition. The damaging effect of pre-ignition is not due to the primary auto-ignition itself but to the resulting extreme knock later in the course of combustion. Primarily, these effects have to be eliminated in order to increase the allowable load level. This means that in the case of the engine which does not show any local concentration of pre-ignitions, the source of lubricant oil droplets must be eliminated (e.g. by an improved spray targeting). In the case of the second engine, spark plug temperature has to be reduced. After that, load can be increased further. Then, perhaps other stochastic effects such as solid particles might become crucial.

Depending on the cause of pre-ignition, enhancing mixture formation is another option to increase load in downsized engines while avoiding pre-ignitions. In order to improve mixture formation, a charge motion system was installed on a downsized engine. Figure 19 shows that in comparison to the base configuration, a certain charge motion improves mixture preparation, which allows an increase in the maximum load without pre-ignitions of up to more than 40%.

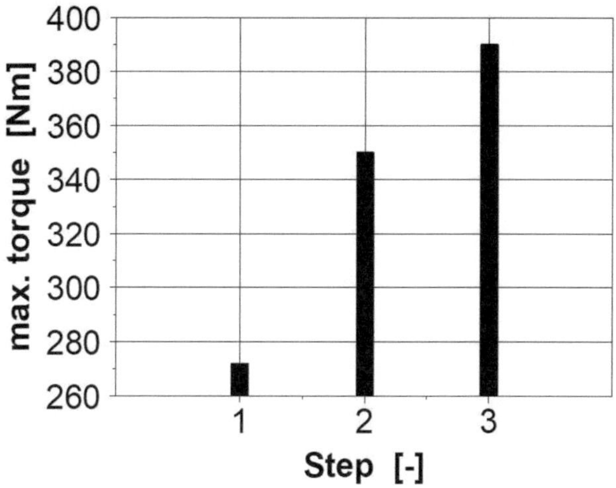

Fig. 19. Measures against pre-ignition occurrence to improve engine behavior

4 Conclusions

The phenomenon of knocking combustion has long been known to engine developers and remains a very important development issue. It has been demonstrated experimentally that knocking occurs by auto-ignition in the end-gas region of spark ignited engines. Fibre optical measurement techniques have been used in a wide variety of engines to determine the locations of knock onset and to improve the understanding of the causes of knocking and mega-knocking.

The development of downsized, heavily boosted engines has led to the discovery of a new combustion irregularity: pre-ignition. Pre-ignition is observed to occur stochastically at low engine speeds and very high engine loads, but is often observed to occur in groups. It is characterized by ignition of the fuel-air mixture before the spark timing and leads to severe knocking. Here too, optical techniques have provided useful information about the locations of the onsets of pre-ignition events and therefore about the possible causes of pre-ignition. The successful development of high-output downsized engines requires extensive research work to provide understanding about the occurrence of pre-ignition, its effects on the engine, and ways to avoid it.

References

1. Ricardo, H.R.: The Internal-Combustion Engine, vol. II. Blackie and Son Limited, Glasgow and Bombay (1923)
2. Withrow, L., Rassweiler, G.M.: Slow motion shows knocking and non-knocking explosions. SAE J. (Trans.) 39(2) (1936)
3. Miller, C.D., Olsen, H.L.: Identification of Knock in NACA High-Speed Photographs of Combustion in a Spark Ignition Engine. U.S. National Advisory Committee for Aeronautics, Report No. 761 (1942)

4. Miller, C.D., Olsen, H.L., et al.: Analysis of Spark-Ignition Engine Knock as Seen in Photographs Taken at 200,000 Frames per Second. U.S. National Advisory Committee for Aeronautics, Report No. 857 (1946)
5. Male, T.: Photographs at 500,000 frames per second of combustion and detonation in a reciprocating engine. In: Third Symposium on Combustion, Flame and Explosion Phenomena (1949)
6. Ricardo, H.R.: The High-Speed Internal Combustion Engine, 4th edn. Blackie and Son Limited, London (1953)
7. Maly, R., Ziegler, G.: Thermal Combustion Modeling – Theoretical and Experimental Investigation of the Knocking Process. SAE-Paper 820759 (1982)
8. Leppard, W.R.: Individual-Cylinder Knock Occurrence and Intensity in Multi-Cylinder Engines. SAE-Paper 820074 (1982)
9. Spicher, U., Kollmeier, H.P.: Detection of Flame Propagation During Knocking Combustion by Optical Fiber Diagnostics. SAE-Paper 861532 (1986)
10. Spicher, U., Kröger, H., Ganser, J.: Detection of Knocking Combustion Using Simultaneously High-Speed Schlieren Cinematography and Multi Optical Fiber Technique. SAE-Paper 912312 (1991)
11. Spicher, U., Krebs, R.: Optical Fiber Technique as a Tool to Improve Combustion Efficiency. SAE-Paper 902138 (1990)
12. Spicher, U., Spiegel, L., Reggelin, B., Heuser, G.: Investigation into the Applicability of an Optical Fiber Sensor for Knock Detection and Knock Control System. SAE Paper 922370, San Fancisco (1992)
13. Rothe, M., Han, K.-M., Spicher, U.: Untersuchungen zur Ursache von Extremklopfen bei Motoren mit Direkteinspritzung. In: 15 Aachener Kolloquium Fahrzeug- und Motorentechnik (2006)
14. Dahnz, C., Han, K.-M., Spicher, U., Magar, M., Schießl, R., Maas, U.: Investigations on Pre-Ignition in Highly Supercharged SI Engines. SAE Technical Paper 2010-01-0355 (2010)
15. Palaveev, S., Kubach, H., Spicher, U., Magar, M., Schießl, R., Maas, U.: Premature Flame Initiation in a Turbocharged DISI Engine – Numerical and Experimental Investigations. SAE-Paper, 2013-01-0252
16. Spicher, U.: History and Diagnostics of Irregular Combustion in Spark Ignition Engines. COMODIA 2012, Fukuoka (Japan) (2012)
17. Dahnz, C., Spicher, U.: Irregular combustion in supercharged spark ignition engines – pre-ignition and other phenomena. Int. J. Engine Res. 11 (2010). JER609

Fuel/Lubricating Oil

Impact of Detailed Fuel Chemistry on Knocking Behaviour in Engines

Roger F. Cracknell[1(✉)], Arjun Prakash[2], Kieran P. Somers[1,4], and Chongming Wang[3,5]

[1] Shell Global Solutions, Manchester, UK
Roger.Cracknell@shell.com
[2] Shell Global Solutions, Houston, USA
[3] Shell Global Solutions, Hamburg, Germany
[4] National University of Ireland, Galway, Ireland
[5] University of Birmingham, Birmingham, UK

Abstract. Demand for more efficient gasoline vehicles has driven the development of downsized, engines, which benefit from higher octane.

Features on modern SI engines such, direct injection, inter-cooling in boosted engines, cooled EGR and Millerisation lead to a much lower temperature for a given pressure in a real engine as compared to the test conditions in the CFR engine used to define the Research Octane Number (RON) and Motor Octane Number (MON) octane rating scales.

Because the end-gas in modern engines experiences a different pressure/temperature history during knocking cycles, as compared to the CFR engine, there is a growing body of evidence to suggest that for a given RON, it may actually be beneficial to have a high octane sensitivity (RON-MON) or in other words a lower MON.

To explore this further, tests have been conducted in a single cylinder DISI engine over the whole speed load map using three different compression ratios, and fuels with two different levels of RON but with three octane sensitivity levels ranging from 5 to 15. These results have been further interpreted by reference to chemical kinetic models for gasoline autoignition, which can be used to rationalise how the influence of sensitivity varies over the speed/load map.

1 Introduction

Driven by increasing market demands for higher fuel efficiency, gasoline engines have drastically increased in complexity over the past two decades to incorporate direct injection (DI), turbocharging, cylinder deactivation, intercooling and EGR. Fuel octane quality can play a significant role in improving engine efficiency, through preventing knock, and hence allowing the engine to operate close to optimal spark timing [1–4].

Octane is defined using two industry standards: RON and MON, which are measured on a single-cylinder standardized naturally aspirated carburettor engine designed in 1929, known as the Cooperative Fuel Research (CFR) engine [5, 6]. The octane quality of an actual fuel is defined in terms of the volume fraction of iso-octane in a Primary

M. Günther and M. Sens (eds.), *Knocking in Gasoline Engines*,
https://doi.org/10.1007/978-3-319-69760-4_14

Reference Fuel (PRF) consisting of iso-octane and n-heptane which has the same knocking characteristics as the real fuel.

The auto-ignition characteristics of PRF fuels are different to real-world gasolines. Figure 1 illustrates the ignition delay time as a function of temperature for a surrogate “real fuel” and two PRF mixtures, where PRF95 denotes a mixture with 95% by volume of iso-octane and 5% by volume of n-heptane, which has an octane number of 95 and PRF85 is defined in an analogous fashion. Such a plot is familiar from experimental techniques such as rapid compression machines and shock tubes.

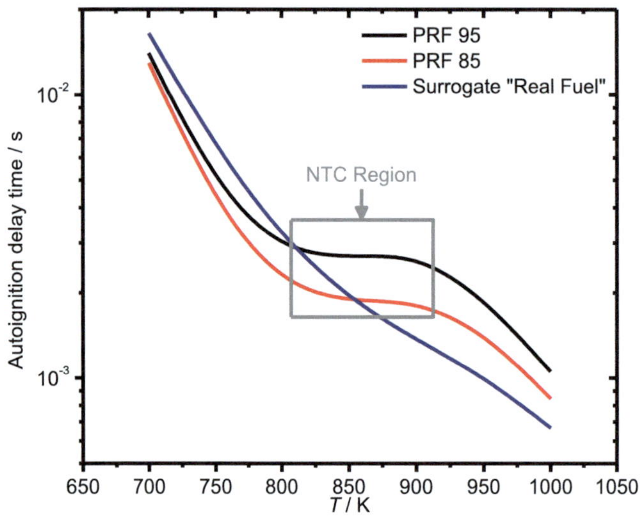

Fig. 1. Calculated autoignition delay times at 40 atm for a surrogate gasoline comprising 62 vol. % toluene, 16 vol. % i-octane, 22 vol. % n-heptane (95 RON, 85 RON), PRF95, and PRF85. Surrogate compositions were computed based on the work of [7]. All simulations were carried out for stoichiometric fuel in air ($\phi = 1$) mixtures using the chemical kinetic mechanism from [8].

For the surrogate “realistic” gasoline, the autoignition delay time decreases monotonically with increasing temperature. By contrast the PRF fuels show a “Negative Temperature Coefficient” (NTC) region in which the ignition delay does not decrease with increasing temperature.

Approaching the NTC region, the line for the surrogate “real fuel” crosses over the curve for PRF 95 and then at higher temperatures it crosses the curve for PRF 85. The temperatures at which the surrogate intersects the PRF 95 and PRF 85 curves respectively can be thought of as representing the conditions in the RON and MON tests because the ignition delay of the real fuel and the PRF defining the octane number are the same. The presence of the pronounced NTC behaviour in the primary reference fuels is the main reason why the RON of a real fuel is typically around 10 numbers higher than the MON. At temperatures, lower than the NTC region, the surrogate “realistic” gasoline is more resistant to autoignition than either PRF 95 or PRF 85. If the engine operates in a regime of pressure -temperature space such that it is more resistant to autoigniton than its RON value, then the engine is said to be operating in the "beyond RON" region [9–11].

A practical approach developed by Kalghatgi [9, 10] to quantify the octane appetite of modern engines is to define an octane index (OI) as a linear combination of RON and MON with an engine condition dependent weighting factor, K, such that

$$OI = K \times MON + (1 - K) \times RON. \quad (1)$$

This can also be written as

$$OI = RON - KS, \quad (2)$$

where S is the sensitivity (RON-MON). The K value is dependent only on the in-cylinder temperature and pressure conditions in the end-gas prior to the onset of auto-ignition.

Most investigations into the effects of octane index have found that modern engines exhibit negative K values when operating at near-knocking conditions [4]. A negative K value means that the engine will be more knock resistant using fuels with a higher RON and a higher sensitivity [4]. It has however been noted that the K value can be different at different regions of the speed load map and where the acceleration follows different trajectories. Many authors have reported that K decreases with increased engine load and intake pressure [9–15]. This can be rationalised in terms of Fig. 1 because higher intake pressure in a boosted engine with intercooling, means that the temperature for a given pressure will be lower, pushing the combustion regime further "beyond RON" in the left of Fig. 1. Other authors have also reported that the value of K increases with engine speed [4, 9, 16].

To explore this further, tests have been conducted in a single cylinder DISI engine over the whole speed load map using three different compression ratios, and fuels with two different levels of RON but with three octane sensitivity levels ranging from 5 to 15. The detailed results of these experiments are reported elsewhere [3, 18].

In this work, these results have been further interpreted by reference to chemical kinetic models for gasoline autoignition combined with a 0D engine model, which can be used to rationalise how the influence of MON varies over the speed/load map.

Table 1. Fuel matrix properties

	RON	MON	Density (kg/m^3)
F1	92	87	718.0
F2	92	82	742.3
F3	92	78	731.1
F4	98	93	698.6
F5	98	88	730.0
F6	98	83	749.0

2 Single Cylinder Engine Testing

2.1 Fuel Matrix

The six fuels tested are reported in Table 1. They all contain approximately 10% ethanol by volume and were blended in Shell Technical Centre Hamburg (STCHA) without any deposit control additives or other performance boosting additives. Given the study's focus on octane sensitivity, the fuels are split into two subsets of constant RON but varying MON. F2 is representative of regular grade gasoline in the North American market while F5 represents premium grade in the same market. The properties of all fuels were tested using standard methods before being transferred to the single cylinder engine.

2.2 Experimental Facilities and Single Cylinder Engine

A 4-stroke single-cylinder AVL DISI research engine was used to carry out all testing. The Compression Ratio (CR) of the engine can be varied between 9.5:1, 10.5:1 and 11.5:1 by inserting metal sheets of varying sizes between the cylinder liner and crank-case; these are denoted CR1, CR2 and CR3 respectively. The 4-valve pent-roof cylinder head with centrally mounted piezo direct injector and Variable Valve Timing (VVT) system for all four valves make up the engine's combustion system. The spark plug was centrally located within the combustion chamber.

An electric dynamometer was used to maintain the engine operating at a constant speed ($\pm$1 RPM) for all power outputs. Real time combustion indication and analysis was achieved using an AVL Indicom system. An IAV FI2RE management system was used to control the engine. A Siemens CATs system controlled the fuel, air, coolant, oil conditioning units and emissions measurement equipment during operation. The CATs system also carried out signal acquisition and recording during the engine's operation.

The in-cylinder pressure was measured using a Kistler pressure transducer, installed in a sleeve on the intake and exhaust bridge. A charge amplifier (ETAS ES630.1) was used to collect pressure measurements. In the range between 30° and 70° CA After Top Dead Centre (ATDC) the data was recorded at a resolution of 0.1° CA, in the remaining range at a resolution of 1° CA.

Boosted air (up to 3 bar) was delivered to the engine using a connected external air handling device. Before entering the engine, the air was dried, filtered and delivered to a 200L conditioning unit where the air pressure and temperature were tightly controlled using a closed-loop approach. The engine was operated with stoichiometric combustion with any fuel volumetric energy content differences compensated through the injector pulse width. An AVL fuel mass flow meter was used to control fuel consumption and individual AVL conditioning systems controlled the fuel, coolant and lubricant during engine operation.

2.3 Experimental Procedure

In the first experimental phase, F2 and CR3 (11.5) were used to determine the engine's optimization parameters such as optimum injection mapping and valve timings. These remained unchanged throughout all testing thereafter. In the second phase F1-F6 were tested for all compression ratios at a number of speed-load conditions, described in Table 2. Tests for F2 CR2 (10.5) were repeated at the start, middle and end of the procedure to verify consistency of results.

Table 2. Speed-Load conditions tested

<table>
<tr><th>Engine Speed (RPM)</th><th>Engine IMEP (bar)</th></tr>
<tr><td>1000</td><td>2, 4, 6.5,8, 9.5, 12, 14, 16</td></tr>
<tr><td>1800</td><td rowspan="3">2, 4, 6.5,8, 9.5, 12, 14, 16, 18, 20, 22</td></tr>
<tr><td>2500</td></tr>
<tr><td>3500</td></tr>
</table>

Before each experimental set-up, the fuel system was flushed with the new sample and the compression ratio was manually adjusted as previously described. An engine warm up period of about 30 min was required to warm up oil and coolant to temperatures of 360 K and 355 K respectively. Engine leakage was monitored and oil dilution checked using hydrocarbon emissions data. Daily engine operating check points were performed to ensure correct operation of all measurement systems and engine, after which the testing protocol begun.

With each testing condition the spark timing was gradually advanced until MBT was reached, estimated by placing MFB50 at 7.5 ± 0.5° ATDC. In some cases the onset of knock prevented reaching MBT advance and the testing was carried out in Knock-Limited Spark Advance (KLSA). At 1800 RPM the tolerated threshold for engine knock was 2 bar cycle-based maximum in-cylinder pressure oscillation. The CATs software was used to take 10 Hz measurements for in-cylinder, exhaust and intake pressures as well as ignition and injection signals, which were recorded onto the AVL Indicom system every 100 cycles.

2.4 Impact of Octane Sensitivity on Thermal Efficiency

Indicated thermal efficiency can be obtained from the indicated specific fuel consumption measurements and an efficiency map plotted. Maps for the 10.5:1 CR case as a function of RON and sensitivity are shown in Fig. 2. The complete set of maps is shown in [3].

The benefits of high sensitivity on efficiency are particularly visible for the 92 RON fuels. At high speed/low load points. Focusing on the region around 1500 rpm/16 bar IMEP, it is found that the efficiency clearly increases with increased sensitivity. There is still an increase in efficiency with sensitivity at higher speeds, suggesting a negative K value across the map, so that higher fuel sensitivity means a higher OI from Eq. 2.

The benefits of higher sensitivity are less apparent for the 98 RON base fuels. It has been found previously that the incremental benefits of higher OI fuels decrease as the OI increases [2].

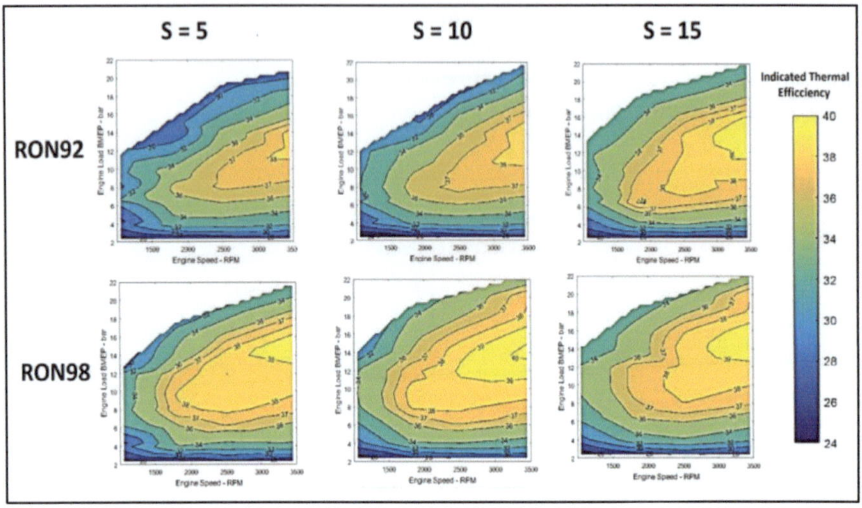

Fig. 2. Indicated thermal efficiency versus speed and load for CR 10.5:1 as a function of RON and sensitivity.

A more detailed analysis of this data set in [18] shows clearly that K value decreases with engine load and also suggests a more negative K value with increased compression ratio. This data does not reveal a clear dependence on engine speed, although as previously noted, several studies have shown that K value increases with engine speed.

3 Chemical Kinetic Interpretation of K Value Dependence on Speed and Load

The chemical kinetic modelling studies shown in Fig. 1 have been extended in order to explore the impact of speed and load on K value. Two major assumptions have been used in this analysis:

1. The pressure and temperature of the end gas (both in the compression stroke before the spark and after the spark) are related to each other by a compression isentrope of constant gamma (we assumed a value of $\gamma = 1.26$). This means that any heat transfer effects are ignored.
2. The point at which auto-ignition of the end-gas occurs can be predicted by the crank angle position at which Livengood-Wu integral [17] reaches unity,

$$\int_{t_i}^{t_f} \frac{1}{\tau(T,p)} dt = 1, \tag{3}$$

where t is time and $\tau(T,p)$ ignition delay time under a constant temperature, T and pressure, p predicted by the chemical kinetic model. The pressure is estimated by a simple 0D engine model which represents the combustion by a Wiebe function.

Figure 3 shows autoignition delay times for the fuels described in Fig. 1, modelled by the Andrae mechanism [8], plotted as a function of pressure, where pressure and temperature are related to each other along a compression isentrope for each fuel. The curve contains similar features to Fig. 1: At the lower p and T region of the TRF 95/85 is more resistant to autoignition than PRF 85. The question of where the LWI will have reached unity is critical: If autoignition occurs before the cross over point of the blue and black lines then the K value is negative, otherwise the K value is positive. Obviously changing the starting conditions will change the isentrope. An initial pressure of 2 atm., and the same intake pressure, will shift the curve to the right. Even though the pressures are higher, the ignition delay times are longer because the temperature for a given pressure is lower.

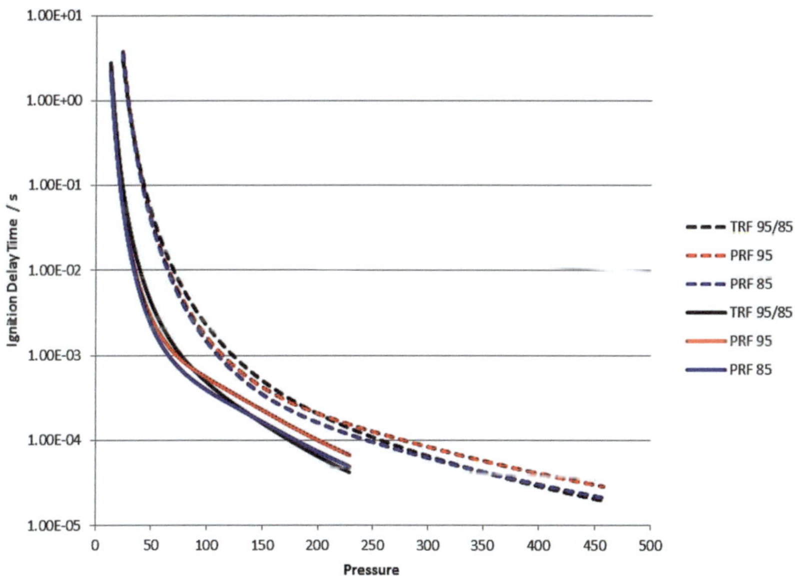

Fig. 3. Ignition delay time plotted against pressure along compression isentropes for each fuel [Air intake pressure = 1 atm (solid lines) 2 atm (dashed lines), Air intake temperature = 300 K, $\phi = 1$].

Changing the engine speed does not alter the isentrope. It is just that with a higher engine speed, the time elapsed per crank angle degree diminishes, and higher values of p and T are reached before the Livengood Wu Integral reaches unity. This is illustrated by Fig. 4, which shows 2 different compression isentropes. The p and T history was related to crank angle using a simple 0D model. The compression ratio (CR) and spark timing (ST) do not affect the locus of the isentrope, but the spark timing will influence the rise of the end gas pressure, which will affect the position on the isentrope where the Livengood Wu integral reaches unity.

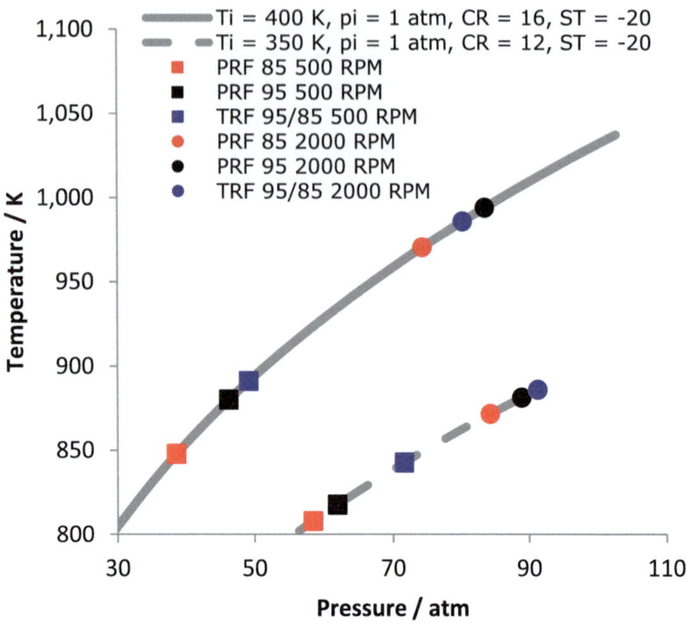

Fig. 4. Values of T and p at which autoignition occurs (defined by when the Livengood-Wu integral reaches unity) at different RPM values and initial conditions. $p_i = 1$ atm. $\phi = 1$. Solid line: $T_i = 400$ K, Dash line: $T_i = 350$ K. Black symbols represent PRF 95, red symbols represent PRF 85 and the blue symbols represent the surrogate "realistic" gasoline with 95 RON and 85 RON with composition described in the legend to Fig. 1.

Looking at the solid line in Fig. 4, representing the initial conditions of 1 atm and 400 K, for the case of 2000 RPM represented by the filled circles, it can be seen that the Livengood Wu integral reaches unity at a higher T and p, for the PRF 95 fuel (black point) than the TRF85/95 which in turn reaches unity at a higher p and T than PRF 85. Thus, the TRF95/85 fuel would knock at a point that is intermediate in crank angle, between the PRF85 and PRF95 fuels. The K value under these conditions would therefore be in the range between 0 and 1. However for the 500 rpm case, represented by the filled squares, the TRF95/85 fuel is actually more resistant to knock than PRF95, suggesting a negative K value. An increase in engine speed appears to be associated with an increase in K value.

The isentrope which passes through 1 atm and 350 K (dashed line) represents the situation with a pressure/temperature trajectory associated with a higher air intake pressure and a higher load. The TRF95/85 fuel is more resistant to auto ignition than the PRF95 fuel at both 500 rpm and 2000 rpm suggesting a negative K value under all these conditions. The extent to which the K value is negative can be visualised by the distance along the isentrope in T, p space between the point at which the TRF 85/95 fuel autoignites (i.e. with a Livengood Wu integral reaching unity) and the point at which the PRF95 fuel autoignites. This suggests both that the K value does indeed increase with rpm and decreases with load.

Note that the model does not include heat loss or friction and is just intended to illustrate trends. The rpm values chosen do not represent a real engine, but do illustrate the trends that would be observed. The effect of heat loss could be approximated by modelling the compression using a polytropic coefficient ($1 < n < \gamma$) and would lead to a lower temperature for a given pressure, which would give a directionally lower K value.

4 Conclusion

Six E10 fuels with two different levels of RON and three octane sensitivity (RON-MON) levels were used to investigate how fuel octane quality can impact efficiency in modern DISI engines. Efficiency gains as a function of octane sensitivity are shown across the speed/load range. The way in which sensitivity affects anti-knock quality can be quantified by a K value, where the octane index is given by OI = RON-K × S. A negative value of K suggests that high sensitivity will be beneficial.

In a complementary study, chemical kinetic modelling, combined with a 0D engine model was used to determine how the K value of an engine will vary with the locus of the p, T isentrope and with engine speed. As expected the model shows that that the cooler the end-gas temperature is (for a given pressure) the more negative the K value. It also shows that K increases with engine speed in agreement with several reported studies.

Acknowledgement. Funding from the European Commission Marie Curie Transfer of Knowledge Scheme (FP7) pursuant to Contract PIAP-GA-2013-610897 GENFUEL is greatly acknowledged.

References

1. Prakash, A., Cracknell, R., Natarajan, V., Doyle, D., Lobato, P., Hinojosa, M.: Understanding the Octane Appetite of Modern Vehicles. SAE Technical Paper 2016-02-0834 (2016)
2. Redmann, J., Beutler, M., Kensler, J., Luebbers, M., et al.: Octane Requirement and Efficiency in a Fleet of Modern Vehicles. SAE Technical Paper 2017-01-0810 (2017). https://doi.org/10.4271/2017-01-0810
3. Prakash, A., Wang, C., Janssen, A., Aradi, A., et al.: Impact of Fuel Sensitivity (RON-MON) on Engine Efficiency. SAE Technical Paper 2017-01-0799 (2017)
4. Remmert, S., Campbell, S., Cracknell, R., Schuetze, A., et al.: Octane appetite: the relevance of a lower limit to the MON specification in a downsized, highly boosted DISI engine. SAE Int. J. Fuels Lubr. **7**(3), 743–755 (2014). https://doi.org/10.4271/2014-01-2718
5. ASTM: D2699 - Standard Test Method for Research Octane Number of Spark-Ignition Engine Fuel. American Society for Testing and Materials (2013)
6. ASTM: D2700 - Standard Test Method for Motor Octane Number of Spark-Ignition Engine Fuel. American Society for Testing and Materials (2011)
7. Morgan, N., Smallbone, A., Bhave, A., Kraft, M., Cracknell, R.F., Kalghatgi, G.K.: Mapping surrogate gasoline compositions into RON/MON space. Combust. Flame **157**, 1122–1131 (2010)

8. Andrae, J.C.G.: Comprehensive chemical kinetic modeling of toluene reference fuels oxidation. Fuel **107**, 740–748 (2013)
9. Kalghatgi, G.: Fuel Anti-Knock Quality - Part I. Engine Studies. SAE Technical Paper 2001-01-3584 (2001)
10. Kalghatgi, G.T.: Fuel anti-knock quality-Part II: Vehicle Studies-how relevant is Motor Octane Number (MON) in modern engines. SAE Technical Paper 2001-01-3585 (2001)
11. Szybist, J.P., Splitter D.A.: Combustion and Flame 177, 49–66
12. Orlebar, C.N., Joedicke, A., Studzinski, W.: The effects of octane, sensitivity and K on the performance and fuel economy of a direct injection spark ignition vehicle. SAE Technical Paper 2014-01-1216 (2014)
13. Mittal, V., Heywood, J.: The shift in relevance of fuel RON and MON to knock onset in modern SI engines over the last 70 years. SAE Int. J. Engines **2**(2), 1–10 (2010)
14. Davies, T., Cracknell, R., Lovett, G., Cruff, L., et al.: Fuel Effects in a Boosted DISI Engine. SAE Technical Paper 2011-01-1985 (2011)
15. Zhou, Z., Yang, Y., Brear, M., Lacey, J., et al.: A comparison of four methods for determining the octane index and K on a modern engine with upstream, port or direct injection. SAE Technical Paper, 2017-01–0666 (2017)
16. Kassai, M., Shiraishi, T., Teraji, A., Wakefield, S., Goh, T., Doyle, D., Wilbrand, K., Shibuya, M.: The Effect of fuel properties on knocking performance of boosted downsized engines. In: Presented at the International Conference on "Advanced Fuels for Sustainable Mobility", 4th–5th November 2014
17. Livengood, J.C., Wu, P.C.: Correlation of autoignition phenomena in internal combustion engines and rapid compression machines. Symp. Combust. Proc. **5**(1), 347–356 (1955)
18. Wang, C., Prakash, A., Aradi, A., Cracknell, R.F., Xu, H.: Significance of RON and MON to a modern DISI engine. Fuel **209**, 172–183 (2017)

Knock Detection

Model Based Knock Detection

Matthias Biehl(✉) and Michael Meister

Robert Bosch GmbH, Stuttgart, Germany
Matthias.Biehl@de.bosch.com

Abstract. The Model Based Knock Detection (MBKD) is an innovative method to identify knocking combustions and pre-ignitions in gasoline engines. The MBKD was a joint development by the specialized calibration team and the function development team for knock detection and control at Robert Bosch GmbH. The overall objective of the MBKD is to increase quality and to reduce the effort required for the parameterization of the individual methods that serve to identify abnormal combustion phenomena.

The first series version of the MBKD is currently calibrated in more than ten series-production projects. In comparison to knock detection methods based on signal filtering in the time domain, it increases the detection quality by reducing the susceptibility for noise interferences. Methods from the field of Artificial Intelligence, Pattern Recognition, statistics, and optimized digital signal processing are combined to an improved approach for detecting knocking combustions and pre-ignitions. Results and experiences from various series calibration projects show the high generalizability and robustness against electrical and mechanical interferences of the new knock detection method. This leads to reduced knock control interventions, less misdetections and therefore to a reduced Real Driving Emissions (RDE) fuel consumption and an optimized drivability.

The MBKD method and software has shown its suitability for mass production due to the usage in various projects with different ECUs and engine variants (number of cylinders and displacement) with high customer satisfaction.

In a current development the first series version of the MBKD functionality is extended by reconstructing the physical knock and pre-ignition pressure characteristics based on signals of structure borne noise sensors. This allows us to predict sensor signal values, which cannot be detected directly or only at high costs in series-production vehicles.

The main benefit is a more specific control behavior in any engine operation point due to a better estimation of the damage potential of single abnormal combustions.

1 Introduction

The ever-growing economical and ecological demands made on modern combustion engines, in particular with regard to the improvement in efficiency and emission performance standards, require permanent new and further development of components, methods, and functions. A vast number of these developments result in increased complexity and in side effects amongst each other, which necessitates adequate continued processing of the calibration.

M. Günther and M. Sens (eds.), *Knocking in Gasoline Engines*,
https://doi.org/10.1007/978-3-319-69760-4_15

2 State of the Art

Since the last International Conference on Knocking the concept of Artificial Intelligence for knock detection has been developed to series-production maturity. This new approach is called Model Based Knock Detection (MBKD) and is presently calibrated in more than ten different series projects.

This new method of knock detection leads to an increased knock detection quality, a reduced amount of knock control interventions, a higher generalizability and an increased robustness against electrical and mechanical interferences. The basis for these improvements is a transfer of the filtering of the structure borne noise measured by the knock sensor from time to frequency domain.

Both methods, MBKD and the method using classical filtering of the knock sensor signal in the time domain (Classic KD), are currently available in ECUs of the Robert Bosch Group. MBKD and Classic KD use the same ECU hardware configuration and the software based components can easily be replaced by each other.

2.1 MBKD Method (with Reference Level)

MBKD in its actual available implementation differs in the following signal processing steps:

- signal filtering
- computation of the knock intensity
- adaption of the knock feature

The differences between MBKD and Classic KD are shown in Fig. 1 and get explained within this chapter.

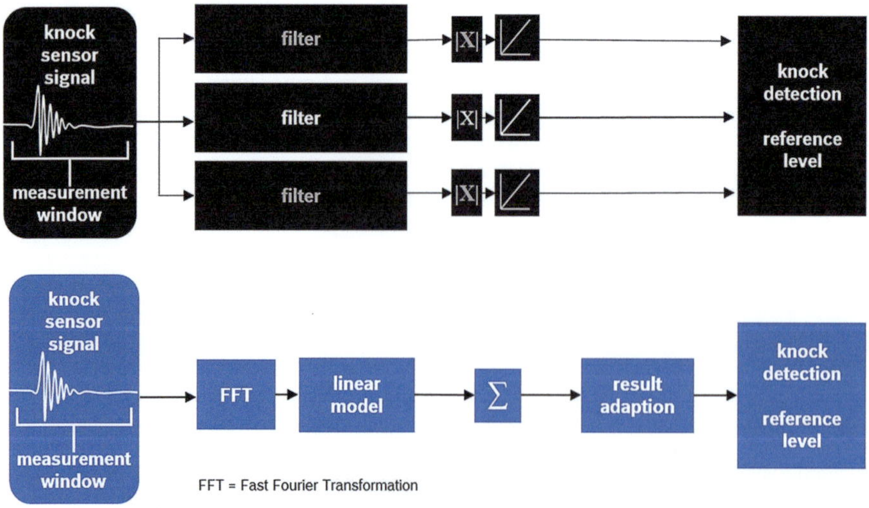

Fig. 1. Comparison of the structure: Classic KD (black) and MBKD (blue)

The knock signal acquisition is kept the same for both methods. Therefore, the measurement and evaluation of the knock sensor signal is time-limited to a calibrated measurement window.

The filtering of the knock sensor signal differs in both methods. In comparison to the signal processing in the time domain of the Classic KD, MBKD uses a Fast Fourier Transformation (FFT) to convert the time signal to the frequency domain. For the extraction of the knock feature from the knock sensor raw signal the classic approach uses standard filtering methods. Therefore, one or up to three different filters can be used simultaneously. Instead MBKD uses frequency models, which are individual weighting factors for each FFT spectral line.

To calculate one knock feature per combustion the Classic KD rectifies and integrates the filtered knock sensor signal. On the other hand, MBKD sums up the weighted spectral lines to get the knock feature of one combustion.

In addition to the Classic KD the MBKD method is extended by a knock feature adaption. This is a multistep optimization of the knock feature to improve the ratio between knocking and non-knocking combustions.

Finally, both methods use a mean value of the previous combustions to calculate a reference level, representing the base noise of the engine. A normalization of each combustion to the reference level is used to compute the final knock intensity. Afterwards a comparison to an adjustable knock detection threshold is used to decide whether this combustion is knocking or not.

2.2 Results

The explained improvements led to an increased correlation between physical knock intensity Peak [bar] measured with pressure indication and the ECU result of the structure borne noise sensor evaluation Rki [].

Table 1 shows an example of this enhanced correlation by comparing the Classic KD with the MBKD method at one specific engine operation point. Both calibrations show a good correlation between Peak [bar] and Rki []. To achieve a stable knock detection quality, the calibration of the Classic KD has to be cautious. For example, higher knock detection thresholds would be more efficient but will lead to an unstable knock detection. With the more specific signal processing in combination with the knock feature adaption, the correlation with MBKD is far more adjustable. Given that the ignition control strategy with both methods is identical, the increased and more robust correlation lead to reduced knock control interventions and therefore to a reduced fuel consumption in real driving situations and an optimized drivability.

The current MBKD method and software has shown its suitability for mass production due to the usage in various projects with different ECUs and engine variants (number of cylinders and displacement) with high customer satisfaction.

Table 1. Comparison of Classic KD vs. MBKD with reference level method

Classic KD	**MBKD with reference level method**
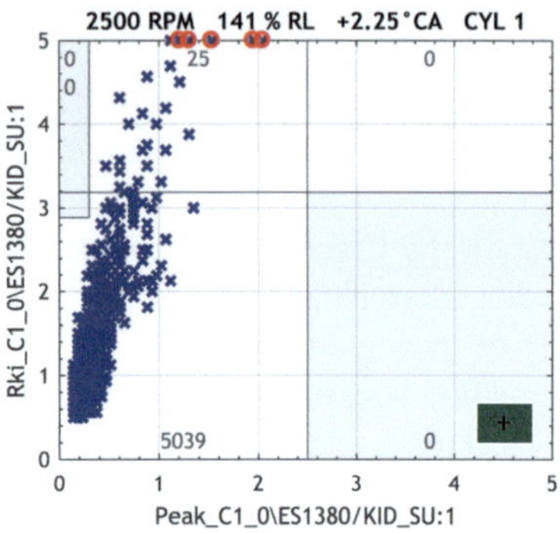	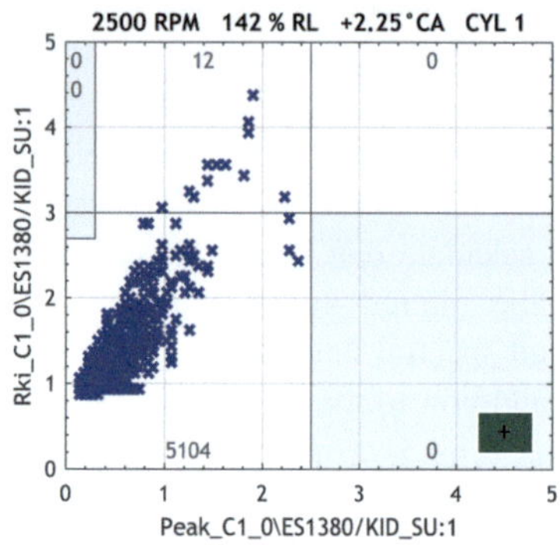
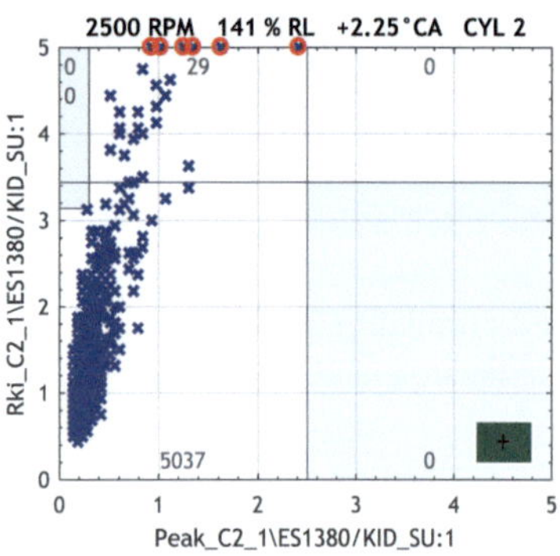	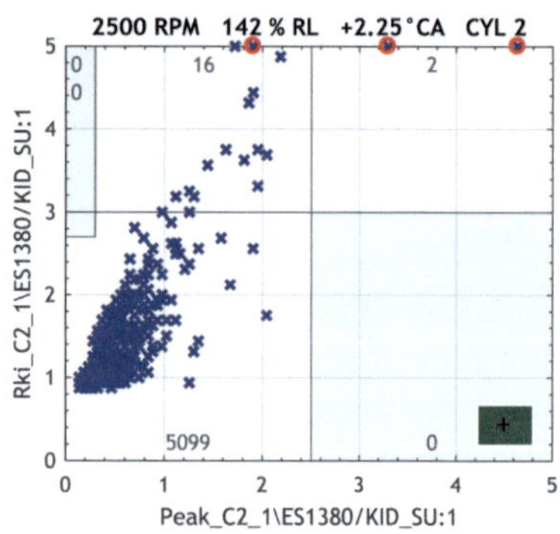
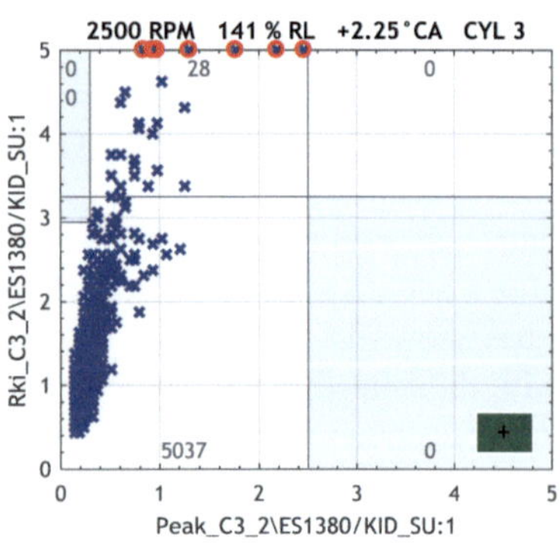	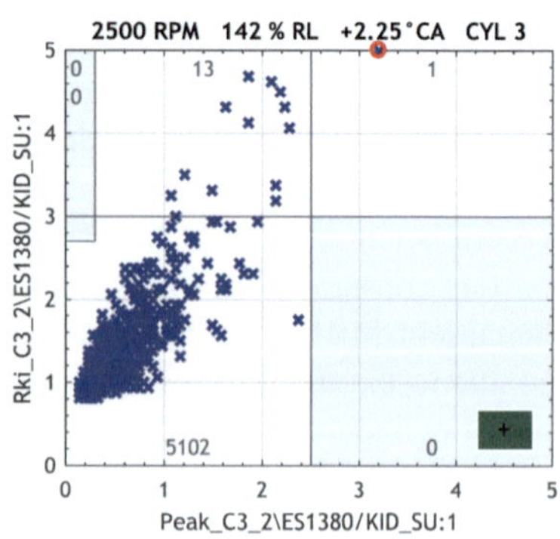

3 Future-Oriented Development

In current developments the MBDK functionality is extended by reconstructing the physical knock and pre-ignition pressure characteristic based on the signal of the structure borne noise sensor. This allows us to predict the sensor signal values, which cannot be detected directly or only at high costs in series-production vehicles. Thereby it is possible to create a common sense among engine developers and calibration engineers by talking of knocking in the unit of physical pressure, bar.

3.1 Pressure Estimation Knock Detection

The MBKD in its future-oriented implementation is expected to differ in the following signal processing steps:

- adjusted calibration of the knock feature adaption
- knock detection with a threshold based on the reconstructed physical knock pressure Peak* [bar]

Figure 2 shows the steps from the structure borne noise sensor to the Peak* value and in the end to the knock decision.

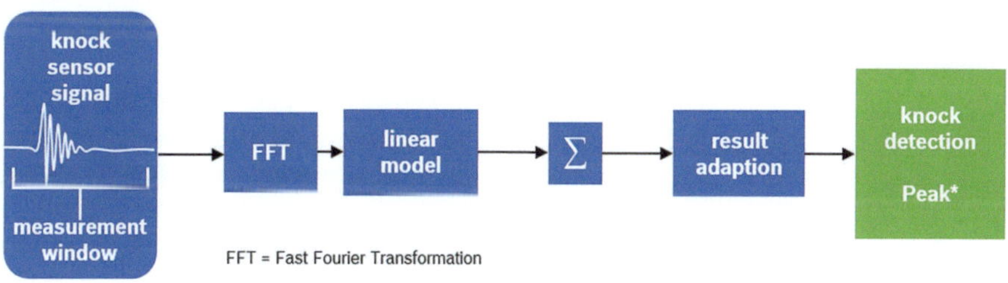

Fig. 2. Structure of MBKD pressure estimation

The knock signal acquisition and the way to one knock feature per combustion, via FFT and the sum of the weighted spectral lines, does not differ from the MBKD method explained in Sect. 2.

The base functionality of the knock feature adaption is equal to the current MBKD method. But the calibration of the different optimization steps may be different in this approach to estimate the pressure.

The decision whether a combustion is classified as knocking or not is now without a direct relation to the engine ground noise. Instead, the knock detection now depends only on a threshold related to the Peak* value. By adjusting this border the intensity when knocking gets detected can be adjusted with respect to hardware capabilities, acoustic behavior and customer preferences.

3.2 Results Knock Detection

Table 2 shows a comparison of the MBKD method with the MBKD method pressure estimation. The measurements show a good correlation between the measured physical knock pressure Peak and the reconstructed Peak*. It is possible to obtain an even further reduced misdetection tendency and less dispersion of the correlation. By adjusting the knock feature adaption the slope of the regression line can be set to 1, so that Peak* almost equals Peak.

Therefore the calibration of the knock detection threshold is not directly comparable anymore. The knock detection threshold now represents the knock pressure amplitude above which the electronic control unit (ECU) detects a knocking combustion.

This calibration results in even fewer knock interventions and therefore reduced fuel consumption.

The results in Table 2 can be misleading because they show even more knock interventions related to the overall measured combustions. The measurements have been done with an optimized ignition control strategy, with changes in the ignition retardation per detected knocking combustion and with a faster adjustment of the ignition back to the base ignition angle. Figure 3 confirms these statements. It shows the average ignition retardation of the single cylinders over the complete engine speed area at a relative load of 140%. You can see an evolution from the Classic KD over the MBKD method with reference level to the MBKD method pressure estimation. In general these development steps lead to less retarded ignition control by the knock detection functionality. With the background knowledge that the ignition control strategy can also be improved without risking an unstable steady state knock detection, the pressure estimation knock detection has obvious advantages in transient vehicle operation too.

3.3 Pressure Estimation Pre-ignition Detection

The MBKD pressure estimation approach does not stop at the knock detection. At least as interesting as an estimation of the knock pressure is the estimation of the maximum pressure P^*_{max} in the combustion chamber during a pre-ignition event.

Compared to the knock detection approach explained in Sect. 3.1 you also have to estimate not only the high frequency pressure oscillation Peak*, but also the base pressure P_{grund} to get P^*_{max}. Figure 4 shows the explained strategy.

By having an estimation for P_{max} of a pre-ignition event it is possible to classify the pre-ignitions, and as a consequence take actions meeting the demands. Thinkable are different actions for single pre-ignitions for example with low, medium or high damage potential to the engine hardware. The frequency of occurrence should also be considered. As there might be no action for a single pre-ignition with low damage potential, there might be an action for multiple pre-ignitions of this category if only for acoustic reasons.

Table 2. Comparison of MBKD with reference level method vs. peak pressure estimation

MBKD with reference level method	**MBKD Peak Pressure estimation**
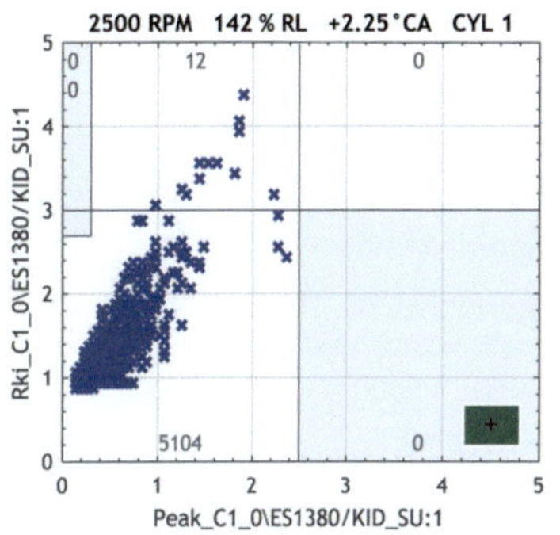	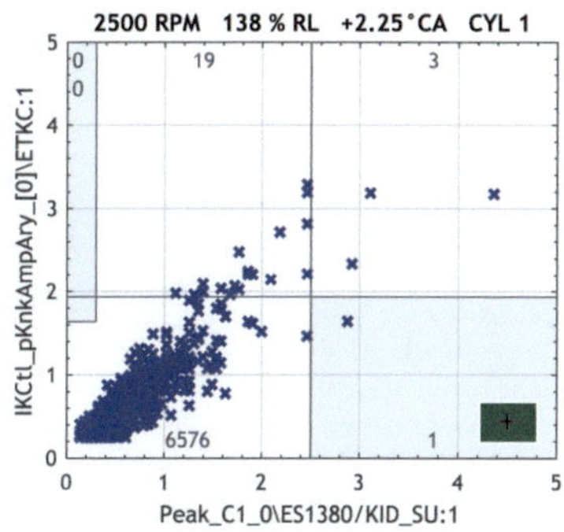
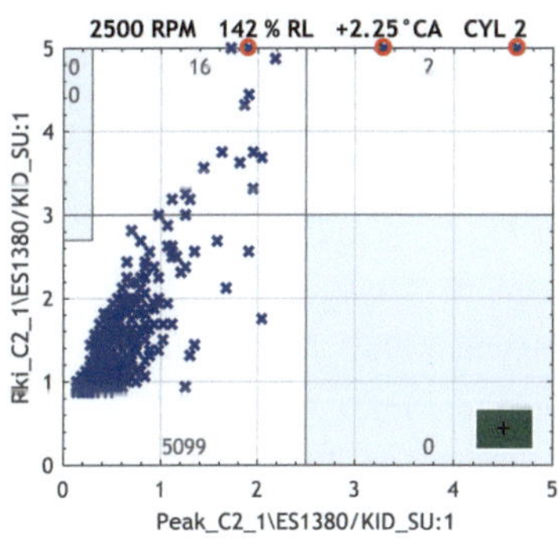	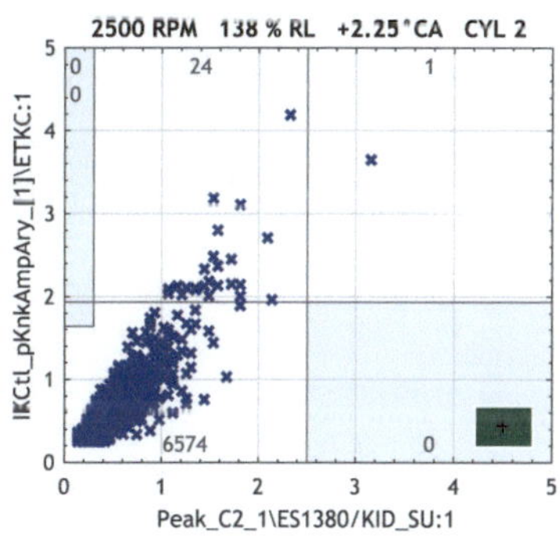
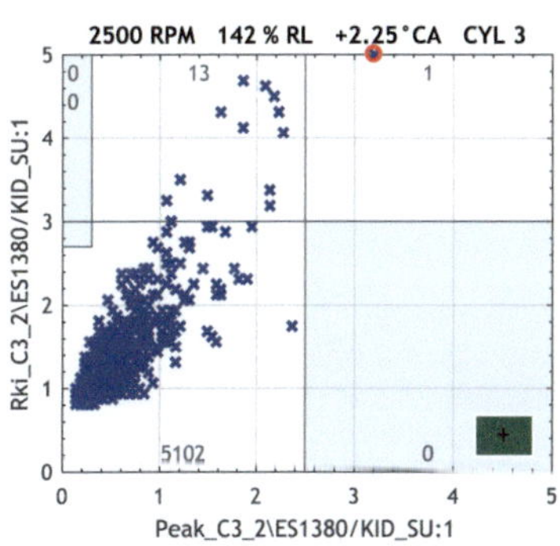	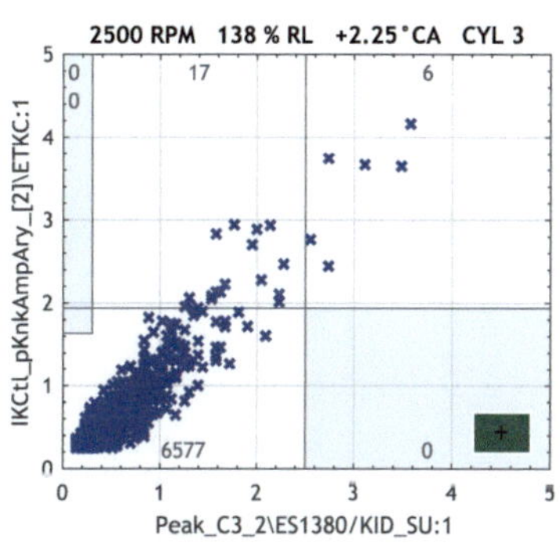

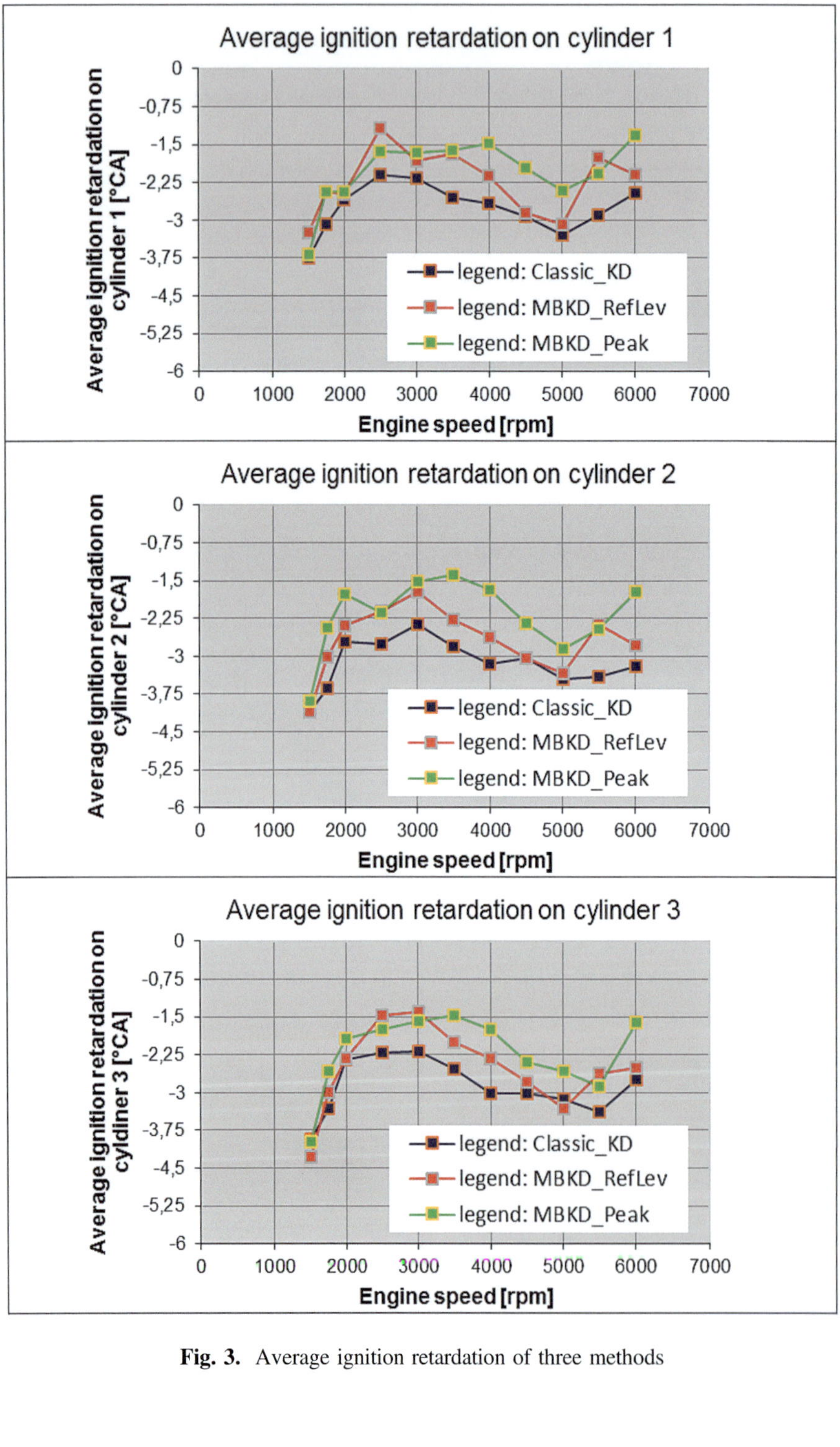

Fig. 3. Average ignition retardation of three methods

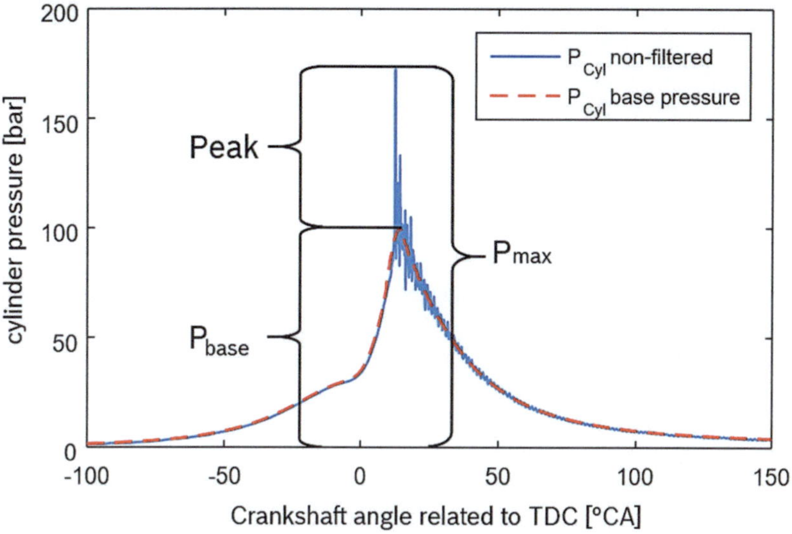

Fig. 4. Pre-ignition pressure estimation

3.4 Results Pressure Estimation Pre-ignition Detection

During simulations, the current algorithm has proved its suitability to get an estimation of P^*_{max} of a pre-ignition event out of the signal of the structure borne noise sensor.

Figure 5 shows multiple abnormal combustions events with partially high maximum pressures in the combustion chamber. Within the simulation, a very good

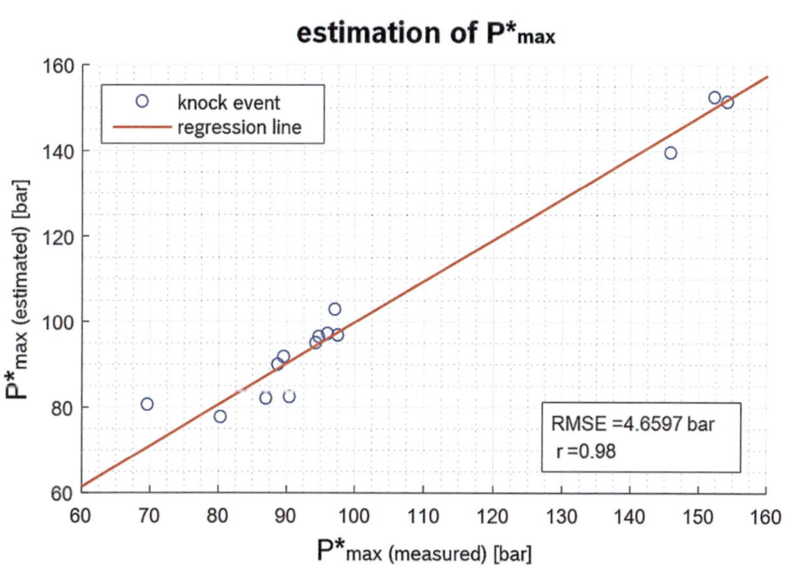

Fig. 5. Results of the pre-ignition pressure estimation

correlation between the measured maximum pressure P_{max} and the estimated maximum pressure P^*_{max} could be achieved. The root mean square error (RMSE) is just below 5 bar pressure. At P_{max} = 150 bar it would be an error of 3.5%. This would be a sufficient accuracy to classify pre-ignitions.

4 Conclusion

The experience from various mass production projects with different ECUs and engine variants (number of cylinders and displacement) shows that the knock detection quality can be clearly improved by using the MBKD method.

With the signal processing via Fast Fourier Transformation (FFT), knock sensor signal filtering in the frequency domain (frequency models) and the knock feature adaption, a reduction of interferences and an optimization of selectivity of knocking and non-knocking combustions can be achieved.

The optimized knock and pre-ignition quality result in an improved robustness against interferences, reduced misdetection tendency and a reduced amount of knock interventions also regarding the depth and duration of the ignition retardations. Additionally to the steady state improvements, the dynamic behavior is also more robust with MBKD. In particular the reduced real drive fuel consumption and the optimized drivability led to a high customer satisfaction whenever the MBKD method was used.

First measurements on the engine with the further development of the MBKD with pressure estimation again show an improvement of the knock detection quality. Prospectively it will be possible to react to knocking combustions and pre-ignitions meeting the demands more precisely regarding to the damage potential of abnormal combustions.

In the end the new method creates a common sense and understanding among engine developers and calibration engineers by talking of knocking and pre-ignitions in physical pressure.

5 Prospect

The development and calibration of the MBKD method pressure estimation will be forced. The new approach has to prove its suitability at different engine projects to secure the methods eligibility under mass production conditions and series spill.

In parallel a tool chain has to be developed to handle the evaluation of the measurements and to secure a uniform calibration strategy.

Trustworthy Estimation and Control of Engine Knocking Level for Transient Operation

Maxime Jean(✉), Thomas Leroy, and Fabien Vidal-Naquet

IFP Energies nouvelles, 1-4 Avenue du Bois Préau, 92852 Rueil-Malmaison, France
maxime.jean@ifpen.fr

Abstract. The contribution of the paper is to propose a knock controller based on the estimation of the distribution quantile of the knock intensity measures. Despite the quantile estimation randomness, the corrections undertaken by the controller are moderated by the level of confidence in the quantile estimation. This strategy offers a fast and stable control ensuring a better compromise between the engine efficiency and the prevention of knock phenomenon. The design of the controller stabilizes the engine torque and is suited for engine transient operations.

Keywords: Knock control · Quantile estimation · Confidence interval · Transient operations

1 Introduction

The ever increasing restrictions on CO_2 emissions added to the upcoming reduction of Diesel vehicle fleet reinforce the interest of improving spark-ignition engine efficiency. To that end, one promising approach is the increase of compression ratios. A drawback of high compression ratio, however, is a higher knock sensitivity. The study of engine knock thus addresses a larger part of the whole engine operating range. In order to avoid engine knocking while maintaining the best efficiency, it is essential to develop fast acting and trustworthy knock estimation and control strategy.

The knock control strategy is based on knock intensity (KI) measurement, usually obtained through signal processing of an accelerometer located on the engine block. Classical control strategies compare cycle-by-cycle knock intensity with a mapped threshold and consequently acts on the spark advance in case the measurement is above the threshold. Instead of acting immediately to a punctual high knock intensity measure, it could be more robust to take into consideration a set of preceding cycles. Furthermore, many recent studies have shown that knock intensities can be considered as stochastic quantities [1]. Accordingly, a knocking engine state (operated at steady-state) would exhibit both knocking and non-knocking cycles. An operating point is deemed knocking based on the fraction of knocking cycles; in general, a fraction of knocking cycles exceeding 5% is considered intolerable [2]. The control objective is then to limit the frequency

M. Günther and M. Sens (eds.), *Knocking in Gasoline Engines*,
https://doi.org/10.1007/978-3-319-69760-4_16

of occurrence of knocking cycles. In other words, it is equivalent to control a quantile (ex. 95%) of the knock intensity distribution.

The complexity relies in the estimation of the quantile in real-time with a convergence time of only a few engine cycles. There are two approaches in the literature. On the one hand, some strategies use no assumption about the distribution of the knock intensities [3–5]. This leads to a convergence time of the quantile estimator longer than the time scale of the engine dynamics, which is not acceptable. On the other hand, some strategies assume that knock intensity are log-normally distributed [6,7]. In that case, the on-line estimation of the quantile can be performed much faster. But if computed over few samples -required to follow the engine dynamics-, its inaccuracy and noisy output leads to irrelevant corrections of the spark ignition timing. The consequences of these instabilities is a sub-optimal spark advance and engine torque fluctuations.

The contribution of the paper is to propose a method that gives a fast and trustworthy estimation of the quantile. This paper discusses the estimation of the expected value of the quantile of the knock intensity distribution, and also calculates the confidence intervals of the quantile estimation. The corrections undertaken by the controller can then be moderated by the level of confidence of the quantile estimation. The developed method gives a sufficient precision for the needs of the control with only ten samples. The rapidity of this estimator suits the engine transient regimes.

The paper is organized as follows. In Sect. 2, the design of the stochastic controller is described. Section 3 describes the experimental set-up and the simulation results.

2 Controller Design

2.1 Controller Input and Output

The potentially damaging shockwaves generated by the knocking phenomena excite the vibrational modes of the cylinders. Different methods exist to obtain a measurement of the knock intensity in the form of a single scalar per cycle. The maximum of the rectified filtered signal is adopted in this work and is used as controller input. It is admitted that it constitutes a sufficient measure of the knock intensity [8]. It provides an intensity quantification of the vibrations caused by at least two combined sources: knock and engine noise. A threshold of KI (to identify knocking cycles) is experimentally calibrated at each operating point of the engine.

In general, the spark advance timing is adjusted based on a comparison of the measured knock intensity to the threshold. The spark advance is decreased to reduce knock intensity; however, there is an associated efficiency loss. In the remaining discussions of this paper, the spark advance timing is the single controller output considered.

2.2 Knock Intensity: A Stochastic Process

Deterministic prediction models of knock occurrences are inconceivable due to the stochastic nature of knocking phenomenon [9]. Moreover, considering the complexity of the underlying physico-chemical processes, it is arguable that knock has an intrinsic probabilistic nature. Only a statistical description has emerged as relevant to describe the experimental observations. This suggests to adopt a probabilistic theoretical framework and to consider the knock intensity measures as realisations of random variables [4,10–12].

The distribution of the knock intensities depends on the engine operating point (speed, load) and state (spark advance, temperature, fatigue and other unmeasured exogenous parameters). The evolution of sets of 300 knock intensity measures and average IMEP (Indicated Mean Effective Pressure) as a function of the spark ignition timing for a given operating point is shown in Fig. 1. The experimental set-up will be described in the last section of this paper. It depicts the randomness of the knock intensities and the compromise between fuel efficiency and knocking occurrences.

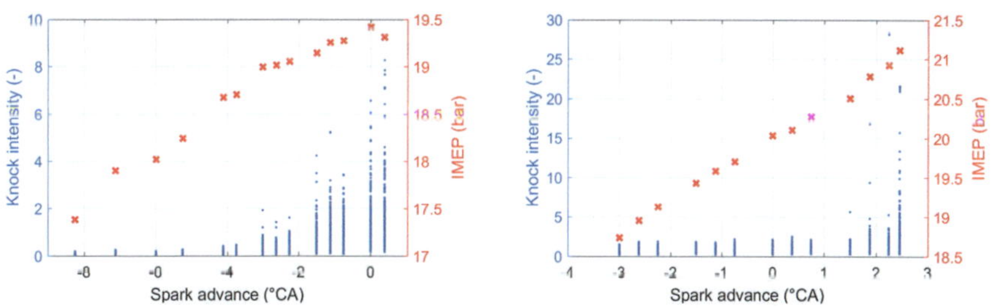

(a) 1750 rpm (b) 5500 rpm

Fig. 1. Knock intensity and IMEP measures for different spark advance timing

When the engine is operating in steady-state over relatively short time scales, the above mentioned parameters stay constant and the knock intensity measures are identically distributed. Futhermore, autocorrelation functions analysed in [1,9] clearly show the cycle to cycle conditional independence of the knock intensities (memoryless Markovian property), given the engine situation. Those two considerations fully support the design of a stochastic controller.

In addition, the knock intensities, given a constant engine state, are assumed to be log-normally distributed. This hypothesis is confirmed in [6] and by the comparison of experimental and fitted log-normal Cumulated Density Functions shown on Fig. 2, computed with the same dataset than previously.

One can assume that there is a known map of the engine speed and torque giving the threshold θ above which the knock intensity measure reveals undergoing knocking phenomena.

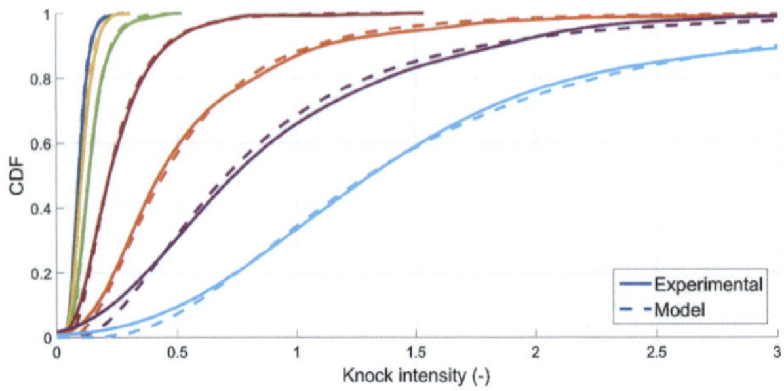

Fig. 2. Knock Intensity Cumulated Density Function

2.3 Acceptable Knocking Margins

In practice, rather than reacting to every single knocking occurrence, a small fraction of knocking cycles is tolerated [7]. The acceptable proportion of knocking cycles such that $KI > \theta$ is considered to be 5%. Note that one could also select another percentage. Rephrasing this condition into a probabilistic perspective is to ensure that KI has a probability of 5% to exceed θ. A sufficient condition to meet this requirement is to maintain the equality between the 95% quantile q of KI and the threshold θ (for a different percentage, the quantile can be adapted).

When $q > \theta$, the engine has an excessive probability of having knocking cycles and SA should be decreased, while the converse assertion $q < \theta$ indicates that the safety limits are respected, possibly with some margin, and SA can be increased.

Only the highest values of the knocking intensities have to be regulated because they represent the knocking cycles. They are fully described by the upper tail of the probability density function. However, our claim is that the possible variations of the distribution for a fixed 95% quantile have a limited impact on upper tail of the distribution, as shown on Fig. 3. Controlling the 95% quantile is then approximately equivalent to completely controlling the upper tail of the distribution.

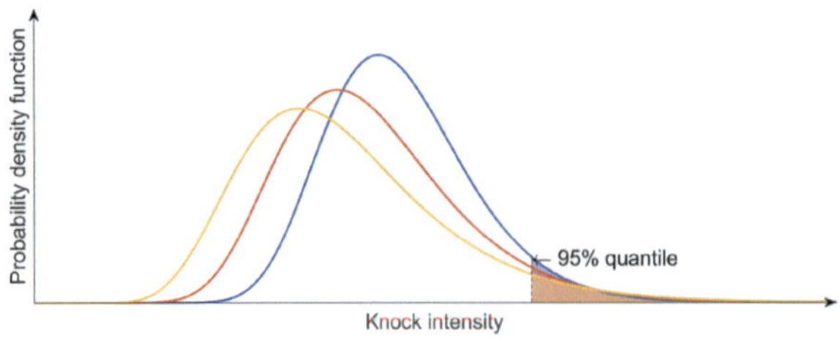

Fig. 3. Log-Normal distributions with the same quantile 95%

A rapidly converging estimator of q can be built, taking into account the log-normal distribution property of KI. An integral controller would be a simple control acting on the spark advance in order to bring back the estimated quantile to the mapped threshold -its setpoint-. However, the intrinsic random nature of KI makes the estimation of q necessarily inaccurate.

2.4 Quantile Confidence Interval Estimator

To overcome the difficulties caused by the randomness of the quantile estimation, the accuracy of this estimation is also evaluated. The corrections undertaken by the controller can then be moderated by the level of confidence one has in their relevance. The proposed algorithm also determines whether the probability that $q > \theta$ or $q < \theta$ exceeds 99%. Note that another high percentage can also be selected. It allows an integral correction to be taken only if one can be sure it is relevant, and to stay unchanged otherwise. The probability to take unnecessary or even counter-productive actions in both directions is 2%.

The algorithm should provide at each cycle m and M defined such that there is a 99% probability that $q < M$ and a 99% probability that $q > m$. In other words, as represented in Fig. 4, $[m, M]$ is the confidence interval of q with a probability of 98%.

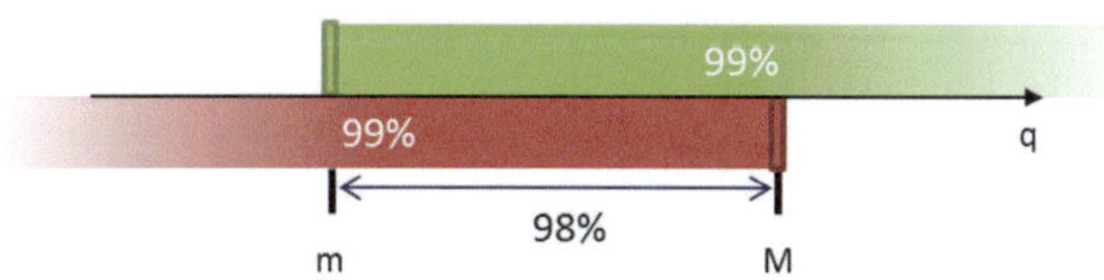

Fig. 4. Confidence interval $[m, M]$

The confidence interval is estimated at each time step, along with the knock intensity expected value $\overline{q}$. This estimation is performed according to the n last measures of knock intensities, which are assumed to be independent, identically and log-normally distributed. During transient regime, the identically distributed hypothesis is approximately respected only during short time scales. The estimation has to be done over a very limited number of measures. Typically, n takes values around 10.

Figure 5 shows three different sets of realisations drawn from the same distribution defined by its unique probability density function. The true quantile of this distribution is shown by the vertical line. The inferred confidence interval of the 95% quantile is represented by the black line segment. The probability to draw a set of realisations that corresponds to the situation represented in the center plot is 98% while this probability is 1% for the situations of the right and the left plots.

The estimator relies on Bayesian inference and other mathematical treatments making it easily implementable thanks to off-line computations summarized in two one-dimensional maps. It requires limited resources:

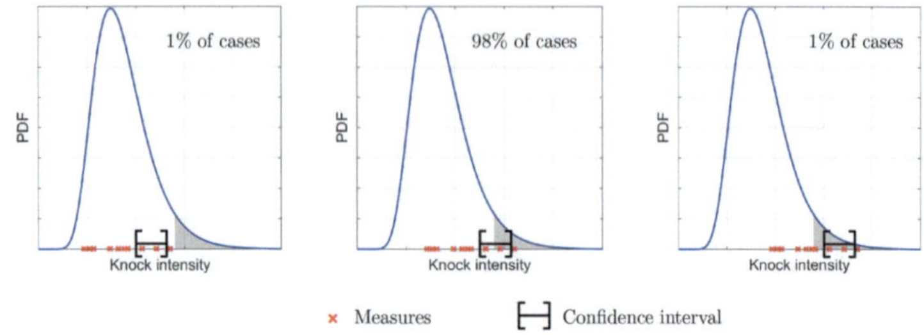

Fig. 5. Quantile confidence interval inferred from realisations of the same distribution

- two one-dimensional maps and a buffer of the n last knock intensity measures: limited memory required;
- basic computations: light implementation in terms of computational power.

2.5 Structure of the Controller

At each cycle, a correction ΔSA proportional to the error between the expected value of the quantile and its target is added to the spark ignition timing of the preceding cycle. However, the correction is inactive if the configuration is such that, according to the confidence interval, one cannot be sure at 99% that a correction should be applied to the spark ignition timing.

$$\Delta SA = \begin{cases} 0 & \text{if } \theta \in [m, M] \\ K(\overline{q} - \theta) & \text{otherwise} \end{cases} \tag{1}$$

where K is a gain to be experimentally adjusted. Indeed, if $\theta < m$, one can be sure at 99% that the spark ignition occurs too soon. Then its setpoint needs a correction to delay the spark ignition timing. The other way around, if $\theta > M$, one can be sure at 99% that the spark ignition occurs too late. Then it needs a correction to advance the spark ignition timing. The structure of the controller is summarized in the block diagram in Fig. 6.

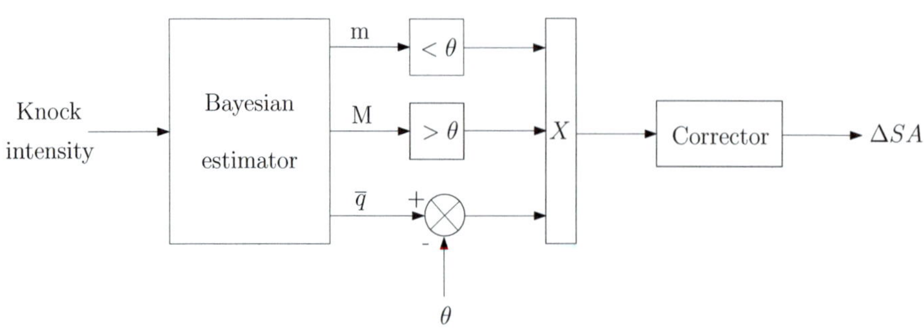

Fig. 6. Controller block diagram

3 Results

3.1 Set-Up

Methodology. The engine under consideration is a 1.2 L three-cylinder turbocharged spark-ignition engine using direct injection technology. For several engine speeds and loads, a spark advance sweep was conducted. On each operating point, several hundred knock intensity values were acquired. The extent of the experimental database is shown in Fig. 7.

As stated before, knock phenomena is stochastic, and still not well understood. Consequently, even under identical conditions, each test is unique. To allow for a more extensive validation and a better performance assessment of the new stochastic controller, based on reproducible results, the behavior of the engine regarding knock is simulated.

In order to obtain knock intensities reflecting real engine operation, the experimental distributions of knock intensities are computed for each operating point, using all cycles values. Knock intensities are then computed with a random number generator whose probability density function varies accordingly, depending on engine speed, torque and spark advance timing. Log-normal hypothesis is then not considered here, since full knock intensities distribution is matched. The methodology is illustrated in Fig. 7, where each point number represents a cycle.

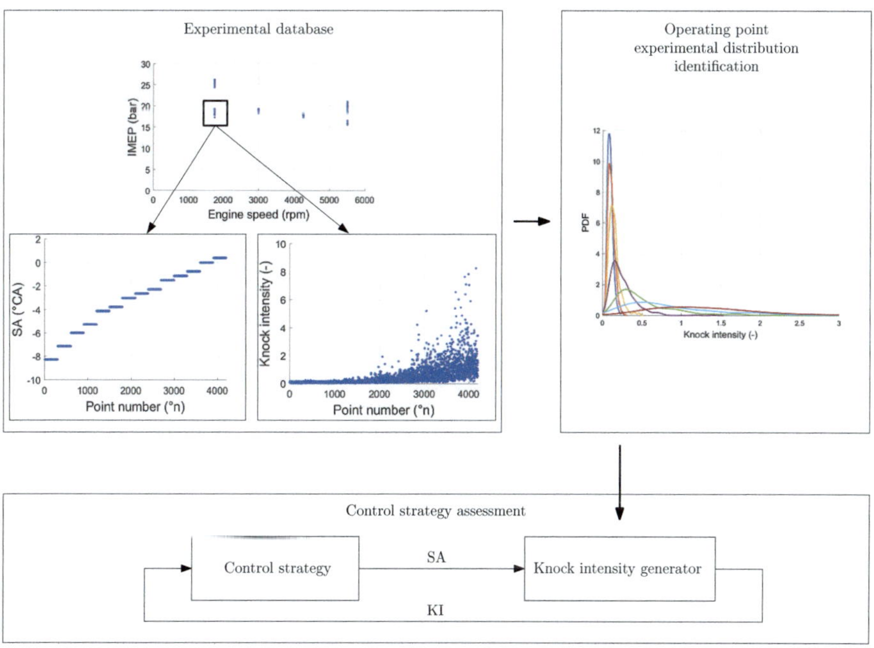

Fig. 7. Methodology

Control Algorithm. For better spark advance control during engine speed and load changes, feedforward design is always adopted. The reference spark advance should be mapped beforehand as a function of these two parameters and then corrected on line by the knock feedback algorithm presented in this paper. However, in order to assess its adaptation abilities, no maps were used in the simulations.

The following parameters were experimentally calibrated: $n = 10$, $K = 0.1$. The map giving the knocking threshold was also built from experimental measurements.

Reference Strategy. This new stochastic knock controller is compared with a reference strategy commonly used. It is a sawtooth shaped spark advance timing control strategy. The spark advance is increased at each time step by a small δSA and suddenly decreased by a large $\varDelta SA$ if the knock measure exceeds a given threshold θ_{st}. Typical values are $\delta SA = 0.01°$ and $\varDelta SA = 3°$. This strategy ensures the shortest reaction time to any occurring knocking cycle, while keeping an average 99.7% of the cycles with knock intensities below the threshold.

3.2 Simulation Results

Steady-State Operation. The proper functioning of both algorithms was first confirmed on stationary operating points. Tests consist in simulating the engine knock in conjunction with a spark advance controller over an extensive period of time. One hundred thousand consecutive cycles are simulated in order to obtain reliable results, which corresponds to more than an hour at 1000 rpm. The evolution of the confidence interval $[m, M]$, i.e. the 98% sure value of q, is shown in Fig. 8 in grey color. If the quantile target θ steps out of this interval, since one is 98% sure that a change needs to be applied on the spark advance, corrective action is taken (ex. at t=876 s and t=880 s). The behaviour of the sawtooth shaped knock controller is shown in Fig. 9. The spark advance timing is continuously increased until a knock intensity measure exceeds the threshold θ_{st}, in which case it is decreased. One can see the high stability of the stochastic controller in comparison with the sawtooth shaped controller.

Efficiency/Knock Compromise. Besides comparing steady-state stability, the level of knock induced by both controllers should be assessed. This can only be done properly by also considering engine efficiency. Indeed, one can only compare knock levels at a given engine IMEP. A comparison between the sawtooth and the stochastic controller is given in Fig. 10. The chosen knock indicator is the proportion of knocking cycles, i.e. with knock intensities reaching the knock area represented in Fig. 11.

Results presented in Fig. 10 are obtained by sweeping the threshold values θ and θ_{st} for stochastic and sawtooth controllers respectively. For each threshold value, a simulation of one hundred thousand consecutive cycles is performed,

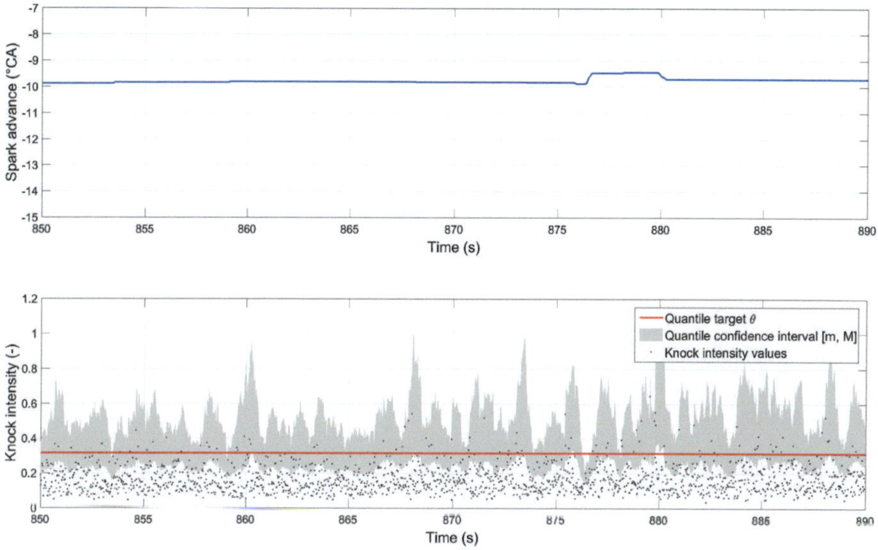

Fig. 8. Stochastic controller

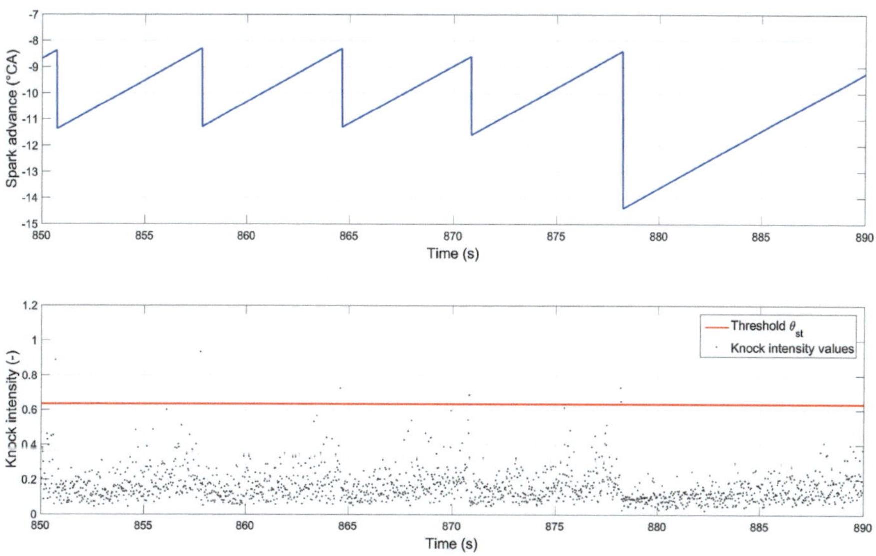

Fig. 9. Sawtooth controller

corresponding to one marker in the figure. As thresholds values increase, IMEP increases at the cost of knock level rises.

To achieve a given IMEP, the occurrence of knocking cycles is inevitable. A spark advance sweep is conducted to determine the smallest percentage of knocking cycles at each IMEP. The set of circle markers in Fig. 10 illustrates these percentages, which consequently represent the best efficiency/knock compromise.

One can then note that in both cases represented in Fig. 11(a) and (b), the stochastic controller virtually reaches the best possible performance regarding the efficiency-knock compromise in steady-state. For a given knock level, the gain of IMEP of the stochastic controller as compared to the sawtooth controller is around 1%.

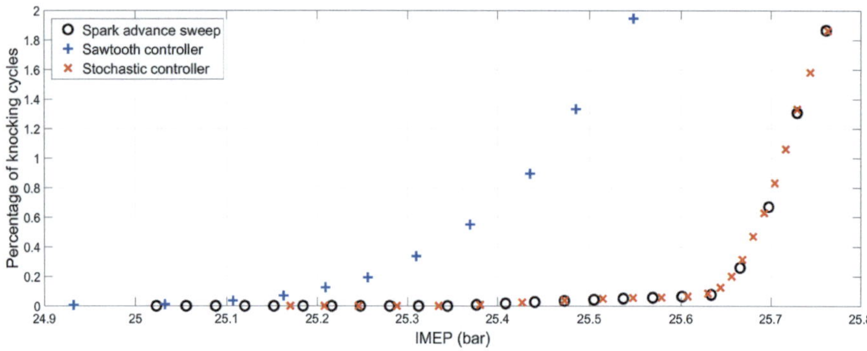

(a) 1750 rpm

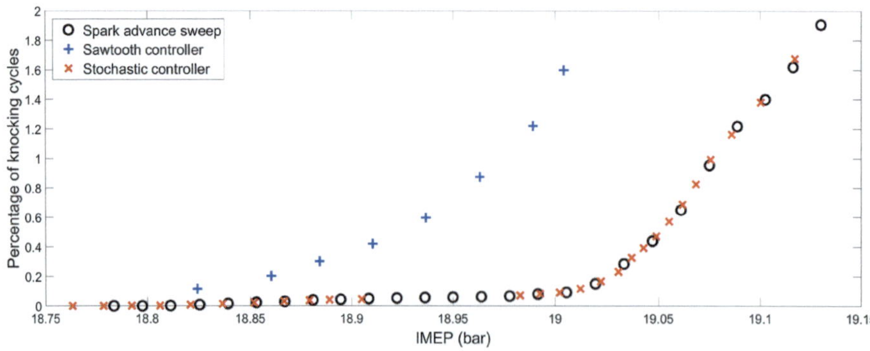

(b) 3000 rpm

Fig. 10. IMEP vs KI

Transient Operation. The behaviour of the controller was simulated on a transient state: a step of charge from 19 bar to 25 bar of IMEP at the constant speed of 1750 rpm, as illustrated in Fig. 12. No spark advance prepositioning is

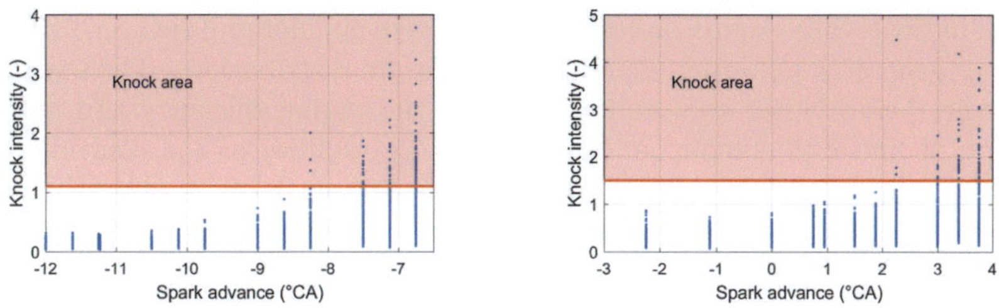

(a) 1750rpm (b) 3000rpm

Fig. 11. Knock area

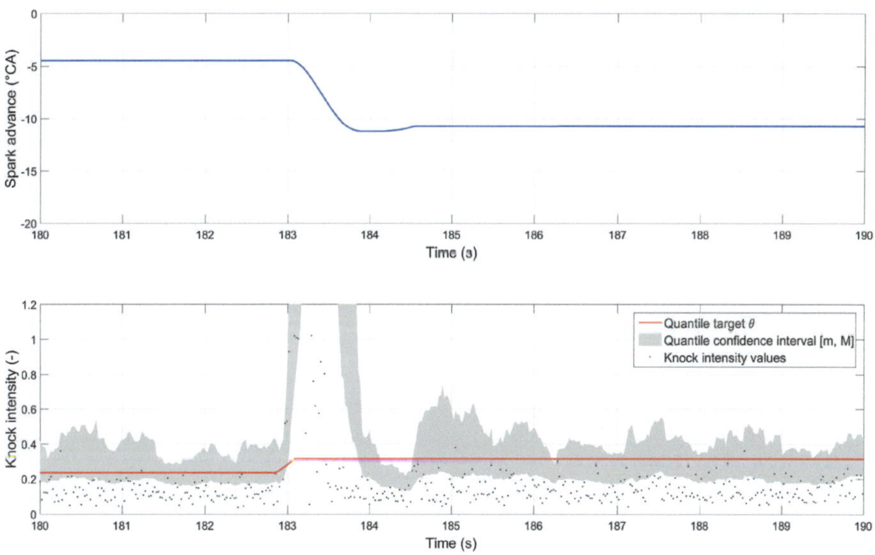

Fig. 12. Transient operation

used, artificially generating a 6° error in spark advance during the transient. The controller immediately starts to decrease the spark advance and reaches its final value within 0.4 s to this excitation. Response time is consistent with the number of cycles used for the knock estimation, i.e. $n = 10$. Note that the controller is satisfyingly reactive in the protection direction.

4 Conclusions

The stochastic controller proposed in this paper estimates the confidence interval of the 95% quantile of the knock intensity distribution. Based on the relative position of the mapped 95% quantile target and the quantile confidence interval, an adapted spark advance correction is undertaken by the controller. This controller was compared with a classic sawtooth shaped spark advance controller in simulations. The engine knock intensities were simulated based on a random number generator following experimental measure distributions. The stochastic controller has a better compromise between the engine efficiency and the prevention of knock phenomenon. In addition, the rapidity of the controller and the requirement of only a few parameter calibration makes it perfectly suited for engine transient operations. The controller is also more stable, which reduces the engine torque variations.

References

1. Spelina, J.M., Jones, J.C.P., Frey, J.: Stochastic simulation and analysis of a classical knock controller. Int. J. Engine Res. **16**(3), 461–473 (2015)
2. Stahr, A., Langfritz, P., Guenther, M., Kratzsch, M.: The delta knocking control - the necessary paradigm shift for engines with high power density. In: 4th International Conference on Knocking in Gasoline Engines (2013)
3. Shi, H., Yang, J., Shen, T., Jones, J.C.P.: A statistical likelihood based knock control scheme. In: Control Conference (CCC), 2013 32nd Chinese, pp. 7768–7773. IEEE (2013)
4. Jones, J.C.P., Frey, J., Muske, K.R.: A statistical likelihood based knock controller. IFAC Proc. Volumes **43**(7), 809–814 (2010)
5. Thomasson, A., Eriksson, L., Lindell, T., Jones, J.C.P., Spelina, J., Frey, J.: Tuning and experimental evaluation of a likelihood-based engine knock controller. In: 2013 IEEE 52nd Annual Conference on Decision and Control (CDC), pp. 6849–6854. IEEE (2013)
6. Spelina, J.M., Jones, J.C.P., Frey, J.: Characterization of knock intensity distributions: Part 2: parametric models. Proc. Inst. Mech. Eng. Part D J. Automob. Eng. **227**(12), 1650–1660 (2013)
7. Lonari, Y., Polonowski, C., Naber, J., Chen, B.: Stochastic knock detection model for spark ignited engines. In: SAE Technical Paper, number 2011-01-1421. SAE International (2011)
8. Naber, J., Blough, J.R., Frankowski, D., Goble, M., Szpytman, J.E.: Analysis of combustion knock metrics in spark-ignition engines. Technical Report 2006–01-0400, SAE Technical Paper (2006)
9. Spelina, J.M., Jones, J.C.P., Frey, J.: Characterization of knock intensity distributions: Part 1: statistical independence and scalar measures. Proc. Inst. Mech. Eng. Part D J. Automob. Eng. **228**(2), 117–128 (2014)
10. Jones, J.C.P., Frey, J., Muske, K.R., Scholl, D.J.: A cumulative-summation-based stochastic knock controller. Proc. Inst. Mech. Eng. Part D J. Automob. Eng. **224**(7), 969–983 (2010)
11. Jones, J.C.P., Spelina, J.M., Frey, J.: Likelihood-based control of engine knock. IEEE Trans. Control Syst. Technol. **21**(6), 2169–2180 (2013)
12. Spelina, J.M., Jones, J.C.P., Frey, J.: Recent advances in knock analysis, simulation, and control. SAE Int. J. Engines **7**(2014–01–1349), 947–955 (2014)

Potential of Series-Compatible In-Cylinder Pressure Sensors for Gasoline Engines Using the Example of Ignition Angle Control

Benedikt van Booven[1(✉)], Harry Schüle[1], Thorben Walder[2], and Hermann Rottengruber[3]

[1] Continental, Siemensstr. 12, 93055 Regensburg, Germany
benedikt.van.booven@continental-corporation.com
[2] Dr. Ing. h.c. F. Porsche AG, Porschestr. 911, 71287 Weissach, Germany
[3] Institut für Mobile Systeme, Otto-von-Guericke-Universität Magdeburg, Universitätsplatz 2, 39106 Magdeburg, Germany

Abstract. To meet increasingly stringent emission legislations, improving the combustion efficiency becomes essential in the development of modern gasoline engines. This paper presents a comprehensive overview of available methods to improve the spark timing for maximum break torque under use of the in-cylinder pressure signal. The suitability of a series-compatible pressure sensor for the use of ignition angle control is summarized and reviewed. Finally, the results of spark timing control in a test vehicle are discussed on base of chassis dyno measurements.

1 Introduction

In addition to the electrification of the powertrain, the development of modern internal combustion engines continues to focus on improving the efficiency - in compliance with the emission limits and without compromising drivability - to meet the stringent legislative requirements. While the use of the in-cylinder pressure signal for the optimization at the gasoline and diesel combustion processes has established itself as a standard during the development process, the use of cylinder pressure sensors in the series environment has become possible only in recent years due to the advances in sensor development and the performance enhancement of modern engine control units.

The use of combustion chamber pressure sensors in the series environment limits itself at the moment exclusively on applications for diesel engines [1,2], where the design of the in-cylinder pressure sensor integrated into the glow plug predominates. Now the development of a series-compatible stand-alone cylinder pressure sensor [4] also makes it possible for the first time to assess the potential of a series pressure sensor for gasoline engines.

The ignition timing particularly affects nearly all the outputs from the combustion process. The precise control of the ignition timing is therefore absolutely necessary for current as well as for future combustion engines. In the automotive industry, open loop controlled ignition systems are currently the standard.

M. Günther and M. Sens (eds.), *Knocking in Gasoline Engines*,
https://doi.org/10.1007/978-3-319-69760-4_17

In this paper, the focus is on closed loop controlled ignition systems. For this purpose, an additional sensor, here a cylinder pressure sensor, is required, which supplies a feedback signal from the combustion process. Such closed loop controlled systems have the potential to minimize the need for extensive calibration, to improve the system robustness with regard to the engine to engine deviations, and to increase the efficiency by means of targeted control interventions in the combustion process.

This paper presents in Sect. 2 a comprehensive overview of methods to improve the spark timing. A number of criteria obtained from the cylinder pressure curve are examined with regard to their suitability for the purpose of ignition angle control, in Sect. 3. Next, the test platform and used measurement equipment is presented in Sect. 4. Then the series-compatible in-cylinder pressure sensor is reviewed and the the pressure signal processing chain for the individual characteristics are examined in Sect. 5. Subsequently, the ignition angle control concept based on a pressure signal is presented in Sect. 6. Finally in Sect. 7, the results and differences between the ignition angle control based on laboratory sensors and those of a series-compatible solution are compared and discussed based on vehicle measurements.

2 Optimal Ignition Angle

In a modern spark-ignition (SI) engine the air is inducted through the intake valves into the cylinder. Inside the cylinder the mixing with residual gas and the injected fuel takes place. After closing the intake valves, the air-fuel mixture is compressed by the upward movement of the piston. At the end of the compression stroke the combustion process is initiated by an electrical discharge at the spark plug. This time is considered as ignition angle. The determination of the engine position where the spark plug shall ignite the air-fuel mixture is done by the spark-advanced control. Since both engine efficiency and emissions are directly affected by the ignition angle, the ignition timing is one of the core control variables for a modern SI engine [7].

In Fig. 1a the cylinder pressure curves in relation to the crank angle position for three different ignition angles are depicted. Under normal operation conditions - no restrictions by drivability, engine knock or catalyst heating strategy - the ignition is initiated $15 - 45°$ crank angle (CA) before the ignition top dead center (iTDC). When the ignition angle is slightly to early, see IGA_1, the combustion and thereby the pressure rise start to early in the compression cycle and counteracts the piston movement. In addition the losses due to heat transfer to the cylinder walls are increased, since the maximum in-cylinder temperature is increased. Therefore more energy will be dissipated during the work cycle of the combustion process.

Further, a late ignition, as depicted by IGA_3, shows a later pressure rise. Although a higher pressure 50 °CA after iTDC can be observed, the overall produced work during the expansion phase is lower compared to the optimal ignition angle IGA_2. The losses in work during the early expansion phase due

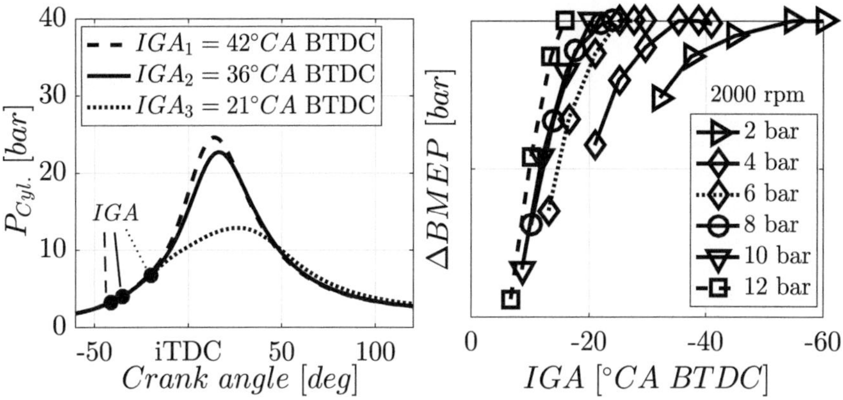

(a) (b)

Fig. 1. Impact of spark timing on cylinder pressure trace and brake mean effective pressure (BMEP). (a) Impact of spark timing on cylinder pressure traces $N = 2000$ rpm and $BMEP = 4$ bar. IGA_1 slightly to early ignition; IGA_2 optimal ignition; IGA_3 to late ignition. (b) Influence of ignition angle variation for several operation points at $N = 2000$ rpm. Δ BMEP refers to maximum BMEP for the given operation point. All other parameters were held constant.

to the reduced work transfer from the cylinder gases to the piston can not be fully compensated [5,7].

The effect of ignition angle on the brake mean effective pressure (BMEP) for the constant engine speed $N = 2000$ rpm and several operation points is shown in Fig. 1b. The optimal ignition angle is defined as the ignition angle that provides the maximum torque, called maximum brake torque (MBT), when all other parameters are held constant. It can be observed that the MBT ignition angle strongly depends on operating point and further engine conditions [7].

2.1 Open Loop Control

Nowadays, open loop controlled ignition systems are the standard in the automotive industry for the non knock limited engine operation. These systems measure or model a number of parameters known to affect the optimal ignition angle and compensate for their effects.

During the development process of a new engine this leads to an extensive testing and calibration effort to ensure an optimal ignition angle under all operation conditions. However, since these systems do not have a direct feedback from the combustion process, they are rather conservatively designed to guarantee a good performance over the entire engine operation range and to be robust for the entire vehicle lifetime and engine series part to part variation [5].

2.2 Closed Loop Control

In contrast to open loop systems, which build up on known or modeled parameters, closed loop systems can in addition directly measure the effect of the applied ignition angle on the combustion process. Thereby, a precise control of the optimal ignition angle is achieved even when non-measured or modeled influence parameters are exposed to large variation. Because of that, a higher system robustness is obtained and the calibration effort can be reduced. Further, as consequence of the increased robustness to external influences and system dispersion the calibration for closed loop systems can be more aggressive compared to open loop systems.

One of the first attempts for a closed loop combustion control was made by Draper and Li [10]. They optimized the ignition angle by using a dithering technique to excite the system. An extention of this work for a four cylinder engine was made by Kohlhase and Isermann [8]. A cylinder individual dithering signal which constantly changed the ignition angle is used first to excite the combustion process. Then in a second step a self-tuning extremum control is used to determine the optimal ignition angle cylinder individually. In [9] Thiel et al. a model free adaptive control approach is used to control the center of combustion.

In addition to that, there exists a group of control strategies which all use a direct measure from the combustion process as a feedback signal. From this signal a criteria is formulated and related to the maximum work. Erikson et al. uses in [11] the feedback from an ion-current measurement. Other related parameters deduced from the in-cylinder pressure are presented in Sect. 3.

3 Empirical Criteria for Optimal Spark Timing

The parameters described in this section are all derived from in-cylinder pressure measurement. Because of their empirical correlation with the MBT all of the three parameters can be used as a feedback signal for the optimal ignition angle control. The three combustion descriptors 50% mass fraction burned (MFB_{50}), pressure ratio (PR) and peak pressure position (PPP) are described below.

3.1 50% Mass Fraction Burned

One of the most common criteria, used to determine the optimal spark advanced during the development and calibration of an engine, is the center of combustion, also called 50% mass fraction burned position (MFB_{50}). According to Heywood [7], an optimal spark timing is achieved when 50% of the air-fuel mixture, trapped inside the cylinder, is burned at about $8 - 10\,^{\circ}$CA after iTDC, regardless of the operating conditions and test engine. This is further supported by the investigations of Bargende in [3].

The MFB_{50} position can be obtained directly by the normalized heat release Q_{norm} from Eq. 1.

$$Q_{norm}(\varphi) = \frac{\int_{\varphi_1}^{\varphi_2} \frac{dQ(\varphi)}{d\varphi} d\varphi}{max(\int_{\varphi_1}^{\varphi_2} \frac{dQ(\varphi)}{d\varphi} d\varphi)} \tag{1}$$

The values of MFB_{50} position included in this paper have been derived by the fast heat release analysis, developed by Hohenberg [12]. According to Hohenberg, the heat release rate can be derived from Eq. 2,

$$\frac{dQ(\varphi)}{d\varphi} = \frac{1}{\gamma - 1} \cdot \left[\gamma \cdot p(\varphi) \cdot \frac{dV(\varphi)}{d\varphi} + V(\varphi) \cdot \frac{dp(\varphi)}{d\varphi}\right] \tag{2}$$

where φ represents the crank angle, p is the combustion chamber pressure, V is the cylinder volume and γ is the polytropic coefficient. The normalized heat release and the according MFB_{50} for three different ignition angles is illustrated in Fig. 2. This method by Hohenberg to determine the heat release based on the in-cylinder pressure measurement has been widely used and the accuracy due to the simplification is well known.

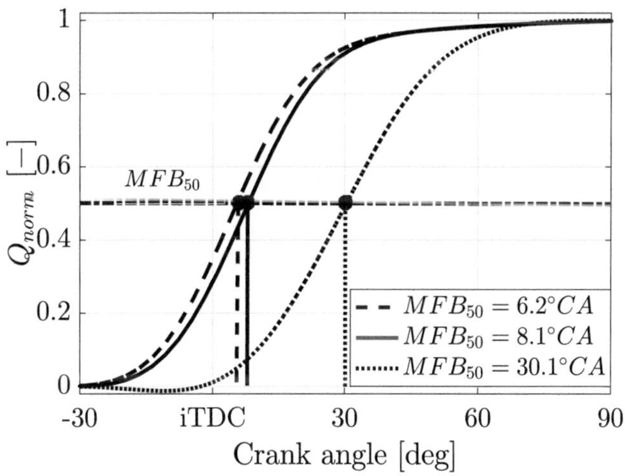

Fig. 2. Normalized heat release rate Q_{norm} with 50% mass fraction burned position for three different spark timings.

3.2 Pressure Ratio

In [13], Matekunas proposed the pressure ratio (PR) as an other descriptor to determine the optimal ignition angle. The PR is defined by the ratio between the pressure of a fired combustion cycle $p_f(\varphi)$ and the pressure from a motored cycle $p_m(\varphi)$ (see Eq. 3). Therefore, the pressure of the motored cycle is estimated by an adiabatic compression and expansion for the entire combustion cycle that follows the ignition angle.

$$PR(\varphi) = \frac{p_f(\varphi)}{p_m(\varphi)} - 1 \tag{3}$$

When the PR is normalized by its maximum, it results in a trace that is very similar to the normalized heat release. Similar to the MFB_{50}, the $PR_{norm} = 0.5$ position can be used as a feedback signal for the ignition angle control.

$$PR_{norm}(\varphi) = \frac{PR(\varphi)}{max(PR(\varphi))} \tag{4}$$

3.3 Peak Pressure Position

The peak pressure position, as a third criteria, is defined as the position, in degree crank angle, where the in-cylinder pressure maximum value is located. The location is determined in relation to the iTDC. In Fig. 3 three pressure curves and their corresponding PPP are depicted.

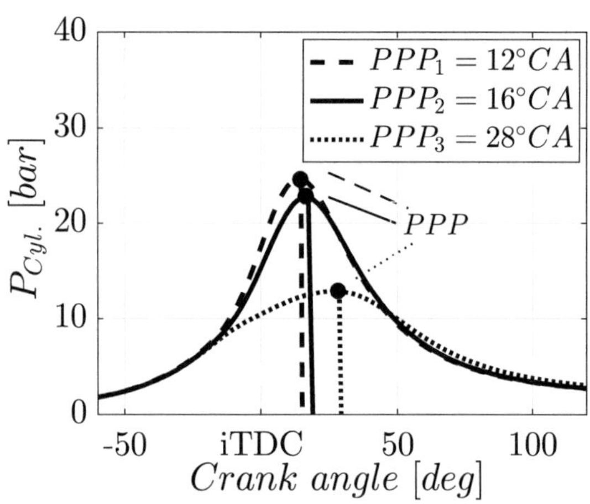

Fig. 3. Peak pressure position for pressure traces with three different spark timings in relation to the crank angle position.

Several applications have shown that the PPP is suitable for the ignition angle control. It was observed that the PPP for a MBT spark timing on a gasoline engine is almost constant at around 16 °CA after iTDC, independend of the operating conditions [5,14,15].

4 Experimental Platform

For the investigations described in this paper, a Sports Utility Vehicle (SUV) with a direct-injected turbocharged 3.6-liter V6 spark ignited engine is used. As a representation of modern gasoline engines, a maximum dethrottling is achieved by the realization of high internal residual gas fractions without compromising the running smoothness of the engine. Therefore, the engine is equipped with continuous variable hydraulic cam phasing actuators on the intake and exhaust side. In addition, an asymmetric lift profile in combination with a two stage valve lift system is installed on one of the valves on the intake side. This improves the combustion stability by an intense charge motion [6, 16].

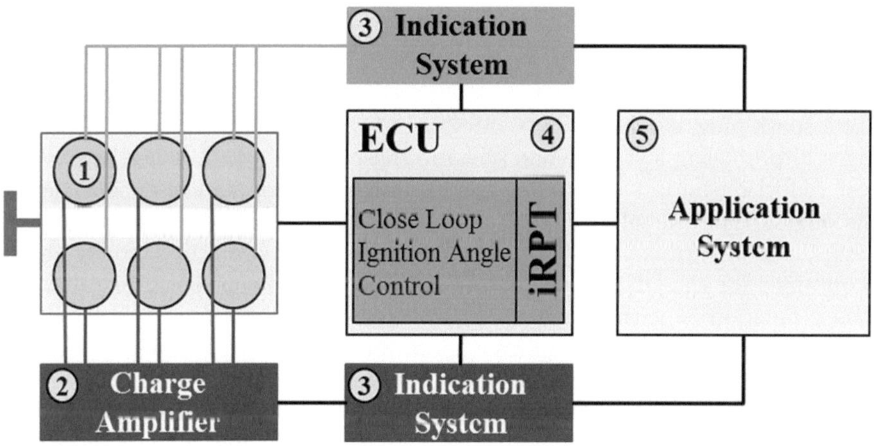

Fig. 4. Measurement setup for closed loop ignition angle control investigation on Porsche Macan test platform.

The measurement equipment required for the validation of the control concept was integrated into a commercial available vehicle. In Fig. 4 a schematic of the used measurement is depicted. In addition to the Continental in-cylinder pressure sensor (see Sect. 5.1), the engine was equipped with piezoelectric pressure transducers integrated into the spark plugs (type 6118BFD35, Kistler) on all six cylinders (1). The spark plug pressure sensors, also called laboratory sensors (Lab.), required unlike the series-compatible sensors a separate charge amplifier (2). Both pressure signals are then separately digitalized and synchronized by the prototyping indication system FI^{2RE} (3). After the combustion descriptors, like MFB_{50} or PPP, have been determined on the indication system, they are transmitted to the ECU (4) by a CAN bus interface. The CAN message received by the ECU, is decoded and subsequently forwarded to the closed loop ignition angle control function. Thereby the control itself, was directly implemented via internal rapid prototyping (iRPT) into the existing series open loop ignition angle control. The ability to monitor the control function and the direct access to all relevant engine control functions was permanently provided by the application system (5).

5 Pressure Sensing and Signal Processing

Cylinder pressure sensors are well known in the field of engine development. In the following the sensing concept of the series-compatible stand-alone pressure sensor (SAPS) of Continental will be presented and the signal process will be analyzed and reviewed with the focus on a series application for a closed loop ignition angle control.

5.1 Stand-Alone Pressure Sensor

Continental ICPS, depicted in Fig. 5a, is based on a stand-alone pressure sensor concept. Although the integration of a pressure sensor into other components, like a spark plug, can be easier to integrate in the engine, there are more advantages in proposing a stand-alone pressure sensor. First of all, it has an obvious cost reduction impact due to the product simplification and in-dependency of the engine type (diesel and gasoline). On the other hand this allows a higher design and integration flexibility. Furthermore, the freedom in sourcing choice is obtained for the OEM's. An additional advantage of the concept is, that the sensor function is neither interrupted by cross dependencies in performance or failures by the combined component in an integrated actuator/sensor concept, such as glow plug, spark plug or injector integration. Moreover, the ICPS has an important cost benefit in after sales, because the sensor does not have to be replaced if the combined component has a defect. All these reasons contribute to the choice to focus the development of a combustion pressure sensor on bases of a stand alone design over the last years [2].

The sensing unit, the core component of the in-cylinder pressure sensor, is illustrated in Fig. 5b The sensor shell and a metallic diaphragm (1) separates the combustion chamber from the piezo-electric ceramic (2) inside of the sensor, which processes the pressure into electrical charges. When a pressure applies force to the diaphragm (4) a part of this force is brought to the piezo ceramic through a "soft transmitter" (3) and compresses the piezo ceramic (5) [2].

A large resistance to high pressure and temperature levels, reached in the combustion chamber, is ensured by the sensing concept through a relatively stiff diaphragm. Further, this design allows an easy adjustment of the mechanical sensitivity of the sensing unit since the ratio of force going to the piezo ceramic is directly linked to the stiffness ratio between the diaphragm and the soft transmitter. In addition to that, the sensing unit is preloaded during the assembly, to ensure constant contact quality between all the mechanical parts for the entire component lifetime. Thereby the sensing concept provides an advantage regarding the ability to sustain fatigue, stress and durability (eg. vibration, mechanical and thermal expansion) [2].

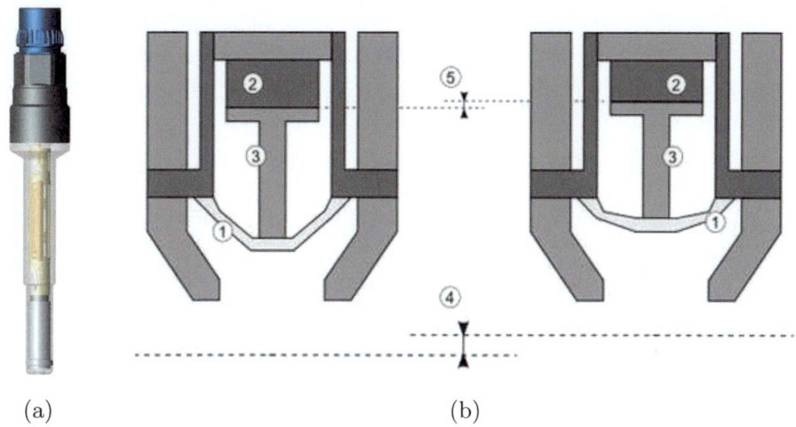

Fig. 5. Series compatible in-cylinder pressure sensor (ICPS) of Continental [2]. (a) Sensor design. (b) Schematic of ICPS sensing unit operation principle.

5.2 Signal Processing

As most of the commercial available combustion pressure sensors, used in the engine development, the Continental ICPS is a piezoelectric based sensor. Whereas piezoresistive sensors are able to do absolute measurement, the ICPS performs a relative pressure sensing. The high sensor performance is ensured by an electronic module directly integrated into the sensor. Thereby, one of the key features of the Continental ICPS, a minimal time delay, is achieved through the internal direct analog amplification and signal processing [2].

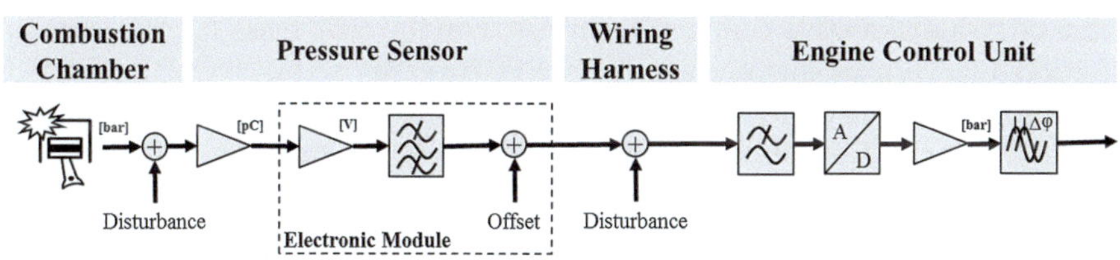

Fig. 6. In-cylinder pressure signal processing from the physical pressure inside the combustion chamber to the digitized pressure trace inside the engine control unit.

In Fig. 6, the signal processing path from the physical pressure inside the combustion chamber, via pressure sensor, wiring harness to the crank angle synchronized and digitized pressure trace at the application software level inside the engine control unit (ECU) is illustrated. In the past decades several publication, like [17–20], have focused on the evaluation of combustion chamber pressure sensors data quality. Although, central topics such as mounting position or sensor design parameters have been covered by these papers, it is crucial for a series application to pay attention to the entire measurement system. The results of sensitivity analyis, derived from the entire signal processing path illustrated in Fig. 6, is presented in [2].

5.3 Comparison of Combustion Descriptors Between Sensors

Below, a Continental in-cylinder pressure sensor is compared with a laboratory sensor (type 6118BFD35, Kistler) with regard to its suitability for ignition angle control. Figure 7 shows the corresponding measurement data for both sensors. The two left diagrams 7(a) and (c) refer to the combustion descriptor 50% mass fraction burned position. The two right Fig. 7(b) and (d) show the comparison of both sensors for the peak pressure position. For all diagrams, the measured value from the first cylinder of the series-compatible sensor from Continental is plotted on the y-axis and the corresponding measured value of the laboratory sensor is plotted on the x-axis.

Figures 7(a) and (b) show the comparison of both criteria at the stationary operating point N=2500 rpm and BMEP=10 bar for 200 consecutive combustion cycles. The measured values of both sensors show a strong correlation for the individual combustion cycles. The mean value, represented by the solid horizontal and vertical line, is slightly lower for the center of combustion, Fig. 7(a). For the peak pressure position Fig. 7(b), the mean value is slightly increased compared to the laboratory sensor. For both features the standard deviation between the sensors are almost the same. Nevertheless, a higher spread for the peak pressure position can be observed in the data.

Figures 7(c) and (d) show load sweeps of stationary operating points between idle and full load for several engine speeds in the range from 1100 rpm to 6000 rpm. Thus, the data included, represent almost the entire engine operating range. The data depicted in the diagram, have been derived by the mean value of 200 consecutive combustion cycles. For the 50% mass fraction burned position Fig. 7(c), as already shown in Fig. 7(a), the serial-compatible sensor shows slightly lower values in the entire engine operation range. The peak pressure position Fig. 7(d), on the other hand, shows larger spread between the two sensors, compared to MFB_{50} criteria.

Therefore, it can be concluded that the center of combustion is better suited for the ignition angle control due to the better reproducibility from the series-compatible sensor point of view. The offset observed between the Continental SAPS and the laboratory sensor, can be taken into account in the control function. This allows the deviations to be compensated with only a minor calibration effort.

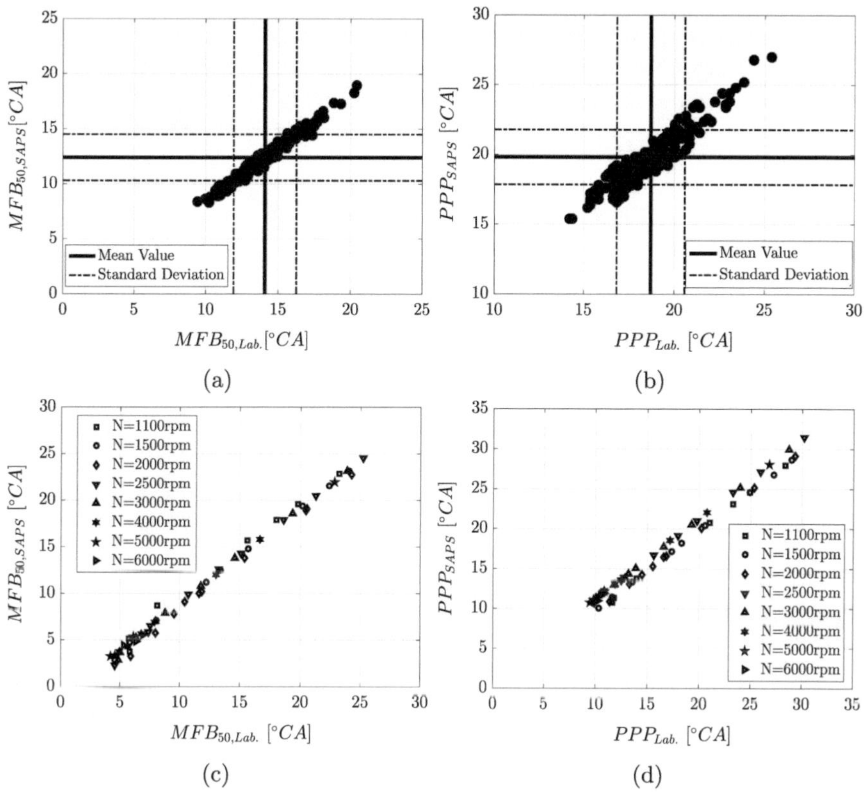

Fig. 7. Comparison of MFB_{50} and PPP derived from the pressure trace of a Continental stand-alone pressure sensor (SAPS) or a Kistler laboratory spark plug pressure transducer (Lab.) of cylinder 1. (a) and (b) At stationary operating point $N = 2500$ rpm and $BMEP = 10$ bar. (c) and (d) Load sweeps in the entire engine operating range from idle to full load for the given engine speeds.

6 Closed Loop Ignition Angle Controller

The control presented in this paper, uses the 50% mass fraction burned position as the combustion feedback signal. Nevertheless, the control could be easily adapted to all three combustion descriptors introduced in Sect. 3.

The results of an ignition angle variation for the loads $BMEP = 2 \ldots 8$ bar at $N = 2000$ rpm is shown in Fig. 8. The y-axis represents the change in brake specific fuel consumption (BSFC) with respect to the $BSFC$ at $MFB_{50} = 8\,°\mathrm{CA}$. By the graph it can be concluded, that a quadratic relation between the 50% mass fraction burned position and the BSFC exists. Further, it can be observed that the center of combustion for all depicted operation points has its minimum at around 8 °CA after iTDC. Therefore, the set point for the ignition angle control is derived, for the application in this paper, as $MFB_{50} = 8\,°\mathrm{CA}$.

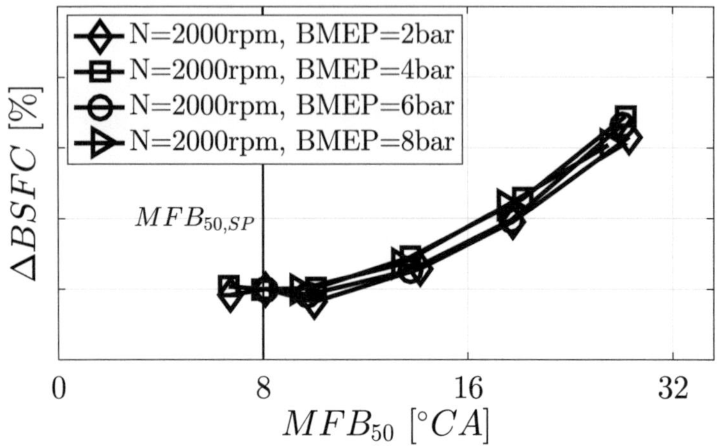

Fig. 8. Ignition angle variation at stationary engine operation. Relative deviation of BSFC with respect to the reference at $MFB_{50} = 8\,°$CA as a function of center of combustion.

The control structure for the closed loop ignition angle control is shown in Fig. 9. The base ignition angle (IGA_{BAS}) is determined, based on the parameters and modeled variables available to the engine control, such as the engine speed (N) and the modeled indicated torque (TQI). Illustrated as pre control, this calculation follows the standard ignition angle control, which is already available. In addition to that, a I-controller was integrated into the ignition angle control. The controller is fed by the difference (e) between the measured MFB_{50} from the actual combustion process and the set point for the desired 50% mass fraction burned position ($MFB_{50,SP}$). As an output, the controller generates an ignition angle offset, here marked as ΔIGA_{CLC}. This offset is added to the base ignition angle, before the next electrical discharge at the spark plug is triggered at the adapted new ignition angle (IGA).

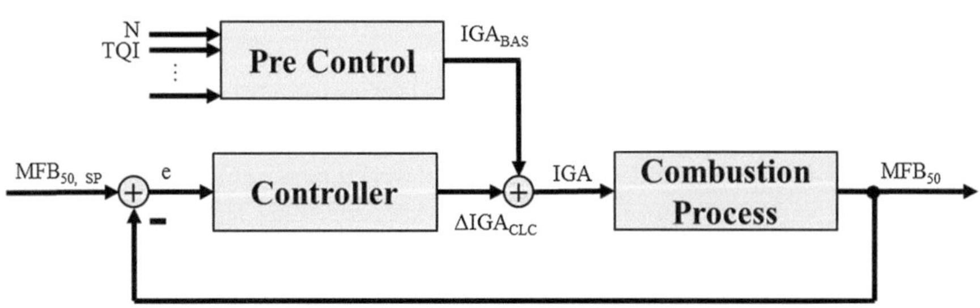

Fig. 9. Structure of closed loop ignition angle control concept.

The I-controller follows the update of the ignition angle offset by the following control law (see Eq. 5),

$$\Delta IGA_{CLC}(n+1) = \Delta IGA_{CLC}(n) + K \cdot [MFB_{50,SP}(n) - MFB_{50}(n)] \quad (5)$$

where $\Delta IGA_{CLCC}(n+1)$ is the new ignition anlge offset, $\Delta IGA_{CLCC}(n)$ is the offset from the past ignition event, $MFB_{50,SP}(n)$ is the set point for the desired center of combustion, $MFB_{50}(n)$ is the actual center of combustion and K is the gain factor of the controller [5].

The gain factor K represents a tuning parameter. From the control point of view, the factor should be selected as large as possible to ensure a fast control behavior. However, due to the characteristic cycle-to-cycle variations of a gasoline engine, the gain factor should be selected in such a way that the variation do not affect the ignition angle to much. Therefore, the gain factor was selected to be $K = 0.125$. Due to the low gain factor, the controller shows a filtering effect on the cyclic fluctuations of the MFB_{50}. However, this causes a decrease in the controller response time. But during transient operation, this effect is compensated by the fast pre control [5].

Furthermore, it must be ensured that a spark advance by the controller is avoided at the occurrence of a knocking combustion. Therefore, the controller is turned off for all knock limited operation points. In addition to the limitation by knock, an intervention of the control for operating points with actively demand spark retard, as a fast torque reserve, must also be handled carefully, in order to maintain a good driveability.

7 Closed Loop Control Demonstration

In this section, the function principle of the closed loop ignition angle control is demonstrated and reviewed using vehicle measurements from a chassis dyno. Therefore, a specific driving cycle for the chassis dyno was defined. The specific driving cycle consists of an urban driving cycle (UDC) followed by several extra urban driving cycles (EUDC). In order to evaluate the benefits of the MFB_{50} control, the control was switched on and off alternately for each of the EUDC cycles. Thereby, both the driver's and test bench's influences were minimized as well as the variation caused by the thermal state of the engine.

An example of the described specific test cycle is illustrated in Fig. 10. In addition, selected measured variables were determined at the positions indicated by letters A to D by the mean value formation in a 5 s interval. These measured variables are summarized in Table 1.

The vehicle speed profile as well as the alternating activation of the ignition angle control are shown in the graph at the top of Fig. 10.

Below, with reference to the right-hand y-axis, the center of combustion for the first cylinder is shown. The dotted line indicates the desired set point at $MFB_{50,SP} = 8\,°CA$. The left axis in the same graph corresponds to the ignition angle offset, determined by the ignition angle control based on the Continental series-compatible SAPS. There a spark advance of up 12 °CA can be observed for

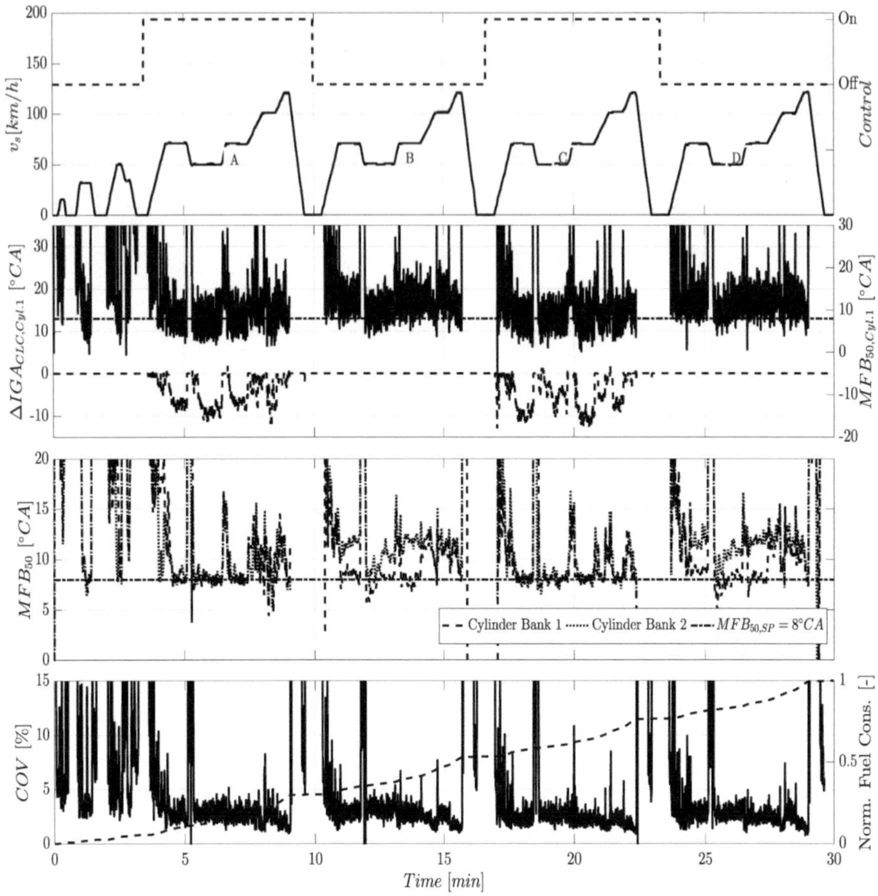

Fig. 10. Vehicle measurements of functional demonstration of closed loop ignition angle control from specific driving cycle on the chassis dyno.

Table 1. Selected measured variables of driving cycle depicted in Fig. 10 from the positions indicated by letters A to D. Values have been derived by mean value formation in a 5 s interval.

Parameter	A	B	C	D	Unit
Control	ON	OFF	ON	OFF	-
$PMI_{HP-Loop}$	4.00	4.00	3.75	3.76	bar
$PMI_{HP-Loop,CBK.1}$	3.94	3.95	3.74	3.78	bar
$PMI_{HP-Loop,CBK.2}$	4.06	4.05	3.74	3.74	bar
COV	2.68	2.77	2.24	2.72	%
$MFB_{50,CBK.1}$	8.02	8.13	7.96	8.86	$°CA$
$MFB_{50,CBK.2}$	8.12	11.98	8.06	11.83	$°CA$
Fuel consumption	−0.88		−1.89		%

the first cylinder. The control interaction in direct relation to the measured 50% mass fraction burned position, shows that the average of $MFB_{50,Cyl.1}$ shows good control behavior. However, the controller is unable to compensate the typical cycle-to-cycle variations of a gasoline engine. Furthermore, the transient response of the control can be observed. First, it can be derived that the control is not permanently turned on due to the limitation mentioned in Sect. 6. In addition, the trace indicates that the controller only response slowly to adapt the necessary spark advance due to the chosen low K factor.

In the third graph, the mean value of the center of combustion is plotted separately, for each of the two cylinder banks. The dashed-dotted line also marks the desired set point of the ignition angle control. The actual 50% mass faction burned position for the two cylinder banks is drifting apart by about 4 °CA when the control is deactivated.

This also has an effect on the engine roughness (COV), which is shown as a moving average for the past 10 combustion cycles in the lowest graph with respect to the left y-axis. By equalizing the actual MFB_{50} for each cylinder separately the variation of MFB_{50} within one cylinder bank and also between the cylinder banks is minimized. Both improvements will lead to an decrease in engine roughness of approximately 0.5%. The last trace illustrated by the dashed line in the bottom graph of Fig. 10, is the fuel consumption, with reference to the right y-axis. Here, the normalized accumulated fuel consumption is used to determine the reduction in fuel consumption by the ignition angle closed loop control. As shown in Table 1, a reduction in fuel consumption for individual operating points of up to 1.9% can be achieved by the control.

In addition to the specific driving cylce for the control validation, several worldwide harmonized light vehicles test cycles (WLTC) and new european driving cycles (NEDC) have been carried out. All test cycles were performed consecutively one after the other. As a result the overall available test time could be reduced. However, since pre conditioning of the vehicle could not be ensured for this test procedure, the results can not be transferred one-to-one to a standardized WLTC or NEDC test procedure.

Figure 11 summarizes the reduction in fuel consumption, that was achieved for the performed driving cycles WLTC and NEDC as described above. The cycles were repeated several time consecutively one after the other. In this process, the control was alternately activated on base of the Continental SAPS or the laboratory sensor, or deactivated to derive the reference fuel consumption for the investigated test platform.

For both test cycles a fuel reduction was detected in the range of 0.5% to 1.5%. There was a minimal difference in the results achieved by the two sensors compared in this paper, which can be attributed to the different calibration of the control function as well as the statistical variation in the results of the tests carried out. Furthermore, it can be observed that a higher fuel reduction for the WLTC was achieved compared to the NEDC. The control thus fulfills the expectation, as stated in Sect. 2.2 for a closed loop control, in the entire operating range of the engine, independent of the prevailing operating conditions.

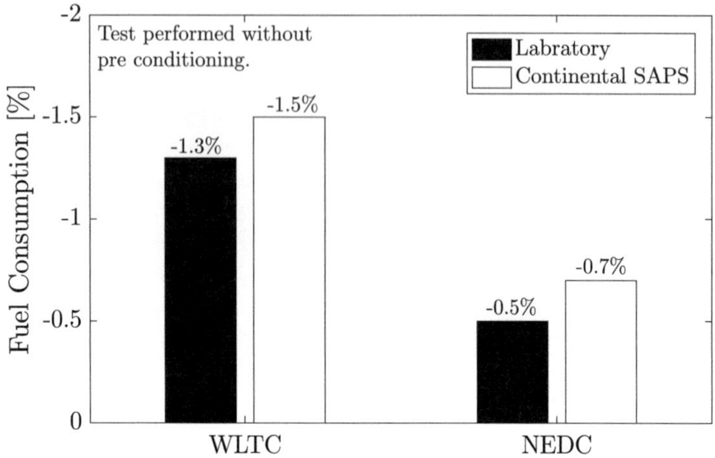

Fig. 11. Reduction in average fuel consumption for non pre conditioned WLTC and NEDC driving cycles.

The optimal ignition angle is ensured by the closed loop ignition angle control and thereby the efficient operation of the internal combustion engine is guaranteed.

8 Conclusion

In the present study, a closed loop ignition angle control with a feedback signal deduced from an in-cylinder pressure sensor on the example of a modern gasoline engine with high variability has been presented. The Continental stand-alone pressure sensor was reviewed and compared to a standard laboratory sensor. Several combustion descriptors where reviewed and analyzed. The feature MFB_{50} showed a significant correlation and signal reliability for the series-compatible sensor in the entire engine operating range and was therefore chosen as the control variable. The control strategy, suitable for a real-time ignition angle control, was implemented for a state of the art V6-SI-engine. The controller was tested and validated at several driving cycles on the chassis dyno. A reduction in fuel consumption depending on the engine operating conditions between 0.5% and 1.9% was demonstrated independently for the Continental SAPS and a laboratory sensor. In addition to that, the variation in center of combustion within the cylinder bank and between the two cylinder banks have been reduced by the control. Thereby, the cylinder individual control of MFB_{50} towards the MBT spark timing showed a decrease of up to 0.5%, at equal operating conditions, with respect to the overall engine roughness. For all the results shown, care was taken during the calibration process of the presented ignition angle controller, that the drivability of the vehicle is not affected.

The evolution of future emission legislation, the expanded operating and testing conditions in view of real driving emissions, especially for cold start condition, and the compliance of CO_2 fleet targets will increase the development effort for future modern SI-engines. Model based controls, which rely on the direct feedback signal derived from the combustion process (e.g. knock control), such as the in-cylinder pressure, can be used as a supplementation or replacement for the existing open loop control functions. Such closed loop controlled systems can address the future challenges by minimize the need for extensive calibration, to improve the system robustness with regard to engine conditions and the engine to engine deviations, and to increase the efficiency by means of targeted control interventions in the combustion process as demonstrated in this paper for a closed loop ignition angle control.

References

1. Neuser, H.-J., Kahrstedt, J., Dorenkamp, R., Jelden, H.: The Euro 6 Engines in the modular diesel engine system of Volkswagen. Motortech. Z. Ausg. **6**(2013), 4–10 (2013)
2. Varoquie, B., Hellemans, A., Bouquey, C., Walker, E., Heunitz, D., van Booven, B.: Combustion control and engine system approach based on cylinder pressure sensor for diesel engines. SIA No. 831, pp. 29–37 (2014)
3. Bargende, M.: Schwerpunkt-Kriterium und automatische Klingelerkennung. Motortech. Z. **56**(10), 632–638 (1995)
4. Hellemans, A., Landrevie, L., Venzal, S., Walker, E.: Direkte Zylinderdruckmessung. Motortech. Z. Ausg. **10**(2011), 770–776 (2011)
5. Eriksson, L.: Spark Advance for Optimal Efficiency. Electronic Engine Controls. SAE Paper No. 1999–01-0548 (1999)
6. Bögl, A., Hemminger, B., Janssen, A., Kerkau, M., Kerner, J., Kronich, A., Schlüter, M.: Die neuen V6 Turbomotoren im Porsche Macan. In: 35, Internationales Wiener Motorensymposium, pp. 197–222 (2014)
7. Heywood, J.B.: Internal Combustion Engine Fundamentals. McGraw-Hill Series in Mechanical Engineering. McGraw-Hill, New York (1989)
8. Kohlhase, M., Isermann, R.: Self-tuning extremum control based on cylinder pressure sensor for direct injection spark ignition engines. Automatisierungstechnik Methoden Anwend. Steuer. Regel. Informationstechnik **57**(1), 23–31. https://doi.org/10.1524/auto.2009.0749. Accessed 17 Sep 2017
9. Thiel, M., Siepmann, M., Jeinsch, T., Schultalbers, M.: Adaptive Regelung und Signalaufbereitung der Verbrennungsschwerpunktlage. In: 7, International Symposium on Automatic Control (2014)
10. Draper, C.S., Li, Y.T.: PrincIples of Optimizing Control Systems and An Application to the Internal Combustion Engine. ASME Publication, New York (1951)
11. Erikson, L., Nielsen, L., Glavenius, M.: Closed loop ignition control by ionization current interpretation. SAE Trans. J. Engines **106**(3), 1216–1223 (1997)
12. Hohenberg, G.: Der Verbrennungsverlauf - ein Weg zur Beurteilung des motorischen Prozesses. In: 4, Wiener Motorensymposium (1982)
13. Matekunas, F.A.: engine combustion control with ignition timing by pressure ratio management. United States Patent 4622939, 18 November 1986
14. Hubbard, M., Dobson, P.D., Powell, J.D.: Close loop control of spark advance using a cylinder pressure sensor. J. Dyn. Syst. Meas. Control **98**, 414–420 (1976)
15. Glaser, I., Powell, J.D.: Optimal closed-loop spark control of an automotive engine. SAE paper No. 810058, pp. 11–21 (1981)

16. van Booven, B.; Schüle, H.; Walder, T.; Kulzer, A. C.; Rottengruber, H.: Real-time assessment of in-cylinder pressure for fresh air mass estimation using the example of modern spark ignited combustion engines. In: 2, Symposium for Combustion Control (2016)
17. Kuratle, R.: Motorenmesstechnik. 1, Auflage Vogle-Fachbuch, Würzburg (1995)
18. Patterson, G., Hartleb, T., Mordelt, G.: Physical techniques for evaluating cylinder pressure transducer data quality. In: 11, International Symposium on Combustion Diagnostics (2014)
19. Randolph, A.: Methods of processing cylinder-pressure transducer signals to maximize data accuracy. SAE Technical Paper 900170 (1990)
20. Randolph, A.: Cylinder-Pressure-Transducer Mounting Techniques to Maximize Data Accuracy. SAE Technical Paper 900171 (1990)

Water Injection

Water Injection for Gasoline Engines - Quo Vadis?

Ingo Hermann[1(✉)], Claus Glahn[1], Matthias Kluin[1], Martin Paroll[1], and Werner Gumprich[2]

[1] Advanced Engineering Gasoline Technologies, Opel Automobile GmbH, Rüsselsheim am Main, Germany
ingo.hermann@opel.com
[2] Engine Analysis, Opel Automobile GmbH, Rüsselsheim am Main, Germany

Abstract. Development on gasoline engines for 2025 and beyond is driven both by enhanced electrification as well as by cost optimized measures for improved engine efficiency. Besides Miller combustion process, cooled external exhaust gas recirculation (EGR) and variable compression ratio, water injection represents one of the relevant technologies. Water injection potential within gasoline engines will be evaluated within this paper.

Basically, the cooling effect of evaporating the injected water can mitigate knocking. This will result in earlier combustion phasing, in less fuel enrichment for component protection and thereby in higher thermodynamic efficiency. Water injection will be systematically investigated on a 1.6 l turbocharged engine for three different concepts: Intake manifold/plenum injection, inlet port injection and the direct injection of a premixed fuel/water emulsion. Complexity and water efficiency are compared with each other for all systems.

Regarding the application strategy, both an efficiency as well as a performance approach are considered. For the efficiency variant, fuel saving potentials will be optimized by increasing the geometric compression ratio. For the performance variant, the increase of stoichiometric power at constant compression ratio/part load consumption will be in focus of the studies.

Understanding the effects of water injection on the working process of the combustion engine is assisted by 3-D-CFD modelling of water distribution to cylinders, mixture formation, evaporation and thermal efficiency. This will be supported by in-situ video records about the water-to-air mixture formation as well as for the injection and evaporation during inlet stroke.

Based on all generated knowledge, the potential of water injection for future combustion engines will be analyzed and pros and cons for the different injection concepts are evaluated. An outlook will be provided.

1 Introduction

The knock propensity of conventional gasoline fuels limits the efficiency of boosted engines. Spark retard as knocking countermeasure leads to late combustion phasing, high exhaust gas temperatures and fuel enrichment for component protection. Thus, both combustion efficiency and stoichiometric operation within the operational engine map are compromised. By the admixture of fluids to the gasoline fuel, the cylinder charge

M. Günther and M. Sens (eds.), *Knocking in Gasoline Engines*,
https://doi.org/10.1007/978-3-319-69760-4_18

will be cooled through the evaporation enthalpy and the knocking tendency is mitigated. As depicted in Fig. 1, ethanol shows a two times higher and water even a 5.5 times higher evaporation enthalpy than regular gasoline RON 95. Admixture of ethanol to gasoline is a prominent measure to reduce knocking. Water with its even higher evaporation enthalpy of 2257 kJ/kg is much more suited to avoid knocking and improve both efficiency and stoichiometric operation. Water is cheap, easy available and guarantees safe handling and therefore appears to be the ideal fluid.

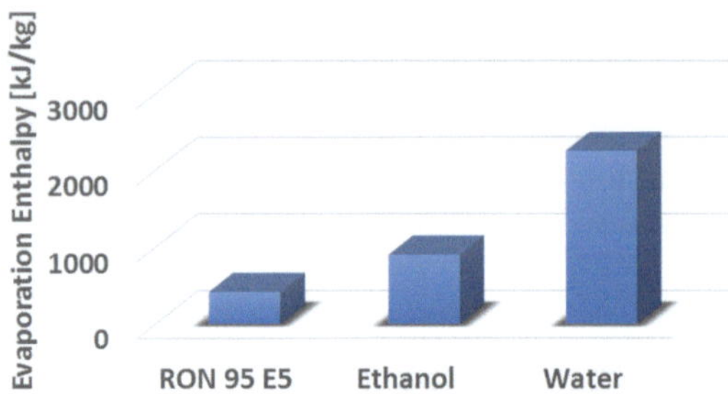

Fig. 1. Evaporation enthalpy of different fluids

Therefore, water injection seems to show a high potential to solve today's main engine development goals, CO_2 reduction and fulfilment of the upcoming stringent emission regulations RDE (Real Driving Emissions).

2 Thermodynamic Principles of Water Injection

Water will be injected into the engine; so the absolute mass flow can be varied depending on the demand within the load/speed map. Figure 2 shows a typical pressure/volume diagram with and without water injection. It can be seen directly, that combustion starts earlier and results in higher peak pressures.

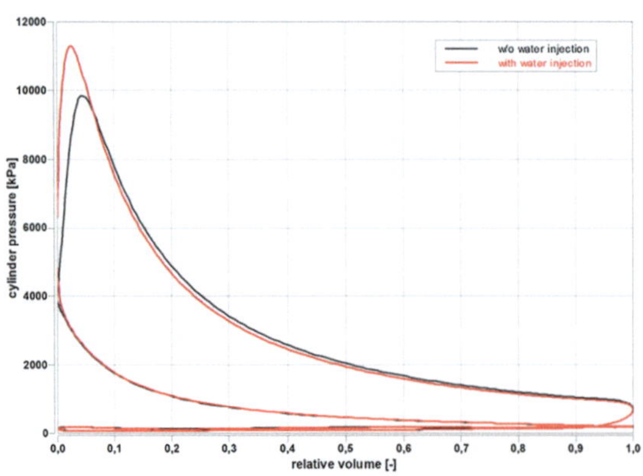

Fig. 2. Pressure/Volume diagram with and without water

The fundamental, thermodynamic principles of water injection are explained in Fig. 3. Regarding the performance concepts, water is injected into the air intake system. Depending on the thermodynamic state (temperature, pressure) of the compressed air, a certain mass fraction of the injected water flow will evaporate and lower the intake air temperature. Consequently, the density of the air charge will rise and the volumetric efficiency will increase. In parallel to the air, fuel flow is also increased by lambda control, leading to a higher torque of the engine. By the lowered intake air temperature, the overall level of the combustion temperature is also reduced which results in knock mitigation.

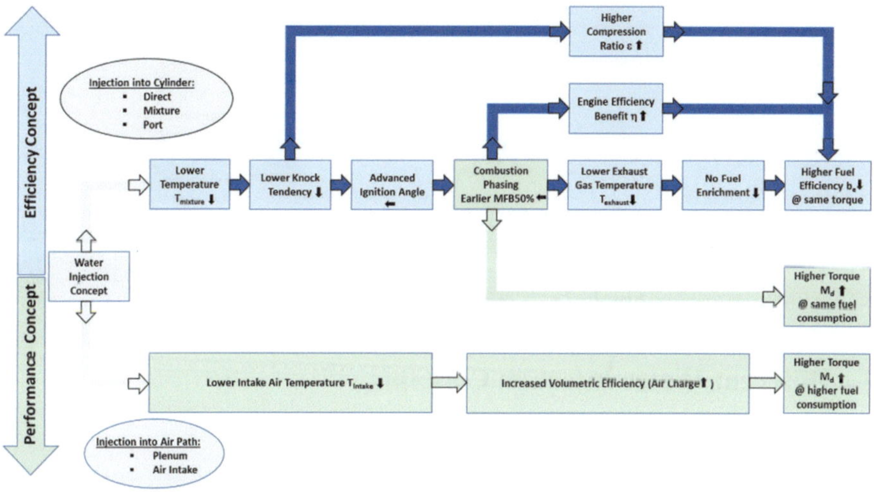

Fig. 3. Thermodynamic principles of water injection

Regarding the efficiency concepts, water is injected into the cylinder. Three different basic water concepts are possible: port injection, mixture injection and direct injection. They will be explained in more detail in the next chapter. During suction and compression stroke of the engine, the injected water will evaporate and lower the temperature of the cylinder charge. Knock tendency is reduced, spark angle can be advanced leading to an improved combustion phasing. Thus, exhaust gas temperature is lowered and fuel enrichment can be reduced or even completely avoided. A stoichiometric load/speed map is achievable ($\lambda = 1$ concept). Fuel efficiency is strongly improved. Earlier combustion phasing leads to a higher internal engine efficiency also contributing to the fuel consumption reduction. For improved part load behavior, which is important for the fuel consumption in drive cycles like the WLTP, the lower knock tendency allows for an increased geometric compression ratio. In literature [1], an increase in ε by 2 points can be referenced, which would result in a WLTP drive cycle efficiency improvement of about 4%.

Earlier combustion phasing can be used either for a better fuel efficiency at constant torque or vice versa for higher engine torque at constant fuel consumption.

3 Demonstrator Engine

Water injection experiments have been executed on a 1.6 l demonstrator engine of the MGE (Midsize Gasoline Engine) family of GM/Opel. By optimizing the integration parameters, the turbocharging concept and the charge air cooler, specific power has been increased by 35%. Figure 4 shows the system build-up.

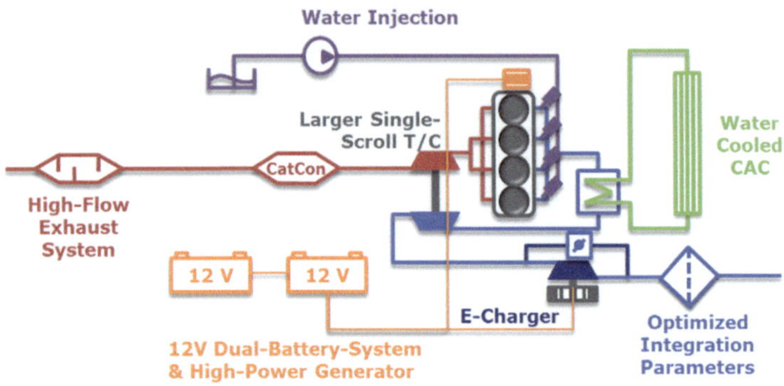

Fig. 4. Demonstrator engine system build-up

4 Different Water Injection Concepts

4.1 Plenum Injection

Water injection takes place in the air intake piping or within the intake manifold, also called plenum.

Figure 5 shows both block diagram and realization on the engine dyno. The water injector is located downstream of the water-cooled charge air cooler. The water droplets have time to evaporate in the long, horizontal part of the intake air piping. Downstream, the air/water vapor mixture falls down to the intake manifold.

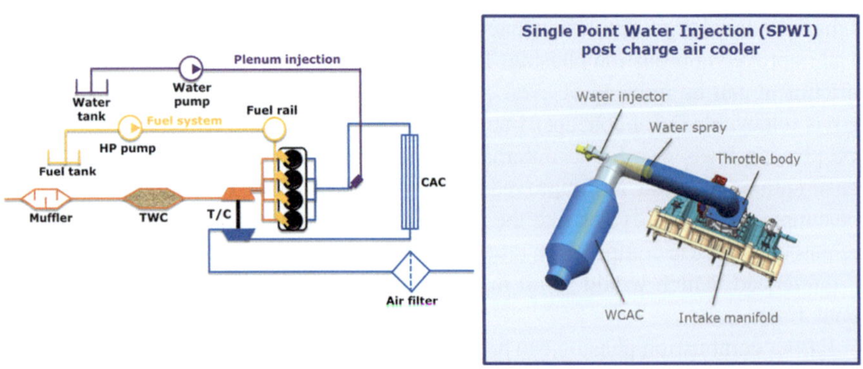

Fig. 5. Plenum injection

4.2 Port Injection

In this case, water injection takes place upstream of each cylinder's inlet channels. There is one water injector per cylinder. The injectors are mounted in a special rig enabling a dedicated positioning of the water injector relative to the two inlet channels. As can be seen from Fig. 6, each water injector has a spray pattern consisting of two cones; each cone is feeding one inlet channel. The spray target of each cone was designed that way, that most of the water spray can enter the cylinder through the open inlet valve. This design has been done purely geometrical for the water injectors; air flow and turbulence has not yet been accounted for in this first configuration. The water injectors are designed for sufficient dynamic mass flow even at the shortest valve opening times.

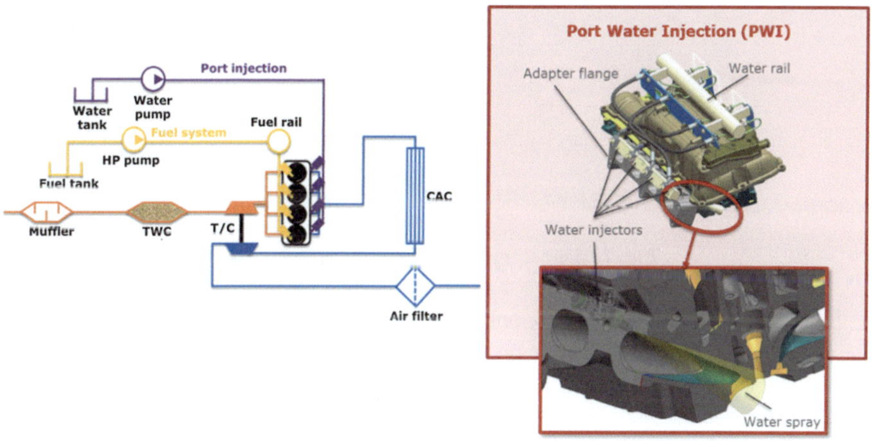

Fig. 6. Port water injection

4.3 Direct Injection of Pre-mixed Fuel/Water

Direct injection of pre-mixed fuel/water is another concept. Within this study, the existing fuel injector in the cylinder roof is used. The injector can handle a mixture of fuel and water within certain quantities and ratios. This enables stationary experiments with this concept. Injection timing of water and fuel takes place in parallel for mixture injection. It will be investigated whether the in-cylinder charge cooling is more effective. That means, that the same thermal efficiency can be gained by less water.

As can be seen in Fig. 7, water is metered into the mixing chamber by two injectors. The mixing chamber itself is filled by the pre-pressurized fuel flow of 4–5 bar. Since water and fuel are not soluble in each other, they are mechanically sheared by a static mixing device especially designed for these conditions. Driving force is the pressure gradient over the mixing device, it is located at the downstream end of the chamber. Further downstream, the high-pressure pump of the engine improves and stabilizes the dispersion of fuel/water to transport it homogeneously mixed through the fuel supply system to the injectors. Emulsifying agents are avoided by purpose, because they have an influence on the combustion which is not desired.

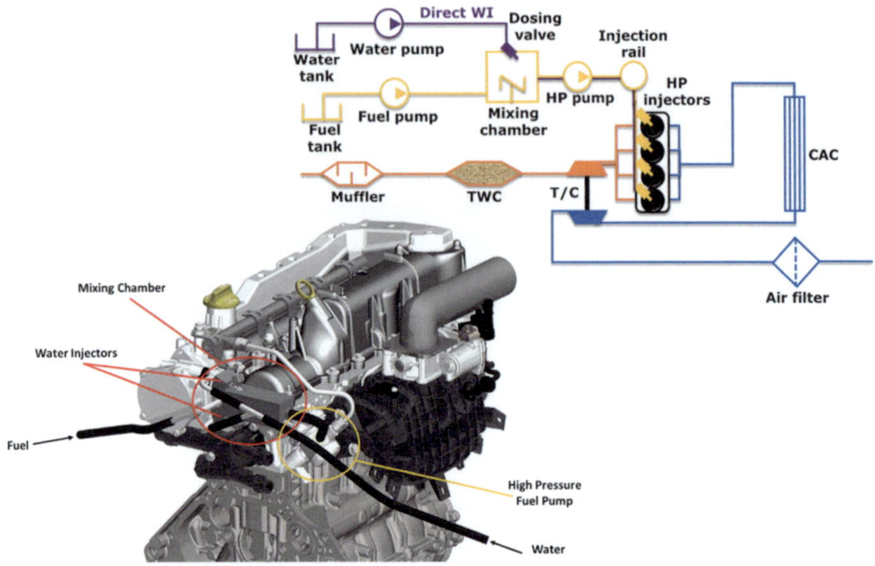

Fig. 7. Direct injection of pre-mixed fuel/water

With this experimental set-up, only stationary operating points are executed. From the results, the principal influence on the combustion shall be concluded. For dynamic conditions within the vehicle, other measures must be undertaken to realize direct mixture injection.

4.4 Direct Water Injection

For direct water injection, there exists a dedicated high-pressure water injector in the cylinder roof. The advantages are obvious; water injection is timely separated from the fuel injection. In contrast to the water injection concepts mentioned so far, direct water injection can take place during the compression stroke after intake valve closing (IVC). At low and medium engine speeds, there may be enough time for evaporation and mixing between intake valve closing (IVC) and spark ignition (SI), which could result in lower compression end temperatures. So, direct water injection might unlock further potentials regarding efficiency.

Figure 8 shows the block diagram and a principal sketch of direct water injection with separate injectors both for water and fuel. In order to avoid two high pressure paths and two injectors in the cylinder roof, a concept with port fuel injection (PFI) and central water injection (DI) seems feasible.

Direct water injection has not been carried out experimentally for this study. Integration efforts are relatively high compared with the other concepts. Though, fuel PFI and water DI are simulated by 3-D-CFD in order to allow a fair comparison between all major water injection systems.

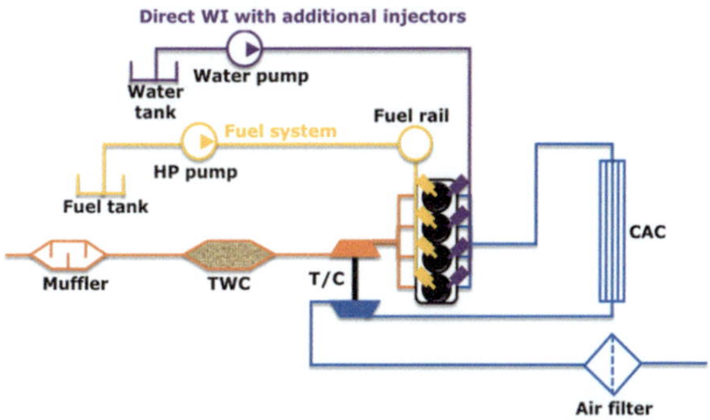

Fig. 8. Direct water injection with dual high-pressure paths

4.5 Experimental Infrastructure for Water Delivery

With the exception of direct water injection, the infrastructure to deliver water with the appropriate inlet conditions to the injectors is shown in Fig. 9. Water needs to be transported with the correct pressure to the water injectors within the different concepts. Therefore, upstream of the water injectors a water rail provides sufficient storage space at an adjustable pressure.

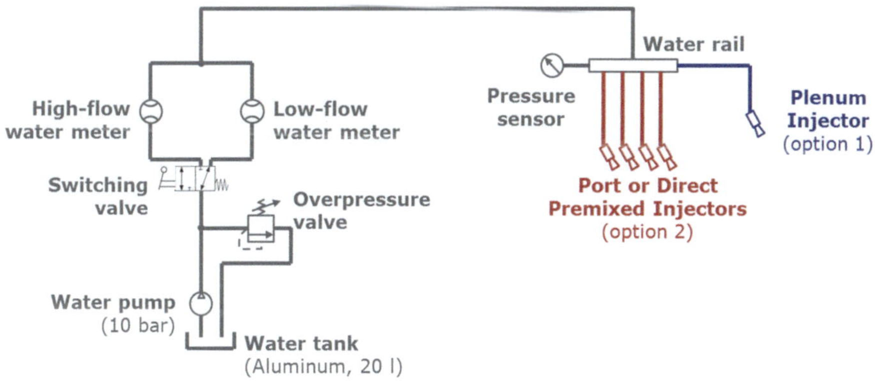

Fig. 9. Water infrastructure on dyno

For testing purposes on the dynamometer, a 20 L canister with de-mineralized water serves as reservoir. An electrical water pump with a capacity of up to 10 bar pressure is sitting within the water. Pressure control is accomplished by an overpressure valve. The pump is running continuously, by narrowing the cross-sectional area in the valve the desired pressure in the rail can be accomplished; excess water is fed back into the canister.

There are two systems for measuring the exact mass flow of the water. Both work according to the displacement principle; one is a turbine flowmeter designed for higher water mass flows, the other one is a gear-wheel flowmeter for low water mass flows.

The water mass flow is an important parameter; in literature the change in thermodynamic behavior of the engine is characterized over the water/fuel ratio, which is defined as the water mass flow divided by the fuel mass flow in percent.

5 Test Results for Water Injection

The test results for water injection on the MGE demonstrator on dyno are discussed within this chapter. Exemplarily, results from port water injection are referenced. Differences between the injection concepts are debated later in Sect. 6. It will be shown, how the factual results from dyno compare to the thermodynamic theory of Sect. 2.

5.1 Stoichiometric Power Increase

The maximum stoichiometric power of the MGE engine is examined without and with water injection and shown in Fig. 10. The black curve represents the base engine. It is run rich ($\lambda < 1$), at higher speeds above the low-end torque knee (LET) for component protection. The stoichiometric specific power of the base engine lies only at 58% of the nominal value. With plenum water injection, stoichiometric power rises up to 66%. That is even topped by port water injection with 92%. Almost within the complete load/speed map of the demonstrator engine, fuel enrichment can be avoided; reason for the missing "upper right corner" in the map was a slight gas slippage in the wastegate of the turbocharger to achieve the high load.

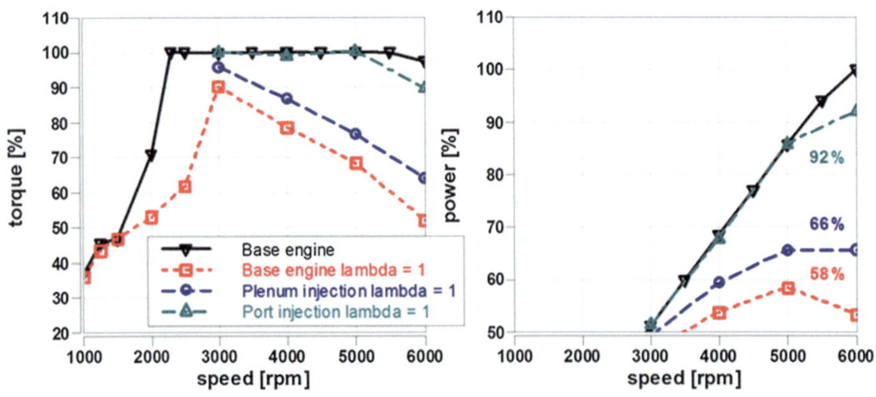

Fig. 10. Stoichiometric power capability

5.2 Thermodynamic Improvements

At maximum engine torque, which is wide open throttle operation, a speed sweep from 3000 rpm to 5000 rpm has been carried out. For each speed, the water/fuel-ratio has been varied.

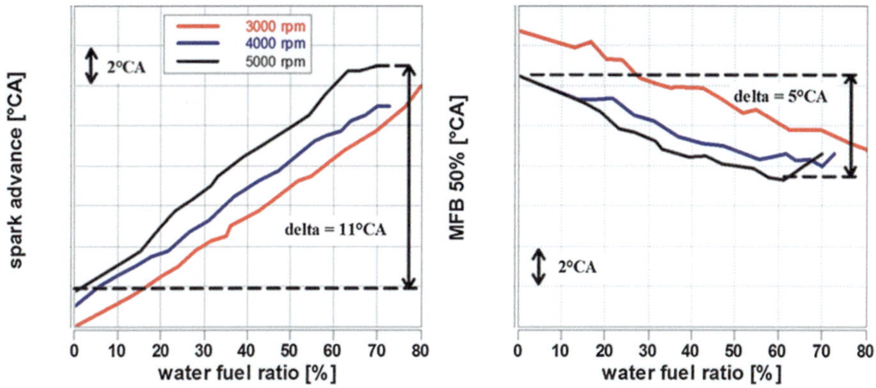

Fig. 11. Spark advance and combustion phasing

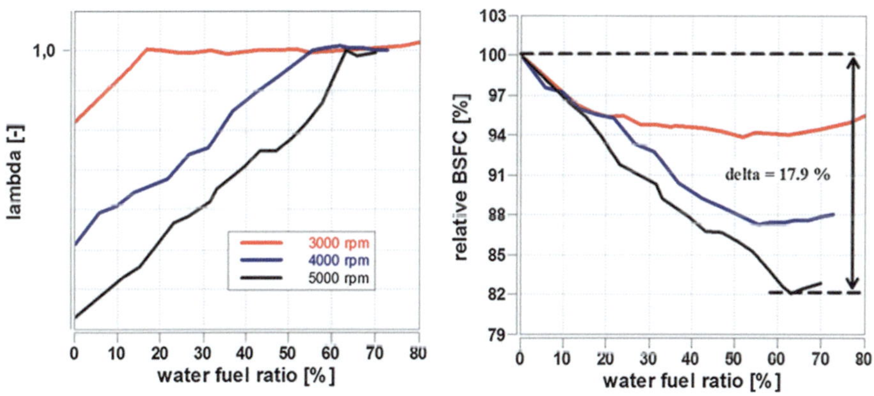

Fig. 12. Lambda and BSFC

As can be seen in Figs. 11 and 12, when there is no water present, enrichment increases with speed and that can be directly correlated with the BSFC values. Without water, higher speeds offer better combustion phasing and some earlier spark. With the addition of water, knocking is mitigated and spark can be advanced. The graphs show a linear correlation between spark advance and water/fuel ratio. At 5000 rpm, a spark advance of 11 °CA could be gained. Since combustion starts earlier, combustion phasing also improves. The center of combustion (MFB50%) also linearly decreases with the water/fuel ratio. At 5000 rpm, MFB50% is enhanced by 5 °CA. It shall be noted, that these mechanisms are independent from λ. The red curve for 3000 rpm reaches λ = 1 at a water/fuel ratio of 15%, but spark advance and earlier combustion phasing are continuing for all measured water rates up to 80%.

By earlier combustion phasing, exhaust gas temperatures drop and lead to an enrichment reduction (Fig. 12). For each speed, stoichiometric conditions are achieved. λ increases linearly with the water/fuel-ratio up to one. At 5000 rpm, the maximum in enrichment reduction could be achieved. Earlier combustion phasing also leads to an improved engine efficiency. Both effects occur at the same time and reduce the specific fuel consumption. BSFC drops linearly with the water/fuel-ratio up to the $\lambda = 1$ value. 17.9% improvement could be achieved at 5000 rpm engine speed.

Cylinder pressure is rising with increasing water/fuel-ratio due to faster combustion as shown in Fig. 13. This must be considered in the design of the base engine.

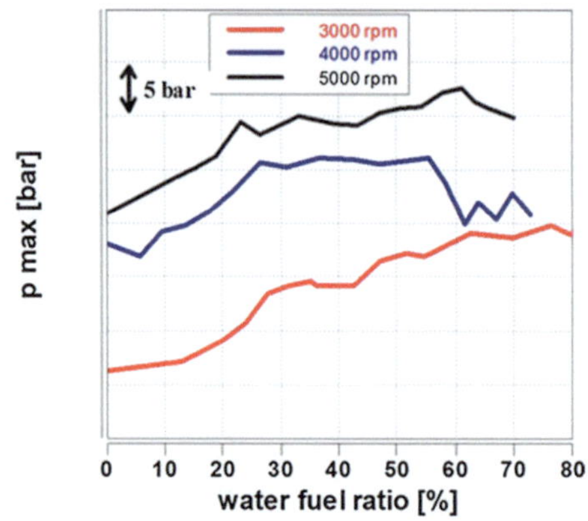

Fig. 13. Maximum in-cylinder pressure P_{max}

As long as the engine is run rich ($\lambda < 1$), both effects, enrichment reduction and enhanced efficiency by earlier MFB50%, lead to fuel savings. If there are already stoichiometric in-cylinder conditions, only the combustion phasing could contribute. Looking at the red curve for 3000 rpm, λ reaches one at 15% water rate. After the linear drop, BSFC then gradually decreases and finally stagnates, although MFB50% is still improving continuously. Overall BSFC advantage is about 1.75%.

Figures 14 and 15 give some insights for the thermodynamic reasons. The exhaust gas temperature EGT before turbine drops when exceeding $\lambda = 1$ from maximum value by 90 °C. All fuel enrichment could be avoided, now the cylinder charge is further cooled down by water evaporation. In fact, there is additional mass in the cylinder by the injected water. The combustion enthalpy is partly used to warm-up this additional mass in the cylinder charge. Overall exhaust gas temperature drops, there is less enthalpy available for generating work at the piston and the overall efficiency drops. There are advantages in the compression stroke, water evaporation enables spark advance and thus improved combustion phasing. But there are disadvantages in the combustion stroke, the additional water mass eats-up a part of the reaction enthalpy and hence reduces the generated power at the piston. These are two opposing effects and they are shown in Fig. 15. This explains well the resulting BSFC curve, which gains only in the 1 digit area when $\lambda = 1$ is overrun. PMEP is not affected, since the additional water mass shifts the operating point in the turbine characteristic towards higher efficiencies. Boost pressure can thus be achieved. With a high water rate, the cylinder charge is diluted; ignition delay linearly increases with the added water vapor mass in the cylinder. Therefore, despite an early spark ignition, fuel mass conversion starts later. The complete burn duration 5–90% lasts longer with higher water/fuel-ratios compared to zero water, though MFB50% is faster. The main advantages of water injection are gained up to $\lambda = 1$. Therefore, water injection is usually executed in the enriched operating regime [2].

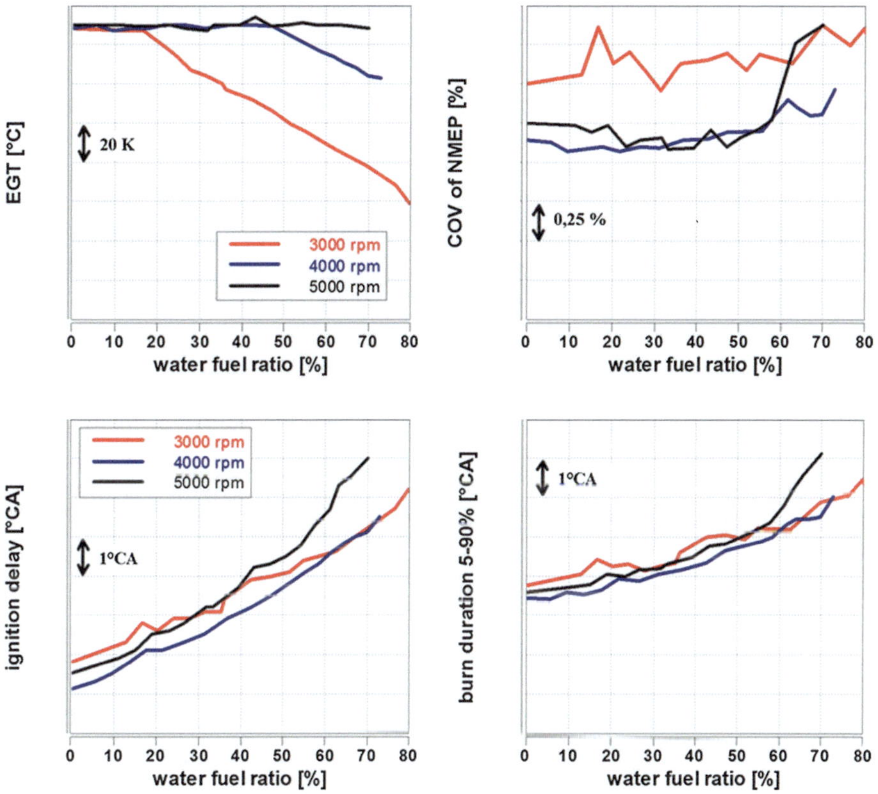

Fig. 14. EGT, COV, ignition delay and burn duration 5–90%

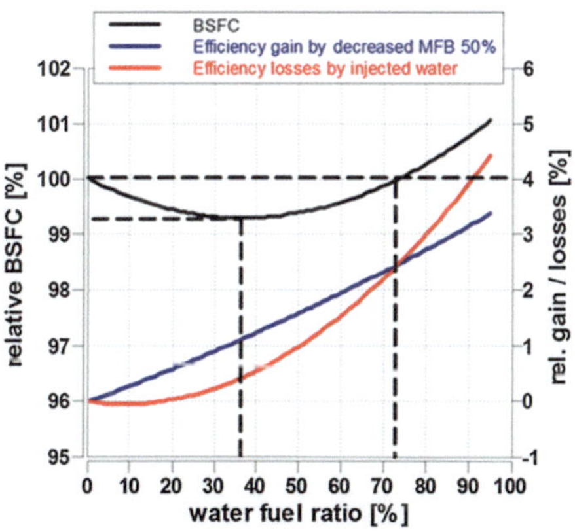

Fig. 15. MFB50% gain and thermal losses

5.3 Gaseous Emission Data

For the speed sweep at wide open throttle, raw emission data upstream of the catalyst are recorded and shown in Fig. 16. With increasing water rate, λ approaches 1, so the combustion is becoming more complete. Consequently, CO emissions are linearly decreasing with the water/fuel-ratio, unburned hydrocarbons HC are also reduced, NOx emissions are rising and excess oxygen O2 is not left over until the water rate for $\lambda = 1$ is reached. When this value for $\lambda = 1$ is exceeded, combustion is becoming incomplete. The high water vapor mass prevents fuel from finding the oxygen molecules for combustion. CO emissions more or less stay constant, unburned hydrocarbons strongly increase and excess oxygen O_2 rises exponentially. The incomplete combustion, when exceeding the water rate for $\lambda = 1$, also contributes to the bad fuel efficiency. Since the water injection enables the stoichiometric operation, all of the gaseous emissions can be converted by the three-way catalyst.

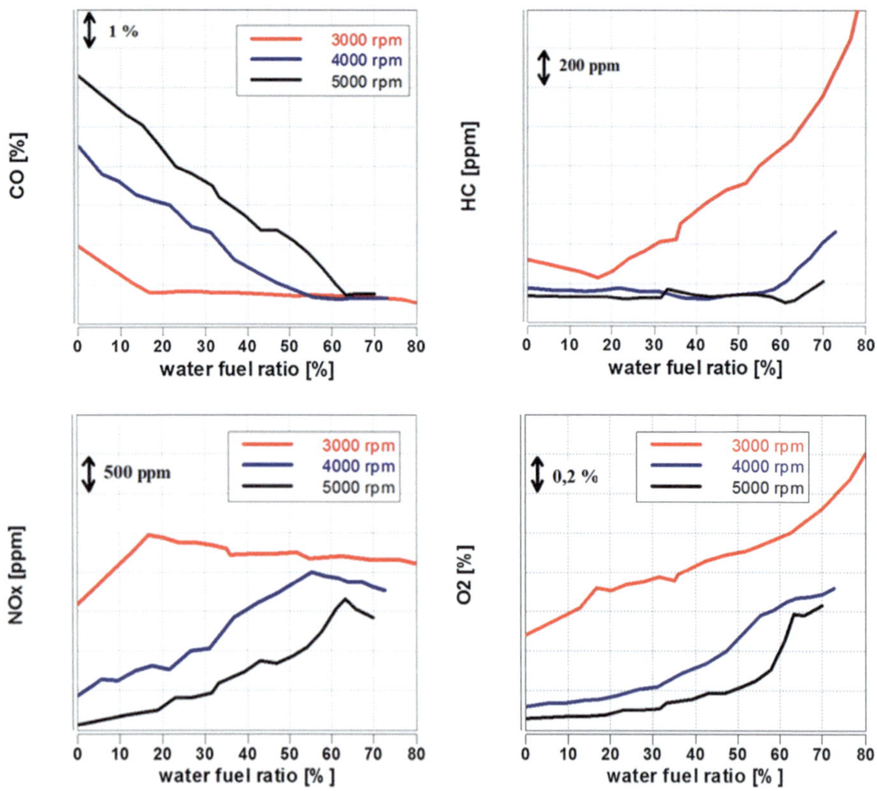

Fig. 16. Raw emissions CO, HC, NO_x and O_2

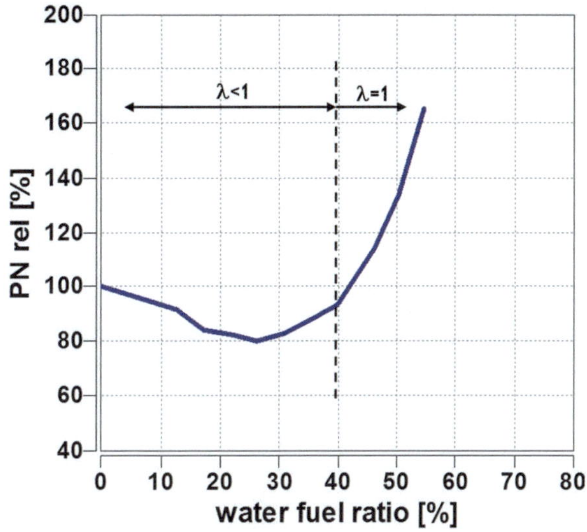

Fig. 17. Particulate emissions

5.4 Particulate Emission Data

Figure 17 shows the relative particulate emissions for a dedicated operation point within the enriched area (5000 rpm/280 Nm). It is an exemplary curve which does look similar for any rich operation point.

A water sweep has been executed. The more water is injected, the less enriched the cylinder charge will be and the relative PN number decreases. A minimum of 20% less PN is reached at 25% water fuel ratio, which is still in the enriched area. With higher water ratios and subsequently also in the stoichiometric region, the particulate number PN strongly increases. The original PN value with no water is exceeded by more than 60%. At first, this appears strange, since we run stoichiometrically. But when the $\lambda = 1$ point is overrun, combustion temperature drops linearly. Additionally, water distribution with port water injection is not even over the cylinder. There may be rich fuel regions, which are surrounded only by water with no air present. Thus, an increase in particulates does seem rational.

6 Evaluation of the Different Water Injection Systems

The different water injection systems are evaluated and compared to each other. The ranking will be done by an evaluation matrix with the following criteria: water distribution to cylinders, water distribution within cylinders, mixture formation, evaporation, thermal efficiency, water efficiency for $\lambda = 1$, transient operation, integration effort and cost.

6.1 Plenum Water Injection

In general, the plenum injection can be realized by different positioning of water injectors somewhere down-stream of the charge air cooler and upstream of the cylinder inlet channels. Plenum injection is characterized by having as few water injectors installed as possible; thus integration effort and costs are minimized. Within this study, one plenum injector is located downstream of the charge air cooler in order to realize a good evaporation, a long mixing time and a minimal wall wetting.

The condition of the charged air at this location has absolute pressures of about 2.3 bar and temperatures of about 25 °C. Thermodynamically, water can be held in the gaseous phase up to the saturation point, which can be calculated for these conditions. Below the saturation point, water stays in the gaseous state and will be distributed homogeneously within the intake manifold to the four cylinders. When the saturation point is overrun, water droplets will begin to condensate. The distribution of liquid droplets within the manifold is not homogeneously; the two inner cylinders will probably get more liquid water than the two outer ones. This is reflected in Fig. 18; the calculated water/fuel-ratio for saturation for the underlying test conditions is 12.5% and is in accurate accordance with the dyno experiments in Fig. 18. The demonstrator engine has pressure transducers within each cylinder; by this indication combustion phasing for each cylinder can be measured.

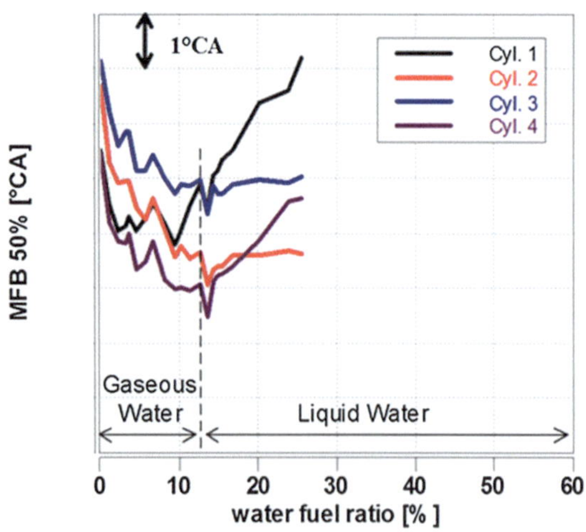

Fig. 18. Uneven distribution for plenum injection

The conclusion for the current configuration is, that water/fuel-ratio is mass-limited for the plenum injection up to the saturation point. Otherwise, water condensation takes place and results in uneven distribution on the cylinders. The only way out could be, to position the plenum injectors within the plastic manifold close to the cylinder head. The water injectors spray target shall hit the inlet channel area of each cylinder equally, so that condensating water droplets are distributed evenly. This would be a quasi-port injection, with several injectors sitting in the plastic manifold. Ideally, the plastic manifold shall support a good flow field to each cylinder and a sweet spot for positioning the injectors.

Pure plenum injection is only suitable for small water mass flows. So, it depends on the application engine. It may be a feasible injection system, if only small amounts of water are either sufficient for stoichiometric operation or if it is combined with another injection system. Below the saturation point, a good mixing with the intake air and a good evaporation takes place. Plenum water injection lowers the intake air temperature. The in-cylinder distribution and mixture formation will happen simultaneously with the air and will show similar quality. Transient operation is good since the dedicated injector can react to air/fuel changes very fast.

6.2 Port Water Injection

For port water injection, each cylinder has its own water injector. Either, the spray of one injector splits into two beams aiming into the two inlet channels of one cylinder (configuration in this study) or there is one dedicated water injector for each single inlet channel. The mechanical effort to integrate the water injectors into the cylinder head rises with their number. Integration effort and costs are higher compared to plenum injection. Nevertheless, distribution per cylinder is excellent as shown in Fig. 19. Transient behavior is excellent as well due to the dedicated injector per cylinder. Port water injection is not water mass flow limited. It is even possible to advance water injection timing into the channels prior to intake valve opening. As the shown test results in Sect. 5 show, stoichiometric operation is always possible.

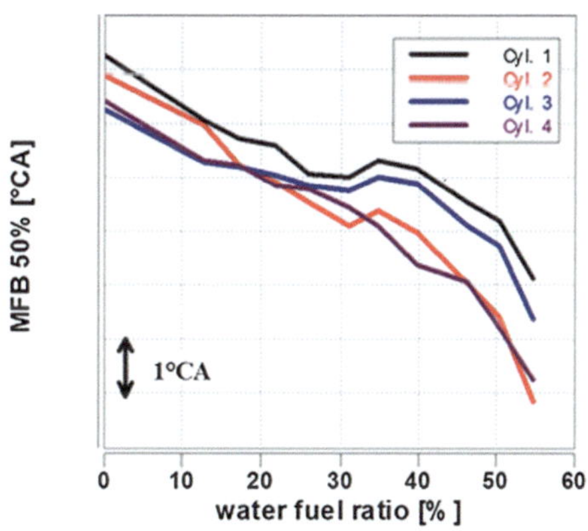

Fig. 19. Homogeneous distribution for port injection

Water and air will partially mix within the inlet channels. Depending on the water rate and wall temperatures, there may be evaporation already or also liquid water staying in the channels. With the intake valve opening, the water/air mixture will flow into the cylinder. It is a complex flow field along the curved shape of the valve into the tumble flow of the engine. For better comprehension, 3-D-CFD simulations have been undertaken as well as in-situ video recording of the air/water/fuel behavior within both intake

Fig. 20. Port water injection (4000 rpm, 250 Nm, 5 ms duration/left side and 7.5 ms duration/ right side)

channels and the cylinder. From the in-cylinder videos in Fig. 20, it can be concluded that mixing will not be perfect. There are zones with more water and there are zones with more gas. At high water rates, the injected water mass just shoots from the intake valve across the cylinder to the opposite walls without being involved in the tumble. In contrast, tumble will be affected in a negative way und thus the mixture preparation. An inhomogeneous mixing will result in more "lost water" wasting the evaporation enthalpy not to the ignitable mixture regions. Mainly wall wetting occurs and cools down the cylinder walls instead of the charged gas.

In Fig. 21, a 3-D-CFD simulation is shown for the evaporation during port injection. The water droplet mass is shown over one engine cycle; three different water/fuel-ratios of one third, two thirds and three thirds are simulated. It is obvious, that liquid water droplets will enter the cylinder; the higher the water rate, the more liquid droplets will go into the cylinder. Nevertheless, complete vaporization of all liquid water droplets will have taken place at crank angles <700°CA, that means prior to ignition top dead center. There is complete evaporation of all water for ratios up to 100%. It can already

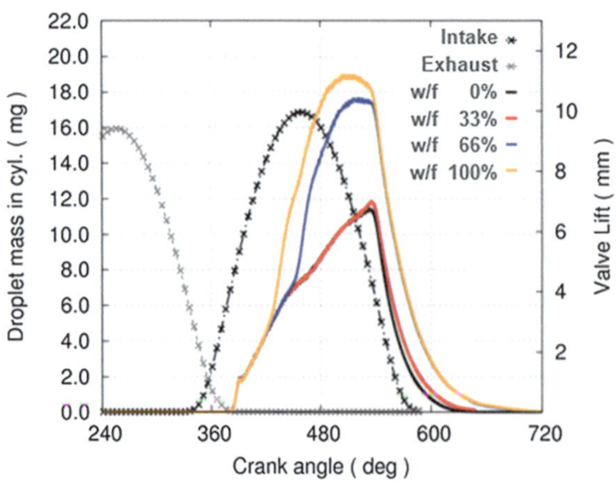

Fig. 21. Evaporation during port water injection

be stated, that for water injection concepts taking place during the intake stroke, there will be complete vaporization prior to compression end.

6.3 Direct Pre-mixed Water Injection

For the direct pre-mixed water injection, there is one dedicated water injector per cylinder. The design must be sufficient to cover both fuel and water mass flow within the complete operation map. In the present study, 350 bar rail pressure is used. By the high pressure, the primary break-up of the water molecules with its high surface tension functions well and a finely dispersed water spray enters each cylinder. Within the in-situ photo from Fig. 22, there is no distinction possible between the fuel and the water, confirming an excellent mixing. Evaporation takes place very fast, the cylinder charge will be directly cooled down homogeneously over a large cylinder volume. All of the water will be used for cooling in a highly efficient way. The temperature profile is overall lower and leads to lowest compression end temperatures. 3-D-CFD simulations and endoscopy video recording are being carried out to give more insight into these phenomena.

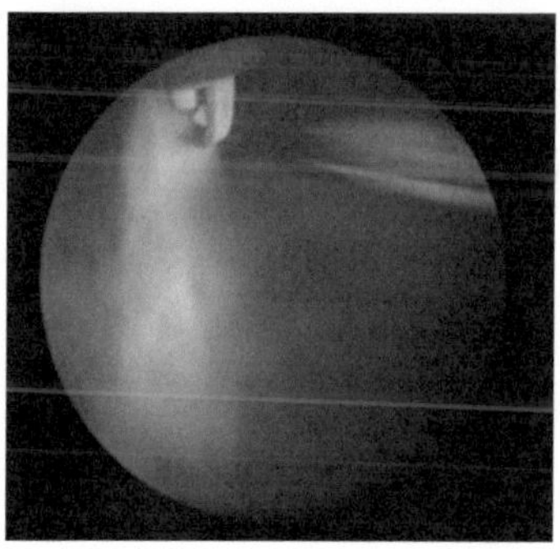

Fig. 22. Direct mixture injection

The handicap of direct mixture injection is the homogeneous mixture distribution to each cylinder. The experimental set-up has been described in Sect. 4.3. Within this concept, there will be turbulence within the mixing chamber, a downstream static mixer which is shearing the two fluids continuously over length and the following high-pressure pump. There is three times a mixing device, so a good emulsion of water/fuel will enter the fuel rail. As shown in Fig. 7, the emulsion is entering the 350 bar rail at one lateral end. It will first serve cylinder 4, then 3, 2 and 1 in sequence. The fuel rail volume has to achieve an equal concentration of water/fuel. There is also a specific flow field in the fuel rail depending on the load point and water/fuel-ratio. For the stationary experiments, the operating point will be run long enough for exchanging the fuel rail volume. Nevertheless, there will be a not homogeneous distribution and also the danger of emulsion separation. The experimental results for an operation point 5000 rpm/280 Nm confirm this assumption. For water rates higher than 15%, the distribution over the four cylinders is not homogeneously and does not follow a clear rule (Fig. 23). As a

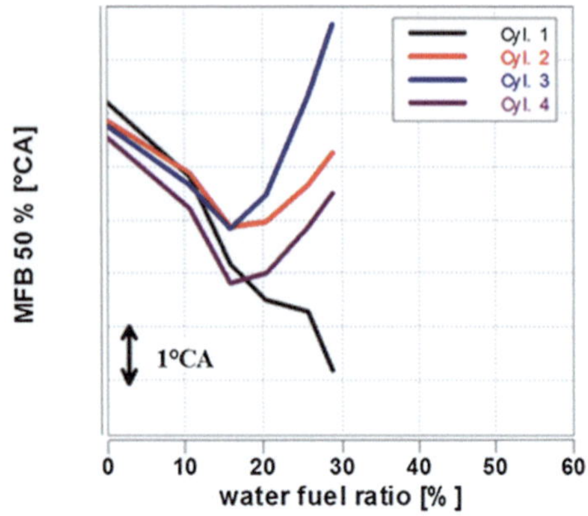

Fig. 23. Uneven distribution for direct mixture injection

conclusion, for mixture injection, piping after the high-pressure pump must have equal length and shortest way to each cylinder. This could improve the issue in order to equalize the cylinder-to-cylinder distribution. It will require a dedicated design of the high-pressure fuel rail for direct mixture injection [4].

Transient operations are even harder to fulfil. The residence time in the rail must be minimized in order to react to dynamic change demands in water/fuel composition. This could be accomplished by a small volume leading to the high-pressure injectors. On the other hand, pressure fluctuations caused by the high-pressure fuel pump and the injector injections must be compensated by a preferably large volume. This is a conflict of objectives. Therefore, a dedicated design for the high-pressure fuel rail is needed.

By calibration of the required water rate over the engine load/speed-map and a good system modelling of the water/fuel supply chain, the time constants for a certain water rate can be simulated and the lag times can be minimized. This results in a very high effort of calibration, modelling and simulation [4]. That is, of course accompanied by high costs.

6.4 Water Consumption for Stoichiometric Operation

From system perspective, the water shall be used highly efficient in order to minimize the water mass and thus the size of the vehicle onboard tank system. Results from measurements at the characteristic operating point 5000 rpm/280 Nm are shown in Fig. 24.

Plenum injection is water rate limited and cannot achieve $\lambda = 1$. It can only be combined with another injection system, e.g. port injection, to achieve this goal. No thermodynamic advantages are visible for plenum injection compared to port injection.

Port injection is capable to reach stoichiometric conditions, but with a relatively high water rate of 40%.

Thermodynamically, direct premixed water injection is by far the most efficient system [3]. More than 10% in water fuel ratio can be saved compared to port injection and $\lambda = 1$ is already achieved at 29% water rate.

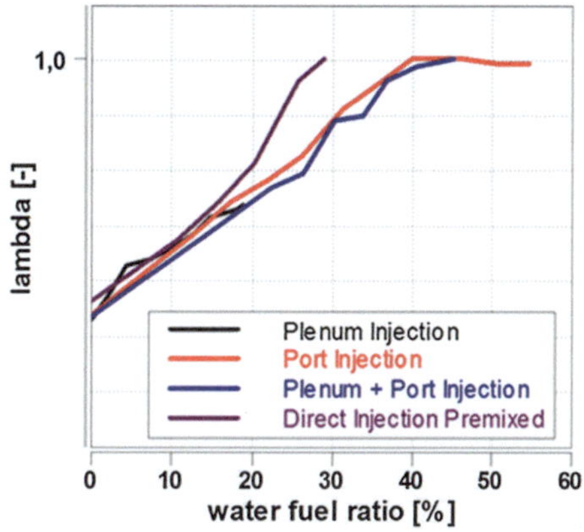

Fig. 24. Water consumption for the tested injection systems

6.5 Direct Separate Water Injection

Direct separate water injection into the cylinder has not been built-up and tested as injection concept within this study. Two high-pressure injectors in the cylinder roof are very volume consuming and very expensive to be integrated. A more feasible variant may be the combination of port fuel injection (MPFI) and high pressure (> 200 bar), direct water injection. This case will be simulated in 3-D-CFD and the results will be compared with the other injection systems. One big advantage of separate direct water injection is the possibility of late water injection timing within the compression stroke. This may unlock further potential to reduce the compression end temperature and further improve combustion phasing and efficiency.

From system perspective, there is one big disadvantage for direct water injection: within the cylinder roof, the water injector tip must be cooled in order to avoid thermal damage. That means, a minimum amount of water needs to be injected during each cycle, even when it would not be needed for the engine thermodynamically. This will reduce the water tank fill. In the extreme case, when the water tank is empty, a power reduction of the engine is required in order to avoid thermal overload of the injectors.

Compared to the direct injection of the mixture, the thermodynamic properties of direct injection into the cylinder are estimated to allow some kind of relative ranking. Thermodynamic properties should be similar, since water break-up by the high pressure is similar and the mixture formation can even be optimized by injection timing. Thermal efficiency of fuel PFI/water DI is assumed to be worse than premixed water DI because the evaporation enthalpy of the fuel is missing at the injection timing. Thus, more water would be needed for stoichiometric operation; also water consumption may be suffering due to thermal injector protection. These assumptions are verified by the 3-D-CFD simulations in the following chapter.

6.6 Evaluation of Water Injection Systems

All arguments pro and con for the different water injection systems are summarized in the table in Fig. 28. Results from 3-D-CFD simulations are shown within section cuts through the cylinder at ignition top dead center which are shown in Fig. 25.

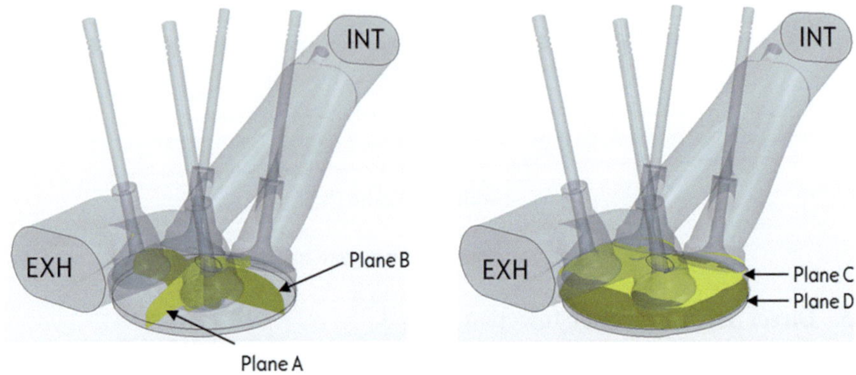

Fig. 25. Cylinder section cuts for 3-D-CFD contour plots

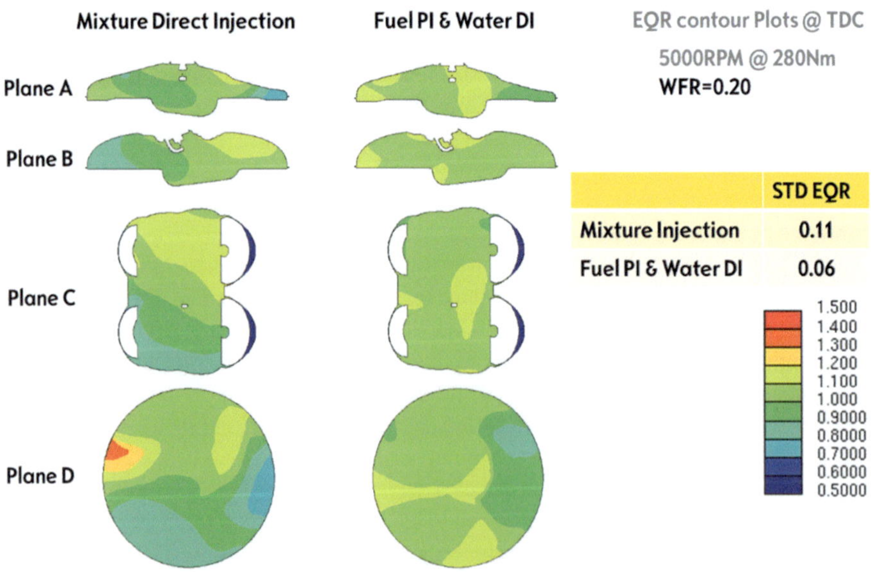

Fig. 26. Mixture formation by EQR contour plots @ TDC

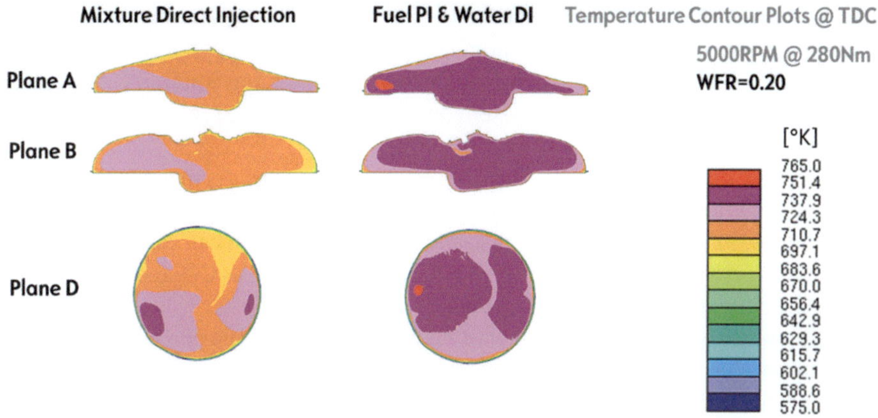

Fig. 27. Temperature distribution mixture versus separate direct water injection

Water Injection	Plenum	Port	Direct Injection of Premixed Water/Fuel	Direct Injection of Water Into Cylinder
Distribution to Cylinders	--	++	+	++
Distribution within Cylinders	--	+	++	++
Mixture Formation	--	+	++	++
Evaporation	-	+	++	+
Thermal Efficiency	--	+	++	+
Water Consumption for λ=1	--	o	++	+
Transient Operation	++	++	--	++
Integration Effort	++	++	--	--
Costs	++	++	--	--
Sum	-5	+12	+5	+7

Fig. 28. Water injection concept comparison

At the characteristic operation point 5000 rpm/280 Nm direct mixture injection is compared with separate direct injection (fuel PFI and water DI). Figure 26 shows the mixing formation exemplarily by contour plots for the standard deviation of the equivalence ratio EQR. The smaller the standard deviation is, the better it is. Separate direct injection exhibits a slight advantage by 0.05 in standard deviation. Injection timing can be optimized for direct separate water injection so that mixture preparation is almost ideal. In Fig. 27, the contour plots of the temperature for the two direct injection concept are shown. It is clearly visible, that mixture injection shows much lower compression end temperatures. Reason is the evaporation of both liquids, fuel and water, at the same injection timing. This fact combined with the excellent mixing explains the lowest water consumption for direct mixture injection.

Plenum injection does not achieve $\lambda = 1$ under all operating conditions. It is ruled out as stand-alone water injection system and could only be used in combination with another injection system.

Port water injection does achieve stoichiometric conditions. As mentioned, mixture formation and thermal efficiency are not optimal; thus higher water rates are required. Nevertheless, integration effort and costs are low. Port injection can easily be adapted to an existing engine architecture. Both $\lambda = 1$ operation as well as CO_2 reduction can be gained in addition to the engine performance.

Direct water injection systems exhibit the best thermodynamic performance. Water consumption can be drastically reduced, especially for the mixture injection. This is of course not for free; direct water injection is closely coupled with the engine development and requires much higher integration and cost effort.

7 Water Injection – Quo Vadis?

Water injection is gaining increased attention within engine engineering of all major OEMs and engineering service providers. Fuel enrichment can be replaced by water injection for component protection which is decreasing the knock tendency and leads to better combustion phasing and fuel efficiency.

There is a trend for higher specific engine powers, a value of 150 kW/l may be stated here. It is desirable to achieve this with the highest degree of commonality with existing engine architectures. Port water injection is the favorable add-on technology to reach this power density without knocking, fuel enrichment and acceptable fuel efficiency. Currently, it is a niche application and the BMW M4 Moto GP Safety Car can serve as the most prominent example for it.

Progressing with time, water injection will be seen up to 2020 mainly in the premium segment; premium vehicles mainly benefit from water injection and have the best capability to cope with the additional costs. In this context, direct water injection may be developed more to profit from the improved water efficiency. It will probably last another five years before water injection enters the mainstream mass market. The technology will be developed by then with reliability and acceptable costs to compete in the mass market with other technologies like VGT (Variable Geometry Turbine), cooled external EGR (Exhaust Gas Recirculation) and VCR (Variable Compression Ratio) for compliance with CO_2 reduction and RDE.

References

1. Pauer, T., Frohnmaier, M., Walther, J., Schenk, P., Hettinger, A., Kampmann, S.: Optimization of gasoline engines by water injection. In: Robert Bosch GmbH, Schwieberdingen, 37. International Engine Congress Vienna (2016)
2. Durst, B., Landerl, C., Poggel, J., Schwarz, C., Kleczka, W., Hußmann, B.: BMW water injection: initial experience and future potentials. In: 38 International Engine Congress Vienna (2017)
3. Durst, B., Unterweger, G., Rubbert, S., Witt, A., Böhm, B.: Thermodynamic effects of water injection on otto cycle engines with different water injection systems. In: BMW AG, 15th Conference "The Working Process of the Internal Combustion Engine", Graz (2015)
4. Böhm, M., Mährle, W., Bartelt, C., Rubbert, S.: Functional Integration of Water Injection into the Engine. BMW AG, MTZ, January 2016

Gasoline Water Direct Injection (GWDI) as a Key Feature for Future Gasoline Engines

Christoph Heinrich(✉), Heinrich Dörksen, Andreas Esch, and Kevin Krämer

Department of Mechanical Engineering, Internal Combustion Engines, Trier University of Applied Sciences, Schneidershof, 54293 Trier, Germany
heinrich@hochschule-trier.de

Abstract. Water injection into internal combustion engines has long been known as a powerful tool to increase either maximum power and/or decrease emissions. Up to now, product costs and application effort have prevented a broad serial introduction of this technology. Competition with electric drive technology and tighter emission legislation make this feature more attractive for serial application in gasoline engines. Therefore, the different benefits of water injection are evaluated in this paper. The direct injection of water into the combustion chamber is realized through the use of an emulsion with a variable water-fuel ratio. Gasoline water direct injection (GWDI) results are contrasted with data obtained by water injection into the intake pipe after the intercooler. Furthermore, investigations regarding knock limited combustion and mitigation of fuel enrichment were performed on two engines from different OEMs, both with turbocharging and direct injection. The study is completed by a detailed testing of the impact of water injection on stochastic (low speed) pre-ignitions (SPI).

1 Introduction

Despite some advances achieved in the area of electric mobility and battery technology internal combustion engines will remain the dominant source of propulsion power for passenger cars in the near future. Considering the tremendous efforts to make diesel engines compliant with upcoming emission legislations (i.e. real driving emissions, RDE) the spotlight is back on gasoline engines, which are superior to diesel engines from an emission perspective due to their clean, stoichiometric, premixed combustion and well established three-way catalyst (TWC) technology. Besides or in combination with other technologies, such as variable compression ratio (VCR), Miller cycling, and cooled EGR, water injection may be a key feature to overcome the fuel consumption deficiencies of gasoline engines.

1.1 Motivation

First applications of water injection were used in the first half of the 20th century in aircraft engines for temporary power increase. Soon after in 1945, it was investigated with regard to efficiency by NACA [1]. Power increase was the driver for applications in racing (Formula 1) in the early eighties. Also the first turbo serial application (SAAB 99)

M. Günther and M. Sens (eds.), *Knocking in Gasoline Engines*,
https://doi.org/10.1007/978-3-319-69760-4_19

had water injection as an option. Currently BMW features a limited serial application (M4 GTS). The list of applications, especially focusing on racing, can be lengthened as required.

Most of these applications were mainly focused on increased power by charge air cooling and knock improvement. Nevertheless, the relevant effects can also be used for efficiency increase. Against the background of tighter CO_2 legislation [2], newer investigations focusing on efficiency and fuel consumption benefits are established. A very good historical overview is given by [3].

The main benefits of water injection into gasoline engines can be highlighted as follows:

1. The charge air and the combustion chamber are cooled by the water's vaporization enthalpy which is several times higher than the fuel's latent heat. On the one hand the knock tendency is retarded by the cooling in favor of the combustion efficiency in the medium and upper load area. On the other hand, a decreased knock tendency could be used for an increase of the overall engine compression ratio.
2. Water injection has the potential to minimize or even completely prevent fuel enrichment which opens the possibility for stoichiometric combustion in all operation modes. Besides fuel reduction full functionality of the emission system (TWC) is achieved which is crucial for the upcoming RDE legislation, subject to high space velocity adapted catalyst volumes.
3. Even if all triggering mechanisms for stochastic (low speed) pre-ignitions (SPI) are not yet fully understood it is expected that water injection will lead to decreasing pre-ignition tendency.

Detailed investigations focused on these three items are carried out in this paper to quantify the benefits expected. Furthermore, vaporized water as inert gas has the same effect like cooled EGR. Besides decreased knock tendency lower peak temperatures and less wall heat losses can be expected. This topic is excluded here due to the more analytic character of this research and is planned to be carried out in a later study.

2 Experimental Setup

The engine laboratory of Trier University is equipped with standard instrumentation for engine research. In-cylinder pressure indication, performance measurement, and emission analyzing tools are completing the picture.

2.1 Engine Specification

Two engines of different OEMs but with comparable technology were used for this investigation. While SPIs (item 3, see above) are investigated on one engine only, a 4-cylinder inline Turbo DI engine, mitigation of knock tendency and fuel enrichment (item 1 and 2 resp., see above) are also studied in addition on a 6-cylinder turbo DI engine from another manufacturer. The reason for this is that different engines do not necessarily show the same behavior because combustion also depends on several contribution factors like charge motion and overall combustion system design.

Characteristics of the two engines are shown in Table 1:

Table 1. Technical data of the engines used

Engine		1	2
Cylinder		4 cylinder inline	6 cylinder inline
Bore	mm	79	84
Stroke	mm	81.5	89.6
Displacement	cm^3	1598	2979
CR		9.5 (10.5)	10.2
Injection		central, 6 hole	piezo
Injection pressure	bar	200	200
VCP		yes	yes
Turbo charged		yes	yes
Max torque	Nm	260 @ 1700–4700/min	400 @ 1300–5000/min
Max power	kW	147 @ 4700/min	225 @ 5800/min

Both engines are equipped with an open access to the ECU to control all parameters needed.

2.2 Water Injection Devices

The effect of water injection is investigated in two ways: Single water injection into the air intake pipe after the intercooler but before the throttle valve (low pressure injection, LPI) and directly as an emulsion of fuel and water by the high pressure fuel rail (gasoline water direct injection (GWDI)).

Low Pressure Injection (LPI)

Water injection into the air intake pipe is performed by a system using pressures between 4 and 8 bar depending on the amount of water requested. Therefore, this way of water dosing is called low pressure injection (LPI) in the following. Since dosing takes place well before the intake ports any control by crank angle timing is not needed. A continuous injection in steady state condition takes place instead. The general setup is shown in Fig. 1:

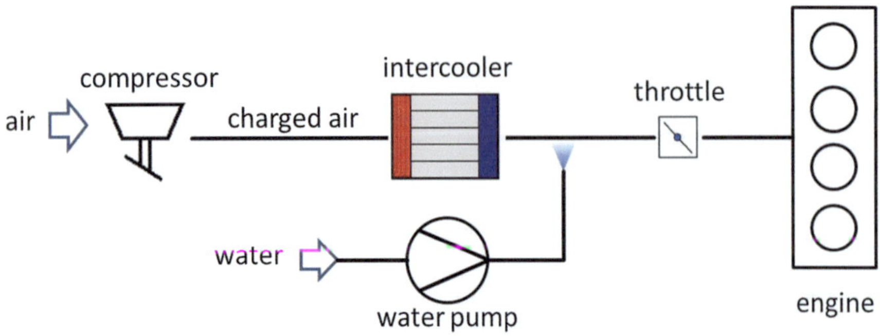

Fig. 1. Setup of low pressure injection (LPI)

The amount of water injection is controlled by pressure variation and nozzle adaptation.

Gasoline Water Direct Injection (GWDI)

In this investigation, direct water injection is performed as a water-fuel emulsion. Some authors prefer separate direct injection of water by an additional lateral injector, see for example [4] and other papers from this group, e.g. [8]. Due to packaging reasons this idea was discarded. Furthermore, direct injection of a water-fuel emulsion leads to better homogenization of the mixture by means of so called micro (steam) explosions [1]. The emulsion approach is also confirmed by results of [9], where the different injection techniques are directly compared.

In fact, the engine laboratory of Trier University has gained comprehensive experience in the lab-scale production of water-fuel emulsions for over a decade. Former investigations were focused on water-fuel emulsions for diesel engines [5–7]. The objective with regard to diesel was focused more on the simultaneous reduction of nitrogen oxides (NOx) and soot than on fuel consumption. Applications using this technology have been realized in inland waterway craft diesel engines, whereas water-diesel emulsion testing for passenger cars struggles with high injection pressure and cavitation effects. Cavitation of gasoline-water emulsions is not expected and has not yet been observed on our engines due to the much lower boiling curve of gasoline which starts below the boiling point of water.

Water-fuel emulsions in general can be produced in two ways. Water and fuel can be mixed by addition of tensides to a length scale of phase domain of 10 nm. These emulsions are called micro emulsions because the length scale is below the wavelength of the visible region and they appear as clear as normal gasoline. Due to the tensides the micro-emulsions are time- and temperature-resistant. These micro emulsions are helpful for lab-scale testing bearing in mind that heat value correction is mandatory. Anyway, micro emulsions are not considered for serial application merely due to cost reasons.

The second way to generate suitable water-fuel emulsions is by the application of pressure and high shear stresses to mix both phases. This leads to short-term time-resistant emulsions of a bigger length scale (~ 10 μm), so called macro emulsions. Macro emulsions appear as a milky opaque liquid, see Fig. 2b. The phases gradually separate over time (in a time frame of minutes) and revert back to two phases, see Fig. 2c. Macro emulsions for this work are generated outside of the engine by means of a modified device derived from the diesel emulsion process, see Fig. 2a.

The serial injection system of the engine is then fed by both types of emulsions. During the investigations no issues concerning the injection system were observed.

Serial applications require emulsion generation very close to the injection nozzle benefiting from the high pressure of today's gasoline direct injection systems. Only a very small dead volume may be allowed due to the fast response to load changes. The emulsion generation will rely on the exact mass dosing of water and fuel. Measures against water residuals after engine stop will have to be taken by defined shut down

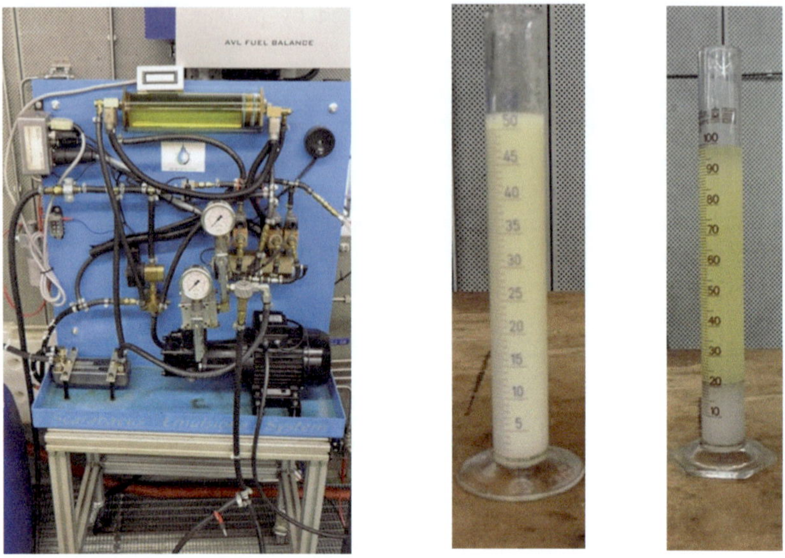

a) b) c)

Fig. 2. Emulsion device (a), macro emulsion (b), and residuals after separation (c)

procedures. Also precautions against the freezing of water storage and supply pipes have to be installed but these are beyond the scope of this investigation.

3 Experimental Results

The experimental testing is focused on the three benefits listed above in Sect. 1.1. As already mentioned, the mitigation of knock tendency and fuel enrichment is tested on the two engines, while the impact on stochastic pre-ignitions is tested only on the 4-cylinder engine.

3.1 Impact on Knock Tendency

The effect of water injection on the knock tendency is based on the cooling of the charged air and, in case of direct water injection, on the direct cooling of the combustion chamber by the vaporization enthalpy of the injected water. On the downside, the vaporized water is an additional inert gas and may lead to an increase of the combustion duration due to the reduced flame speed. So the trade-off between the two effects has to be evaluated and is expected to be engine specific. Therefore, literature does not give a uniform picture of the efficiency gain due to the knock mitigation. In [8], a fuel consumption reduction up to 3.5% is evaluated. In other papers the pure anti-knock benefit is overlapped by other effects like the mitigation of the fuel enrichment [9].

Therefore, investigations were performed on two engines available at the engine laboratory of Trier University, see Table 1. As a preliminary study water addition in the form of LPI was performed on the 6-cylinder engine described in Table 1. The engine was run WOT at 4000 rpm and the air fuel equivalence ratio was first set to stoichiometric ($\lambda = 1$) to inhibit effects of fuel enrichment. As shown in Fig. 3, the anti-knock benefit lead to a gain in the brake specific fuel consumption of up to 2% at a water-fuel ratio of nearly 20%. The ignition timing could start earlier by 3° crank angle.

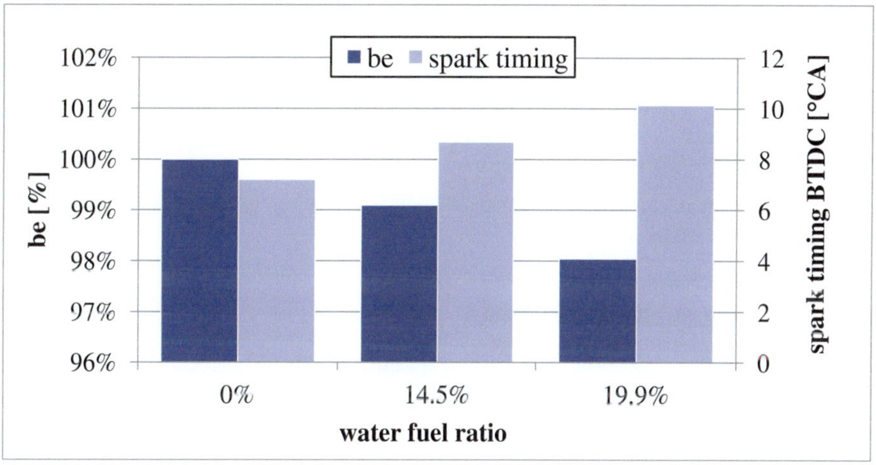

Fig. 3. Specific fuel consumption (b_e) and ignition timing with LPI on the 6-cyl. engine (WOT@4000 rpm, stoichiometric operation)

Unfortunately, GWDI was not successful on the 6-cylinder engine due to oil dilution effects after a short run time.

On the 4-cylinder engine, the anti-knock benefit has to be separated from the fuel enrichment reduction benefit. This engine is not significantly knock limited until the fuel enrichment ($\lambda < 1$) area starts to protect exhaust components. Therefore, investigations on the 4-cylinder engine were controlled by the exhaust temperature in front of the turbine. First, fuel enrichment was reduced and afterwards ignition timing was adjusted. As it will be outlined in Sect. 3.2, fuel saving by reduction of enrichment is superior to anti-knock benefit. Figure 4 illustrates both effects:

Again, the pure spark advance benefit seems to be around 2 to 3%, which is similar to the one on the 6-cylinder engine. Nevertheless, it is worth to mention that other engines with knock limitation in the stoichiometric area might show higher anti-knock gain due to water injection.

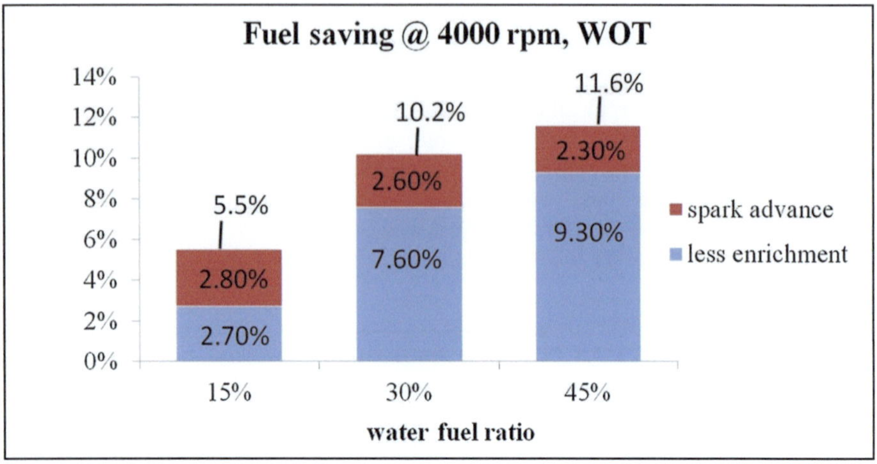

Fig. 4. Fuel saving potential divided in reduction of enrichment (blue) and spark advance/others (red) on the 4-cyl. engine with GWDI

3.2 Mitigation of Fuel Enrichment

The ability of water injection to mitigate fuel enrichment might be the most valuable contribution of this technology to future gasoline engine sustainability. The preliminary study on the 6-cylinder engine demonstrated a fuel saving of 8.3% reached with a water-fuel ratio (LPI) of 19.9% (Fig. 5). In Fig. 5, the base configuration (water-fuel ratio 0%) is tested with the equivalence ratio of base calibration. This is in contrast to test data shown in Fig. 3, where the base configuration (water-fuel ratio 0%) is stoichiometric for all points to work out only the anti-knock benefit.

To investigate this potential systematically with variable water-fuel ratios, several full load points of the 4-cylinder engine are tested with both low-pressure (LPI) and direct injection (GWDI). Since the engine was run with a compression ratio that was one unit higher than in the serial application, an additional safety distance to temperature limits of exhaust parts led to even lower equivalence ratios. The temperature limit of the exhaust gas operates as a control variable and was kept constant during all tests. First, the engine was run with base settings and no water injection. Then water injection was increased and fuel enrichment decreased until the maximum allowed exhaust temperature was reached. Subsequently the ignition was advanced until the knock limit was reached.

Low Pressure Injection (LPI) Results

Because of the simple setup and continuous injection, the water-fuel ratio for the low pressure injection could easily be varied in steps of 0, 15, 30, 45% through pressure adjustment and nozzle variation. After the start of water injection the equivalence ratio was increased in each case until the exhaust temperature reached its limit. Afterwards the ignition timing was advanced until knock limit. The results are shown in Fig. 6. As it can be seen, fuel consumption benefits of 6 to 12% are reached.

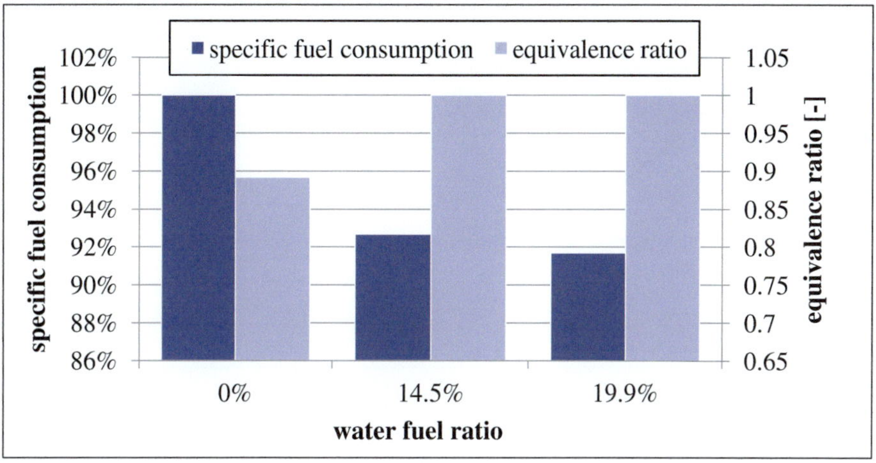

Fig. 5. LPI: Specific fuel consumption benefit (b_e) and equivalence ratio (λ) over water-fuel ratio on 6-cyl. engine (WOT@4000 rpm)

As already pointed out in Sect. 3.1, the predominant part of the savings is due to a decreased fuel enrichment. Figure 7 presents the corresponding equivalence ratio.

Through the analysis of the heat release it can be pointed out that the first step of the water injection (15%) leads to a significant improvement of the spark advance, while the 30%- and 45%-points are nearly negligible with regard to the spark timing improvement. At 4750 and 5500 rpm and 45% water-fuel ratio, the combustion period is extended so far that the effect of the reduced enrichment is overcompensated. This leads to a fuel consumption increase from 30 to 45% water-fuel ratio at these operating points (Fig. 6). In general, it can be stated that over all engine speeds the maximum benefit of low pressure water injection is already reached at a water-fuel ratio of 30% for this engine configuration.

Gasoline Water Direct Injection (GWDI) Results

As mentioned above, GWDI was performed in two ways: by the use of macro and micro emulsions.

Micro Emulsions

First results were performed with micro emulsions following recipes for a water-fuel ratio of 11%. As mentioned above, a heat value correction has to be carried out because tensides are mainly composed of hydrocarbons themselves. Tests with these micro emulsions lead to corresponding results (see Fig. 8) as with macro emulsions (except for the result with an engine speed of 5500 rpm which seems to be an outlier). It should be noted that pretesting with gasoline-tensides mix only lead to a higher knock tendency than pure gasoline. This yields the conclusion that the tensides used here might not be helpful for further spark advance. Since results with micro emulsions at 11% water-fuel ratio fit perfectly to those achieved with macro emulsions (except for the 5500 rpm operating point) further testing with micro emulsions was superfluous. Once again it is worth mentioning that micro emulsions are not taken into consideration for

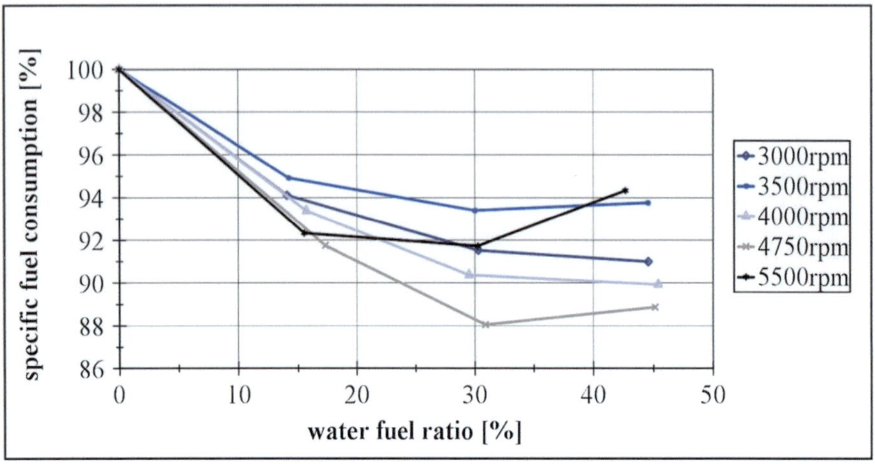

Fig. 6. LPI: Specific fuel consumption benefit (b_e) over water-fuel ratio on 4-cyl. engine (WOT) [15]

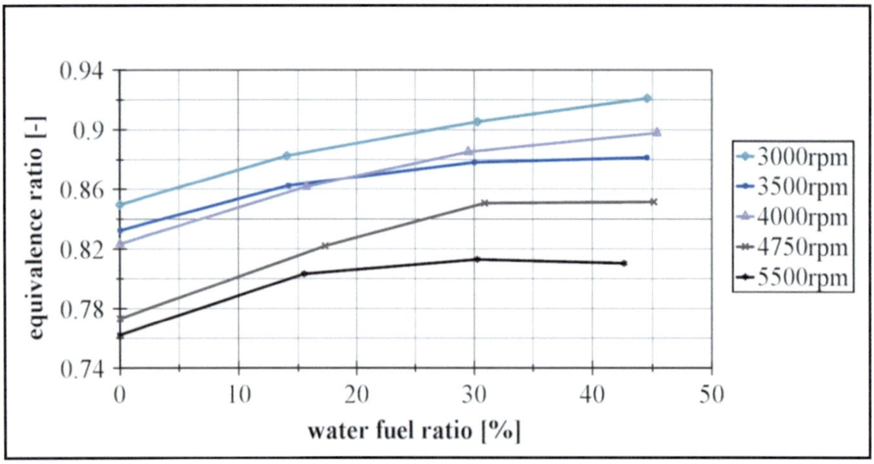

Fig. 7. LPI: Air fuel equivalence ratio (λ) over water-fuel ratio on 4-cyl. engine (WOT) [15]

serial application due to costs reasons but can be used for initial testing if no hardware for mechanical production of macro emulsions is available.

Macro Emulsions

Testing with macro emulsions lead to the results also shown in Fig. 8. Like the results with LPI, the water-fuel ratio is varied from 0, 15, 30 to 45%. First of all, results at an engine speed of 5500 rpm seem to be an outlier. The fuel saving potential is not in range of the other engine speeds. Apart any measurement errors this might be due to a too short time frame for homogenization. Anyway, at the other engine speeds testing with macro emulsions shows a fuel saving potential up to 12%. Since highest fuel

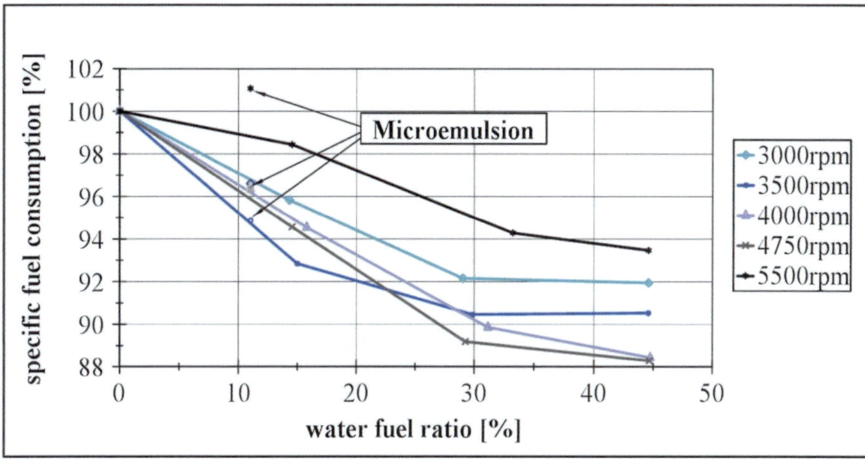

Fig. 8. GWDI: Specific fuel consumption benefit (b_e) over water-fuel ratio on 4-cyl. engine (WOT) [15]

savings can be found at a water-fuel ratio of 45% it can be assumed that even higher water fuel ratios might lead to additional savings which is also stated in [9]. The maximum water-fuel ratio of 45% is limited by the separate chamber volumes of the emulsion device shown above (Fig. 2).

Figure 9 shows the corresponding curves for the increasing equivalence ratio over water-fuel ratio. It is conspicuous that the fuel enrichment decrease, i.e. the increase of the equivalence ratio ($\Delta\lambda \sim 0.09$.) is achieved by a disproportionately high amount of water injection (water-fuel ratio of 0.45). This is remarkable because the evaporation enthalpy of water is several times higher than the one of fuel. A detailed in-cylinder heat balance supported by simulation would bring more clarification.

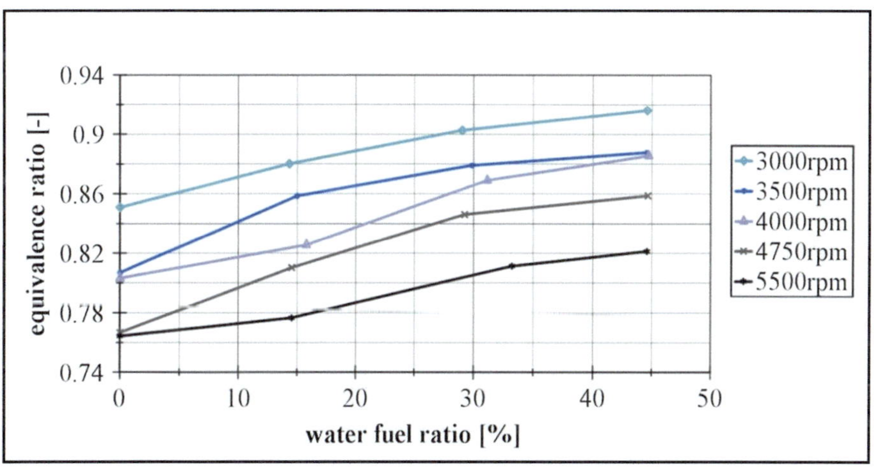

Fig. 9. GWDI: Air fuel equivalence ratio (λ) over water-fuel ratio on 4-cyl. engine (WOT) [15]

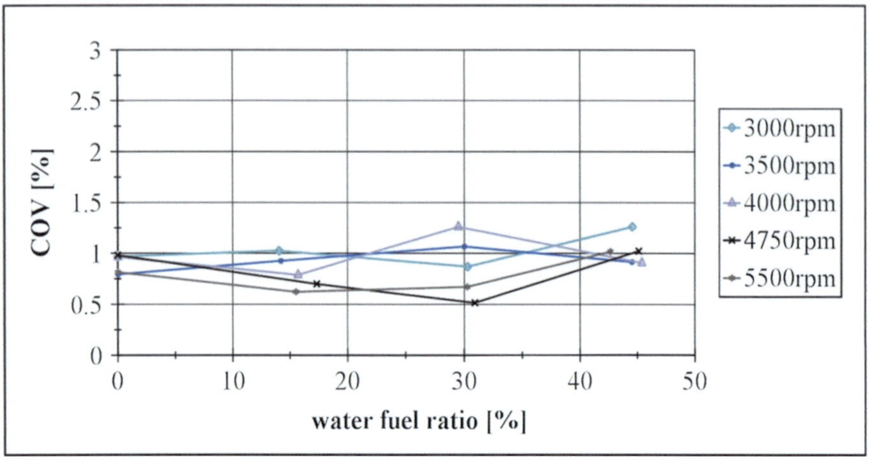

Fig. 10. GWDI: COV over water-fuel ratio [15]

Furthermore, it is remarkable that combustion stability seems to be unaffected over the whole range of the water-fuel ratio. Even when injecting a water-fuel ratio of 45% no significant increase of the COV (covariance of $IMEP_{720}$) is detectable, see Fig. 10. The equivalence ratio increase might lead to an increasing combustion stability here. Nevertheless, it is noteworthy that COV even with high water-fuel ratio is considerably below the limit for serial application release.

3.3 Impact on Stochastic Pre-ignitions (SPI)

In recent years, major steps towards a competitive CO_2 reduction of gasoline engines compared to other propulsion systems have been made by engine downsizing. Its potential is limited by pre-ignition phenomena which seem to occur randomly at full load and low engine speed. In literature these events are called stochastic (low speed) pre-ignitions (SPI or LSPI). SPI can occur rarely but massively which impedes the development of a safety clearance. In many publications liquid droplets are suspected to initialize SPI, see [10–12]. In [13] a detailed study concludes that solid particles are the trigger for SPI which moreover delivers a comprehensible theory for alternating occurrences. Nevertheless, the following properties are considered as promoting factors for SPI:

- high (effective) compression ratio
- high BMEP at low engine speed
- late ignition
- higher amount of slow-evaporating hydrocarbon compounds.

Even if root causes and mechanisms to trigger pre-ignitions are not yet fully understood, the cooling effect of water injection can be assumed to prevent the SPI occurrence. Consequently, investigations are carried out to determine the impact of water injection on the occurrence of SPI. The 4-cylinder engine used before is equipped with a compression ratio that is one unit higher than the serial application to increase the number of SPI occurrences.

The procedure used in this investigation is based on the approach used in [13]. One test comprises three sections of high load running at low speed (2000 rpm) for 15 min each, and the sections are interrupted by conditioning phases (5 min idling), see Fig. 11. Using a 4-cylinder engine we accumulate 180.000 combustion cycles per test in which SPI might occur. The brake mean effective pressure (BMEP) was set to 20 bar which leads to an acceptable number of SPI bearing in mind the higher compression ratio. This turned out to be a good compromise between the incidence of SPI events and the safety clearance to hardware damage.

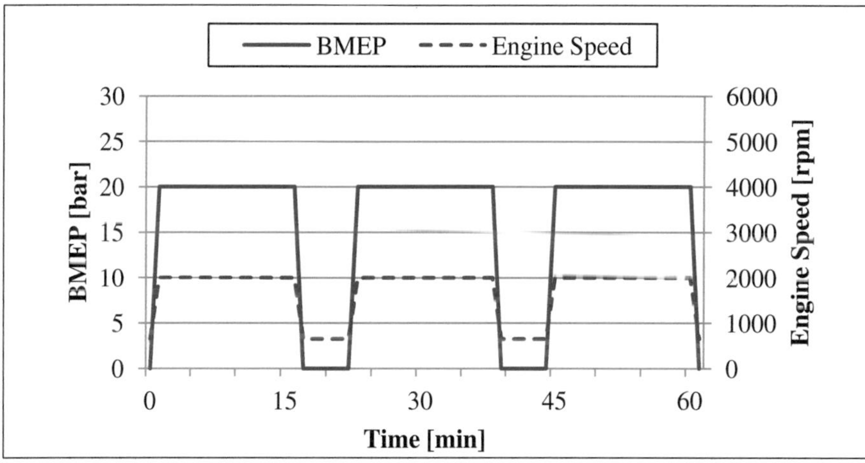

Fig. 11. Procedure for SPI testing

It should be noted that there is no unique definition of a SPI event in literature. Some authors define a fixed pressure threshold (e.g. 105 bar [12] or 130 bar [14]) in combination with an earlier pressure increase. Others use a standard deviation approach for the peak pressure and 2%-heat release [13]. In [14] the maximum pressure rise (threshold: 280% of the NVH target) and the peak to peak knock amplitude (threshold: 830% of the permitted knock amplitude) are also taken into account.

In this investigation, combustion abnormalities are counted as SPI if the indicated pressure curve reports an earlier combustion as spark timing implies and a peak pressure of 120 bar is exceeded. As it turned out, both conditions only occurred together.

Five configurations were tested intensively:

- The base configuration with an equivalence ratio as given in the base calibration ($\lambda = 0.94$) and without any water injection.
- LPI (water-fuel ratio of 45%) and equivalence ratio as given ($\lambda = 0.94$).
- LPI (water-fuel ratio of 45%) and stoichiometric operation ($\lambda = 1$).
- GWDI (water-fuel ratio of 45%) and equivalence ratio as given ($\lambda = 0.94$).
- GWDI (water-fuel ratio of 45%) and stoichiometric operation ($\lambda = 1$).

Since the statistical approach is fairly weak – only a handful of events occur per 180.000 cycles – every configuration was run tree times on different days. The results are illustrated in Fig. 12:

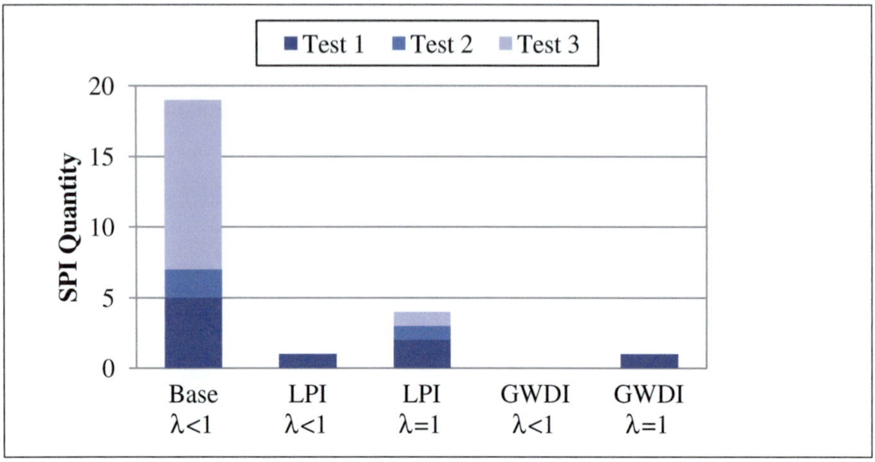

Fig. 12. SPI quantity of test configurations [16]

As expected the base configuration led to the highest count of SPI. Low pressure injection (LPI) reduced SPI occurrence significantly: only one SPI-event was observed in three full tests with enrichment as base configuration. The same configuration under stoichiometric ($\lambda = 1$) conditions led to four events in three tests. With regard to direct water injection it has to be noted that GWDI was able to nearly fully suppress SPI events. No SPI event at all occurred with base enrichment and only one with

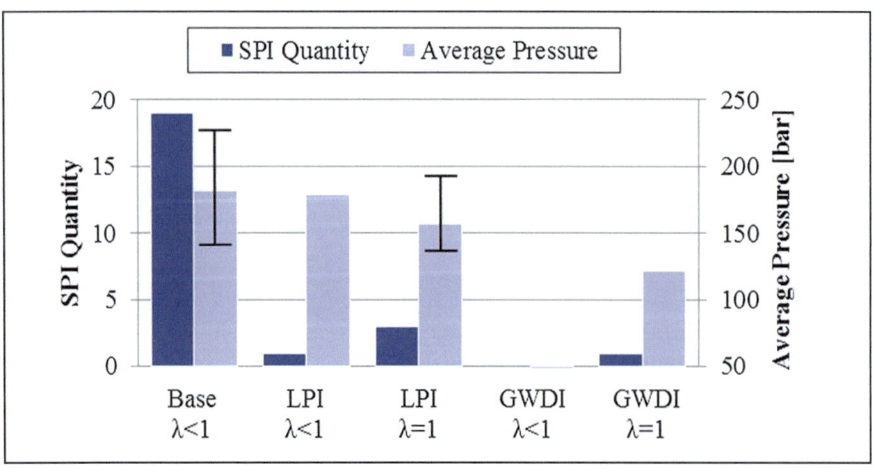

Fig. 13. Peak in-cylinder pressure during SPI events

stoichiometric operation. Furthermore, it is noteworthy that not only the counts but also the peak pressures of SPI cycles were considerably decreasing with water injection (see Fig. 13). This means that the risk of severe engine damage is nearly eliminated through the use of GWDI.

4 Summary and Outlook

Several benefits of water injection into gasoline engines have been demonstrated. The anti-knock impact on fuel consumption is engine related and depends on many influencing factors. In general, there is a trade-off between the earlier ignition and the decreasing flame speed due to the additional inert gaseous water. On the engines tested a fuel consumption benefit of 2 to 3% was observed while other authors reported gains up to 3.5% [4]. On the other hand, water injection can enable engine design with a higher compression ratio which leads to fuel savings throughout the operation map.

The most valuable contribution to future gasoline engines might be the potential to run engines with stoichiometric operation mode even in the high and full load area. Fuel enrichment can be avoided by water injection. Apart from the fuel consumption benefit, a stoichiometric exhaust gas can nearly completely be cleaned by state-of-the-art TWC technology which offers significant advantages with regard to the upcoming RDE legislation.

Furthermore, direct water injection is able to eliminate SPI issues by minimizing SPI occurrence and peak pressure values as shown.

None of the three benefits alone may justify the introduction of a gasoline water direct injection (GWDI) system but the sum of all benefits makes it a key feature for future gasoline engines.

A future water injection system for serial application needs to be developed considering open questions like anti-freeze protection, necessary level of water purity, range of water tank, etc. Nevertheless, from a technical point of view these issues can be resolved. Furthermore, a gasoline water direct injection system (GWDI) will need to produce the required water-fuel emulsion on time, and very close on or in the injector while benefiting from the high injection pressure. Hence future work will focus on so-called in-injector systems (Fig. 14):

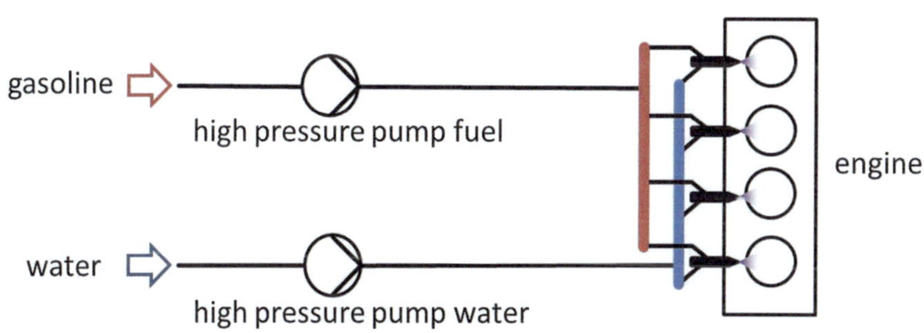

Fig. 14. In-Injector system for fuel water emulsion

Abbreviations

BMEP	Brake mean effective pressure
BTDC	Before top dead center
COV	Covariance of $IMEP_{720}$
CR	Compression ratio
ECU	Engine control unit
EGR	Exhaust gas recirculation
GWDI	Gasoline water direct injection
$IMEP_{720}$	Indicated mean effective pressure (720 CA)
LPI	Low pressure injection (air induction)
NVH	Noise, vibration and harshness
RDE	Real driving emissions
SPI	Stochastic pre-ignition(s)
TWC	Three-way catalyst
VCP	Variable cam phasing
VCR	Variable compression ratio
WOT	Wide open throttle

References

1. Vandeman, J., Heinicke, O.: Effect of water alcohol injection and maximum-economy spark advance on knock limited performance and fuel economy of a large air-cooled cylinder. NACA (National Advisory Committee for Aeronautics) (1945)
2. Regulation (EC) No 443/2009 of the European Parliament and of the Council of 23 April 2009
3. Rohi, A., Satpathy, S., Choi, J., Hoard, J., et al.: Literature survey of water injection benefits on boosted spark ignited engines. SAE Technical Paper, 2017-01-0658 (2017)
4. Thewes, M., Baumgarten, H., Scharf, J., Birmes, G., Balazs, A., Lehrheuer, B., Hoppe, F.: Water injection – high power and high efficiency combined. In: 25th Aachen Colloquium Automobile (2016)
5. Simon, C., Dörksen, H., Will, B., Mengel, C.: Generation and injection of diesel-water emulsions. MTZ Worldw. **71**(7–8), 46–52 (2010)
6. Simon, C., Dörksen, H., Dornbusch, H.: Pollution reduction by diesel-water emulsions. MTZ Worlw. **74**(1), 48–53 (2013)
7. Dittmann, P., Dörksen, H., Steiding, D., Kremer, F.: Influence of micro emulsions on diesel engine combustion. MTZ Worlw. **76**(4), 38–44 (2015)
8. Thewes, M., Birmes, G., Baumgarten, H., Seibel, J.: Water injection for gasoline combustion systems. MTZ Worldw. **76**(2), 10–14 (2015)
9. Durst, B., Unterweger, G., Rubbert, S., Witt, A., Böhm, M.: Thermodynamic effects of water injection on Otto Cycle engines with different water injection systems. In: 15th Conference on the Working Process of the Internal Combustion Engine, September 2015
10. Dahnz, C., Han, K., Spicher, U., Magar, M., Schießl, R., Maas, U.: Investigations on pre-ignition in highly supercharged SI engines. SAE International, 2010-01-0355 (2010)

11. Willand, J., Daniel, M., Montefrancesco, E., Geringer, B., Hofmann, P., Kieberger, M.: Limits on downsizing in spark ignition engines due to pre-ignition. MTZ Worldw. **70**(5), 56–61 (2009)
12. Zahdeh, A., Rothenberger, P., Nguyen, W., Anbarasu, M., Schmuck-Soldan, S., Schaefer, J., Goebel, T.: Fundamental approach to investigate pre-ignition in boosted SI engines. SAE International, 2011-01-0340 (2011)
13. Döhler, A., Pritze, S.: A contribution to better understanding the pre-ignition phenomenon in highly charged internal combustion engines with direct fuel injection. In: 4th International Conference on Knocking in Gasoline Engines, Berlin (2013)
14. Haenel, P., Seyfried, P., Kleeberg, H., Tomazic, D.: Systematic approach to analyze and characterize pre-ignition events in turbocharged direct-injected gasoline engines. SAE International, 2011-01-0343 (2011)
15. Heinrich, C., Dörksen, H., Tölkes, E., Esch, A.: Gasoline water direct injection: fuel saving potential and technical requirements (Direkte Benzin-Wasser-Einspritzung: CO_2-Potentiale und technische Anforderungen). In: 10th Conference Diesel and Gasoline Direct Injection, Berlin (2016)
16. Heinrich, C., Dörksen, H., Krämer, K., Carls, M.: The Impact of Gasoline Water Direct Injection on the Stochastic Pre-Ignitions in Spark Ignited Engines (Einfluss der direkten Benzin-Wasser-Einspritzung auf die stochastischen Vorentflammungen eines Ottomotors). 28. Deutscher Flammentag, Darmstadt (2017)

Potential of Direct Water Injection to Reduce Knocking and Increase the Efficiency of Gasoline Engines

Matthias Hunger[1], Tobias Böcking[2(✉)], Ulrich Walther[1], Michael Günther[1], Normann Freisinger[2], and Günter Karl[2]

[1] IAV, Berlin, Germany
[2] Daimler AG, Stuttgart, Germany
tobias.boecking@daimler.com

Abstract. Given the need to further reduce CO_2 s from passenger cars powered by internal combustion engines and in the light of future tighter exhaust emission regulations, further improvements are necessary to enhance the efficiency and emission behavior of gasoline engines. One major constraint encountered in the gasoline engine is knocking combustion. Its negative effects on efficiency and emission behavior can be counteracted by reducing knock tendency.

One approach to reducing knocking tendency lies in injecting water. This is capable of extracting heat from the cylinder charge and, on this basis, reduce final compression temperature. It also increases charge mass in the cylinder and ultimately improves efficiency as a result of the lower tendency to knock.

With a view to implementing water injection, attention initially turns to the thermodynamic aspects of water injection in the gasoline engine and to comparing and evaluating the water manifold and direct water injection concepts. With the aim of conducting fundamental studies on a test engine with direct water injection, discourse then presents the design-related aspects involved in the implementation of such. The engine tests conducted focus on identifying the main water injection parameters influencing the gasoline engine's working cycle. Selected test results provide the basis for illustrating them and presenting the key underlying mechanisms. The potential for increasing the compression ratio can be determined in relation to the effect of water injection on reducing knock. The resultant higher level of efficiency makes it possible to reduce the emission of CO_2.

Finally, consideration is given to the challenges involved in injecting water, with further concept approaches being presented for water injection.

1 Introduction

The EU Commission's ambitious targets for reducing CO_2 emission will place far higher demands on tomorrow's drive concepts. With today's market penetration of gasoline engine cars, the planned CO_2 fleet emission of 95 g CO_2/km will call for much greater steps to increase efficiency, especially for this drive concept.

In the case of the gasoline engine, the particular focus is on improving part-load efficiency as a way of significantly reducing drive cycle consumption with a view to meeting RDE emission legislation. In the course of introducing RDE legislation, the cycle-relevant load population will shift towards higher engine load.

M. Günther and M. Sens (eds.), *Knocking in Gasoline Engines*,
https://doi.org/10.1007/978-3-319-69760-4_20

In the past, downsizing has become established as a way of further enhancing efficiency. Achieving an identical or higher level of torque from the same or reduced displacement involves increasing mean effective pressure. The aim here is to keep the compression ratio as high as possible. This has the consequence that downsized engines cannot be operated with an efficiency-optimized center of combustion in the higher engine-load range as a result of increased knocking tendency. Taking account of this circumstance, the center of combustion is set at the knock limit.

In downsizing concepts, this means that reducing knock tendency provides much potential for improving efficiency. Added to this, the technologies for avoiding knock come with significant benefits in two respects: on the one hand, fuel consumption could be reduced in a customer cycle on reaching ranges of high engine load in the engine map which, until then, have prevented operation at the optimum center of combustion on account of knock. The potential for cutting fuel consumption from knock-reducing measures could either be achieved with a more favorable center of combustion or by increasing the compression ratio. On the other hand, a lower gas exhaust temperature could reduce or obviate the need for component protection measures in the range of high engine speed. Besides a more favorable center of combustion, a higher air-fuel ratio in this case has a positive effect on fuel consumption and emissions. Alternatively – by designing the turbo-charging unit accordingly – the engine's nominal power output can be increased while leaving the air-fuel ratio unchanged.

Figure 1 shows the engine map of a downsizing concept that illustrates the potential from reducing knock tendency.

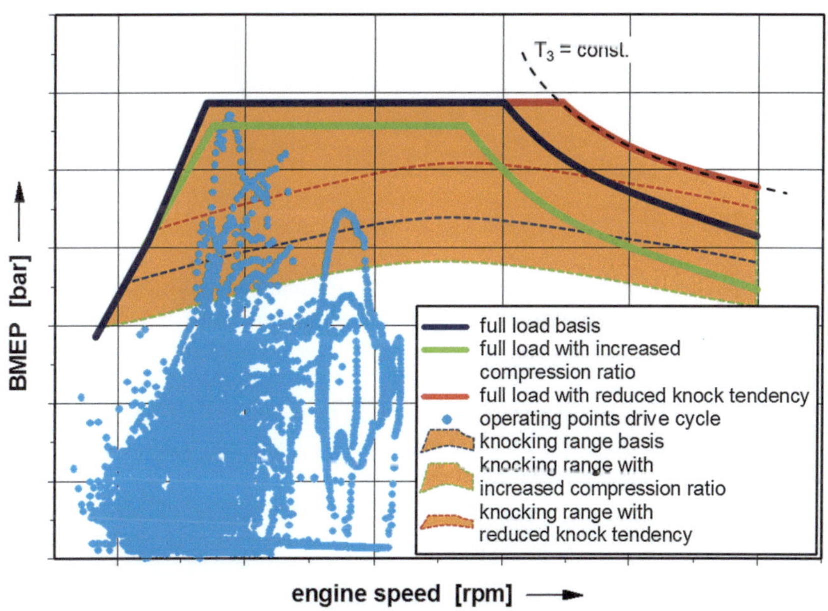

Fig. 1. Map of a gasoline engine

Under consideration of the thermodynamic effects, avoiding knocking combustion and the resultant increase in the high-pressure cycle provide a very effective way of improving efficiency at high engine speed too. Shown to less suitable in a direct comparison is the incorporation of measures, such as cooled exhaust manifold and turbine housing, that merely reduce the temperature of the exhaust gas before reaching temperature-critical components, like the turbine of the exhaust-gas turbocharger or the catalytic converter, or measures that produce a higher permissible exhaust gas temperature, e.g. using materials resistant to high temperature.

Figure 2 summarizes measures and technologies that could help to reduce the tendency of a gasoline engine to knock.

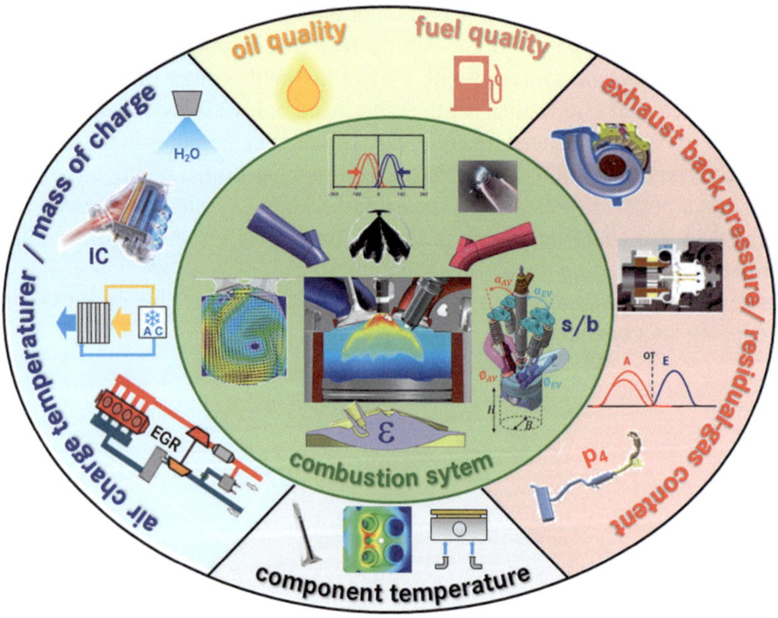

Fig. 2. Measures and technologies for reducing knocking combustion

Apart from the influence of oil and fuel quality, the majority of measures here results in a lower working-gas temperature during compression and combustion. Reducing final compression temperature is regarded as an effective measure for avoiding gasoline engine knock.

For the purposes of this paper, examination in this context focused on the thermodynamic potential for reducing knock tendency by injecting water into the combustion air with the aim of extracting heat from the working gas.

Here, the direct injection of water provides the greatest degrees of freedom for identifying the key parameters influencing the work cycle. For this reason, advance development work involved fundamental experimental and analysis-based investigations in terms of using direct water injection to avoid knock, boost efficiency and reduce exhaust gas temperature.

2 Thermodynamic Aspects of Water Injection in the Gasoline Engine

In the case of the gasoline engine, knock describes the spontaneous ignition and sudden combustion of still unburned mixture components (the so-called tail gas) in the combustion chamber, this mixture being behind the flame front proceeding from the spark plug. Consequently, knock is to a large extent influenced by the temperature of unburned gas and can, for example, be prevented by reducing final compression temperature and combustion temperature. One way of reducing tail gas temperature is to inject water.

Here, water can be introduced into the process in various ways which, however, not only has a major influence on thermodynamics but also on design integration and costs.

Figure 3 shows various options for injecting water.

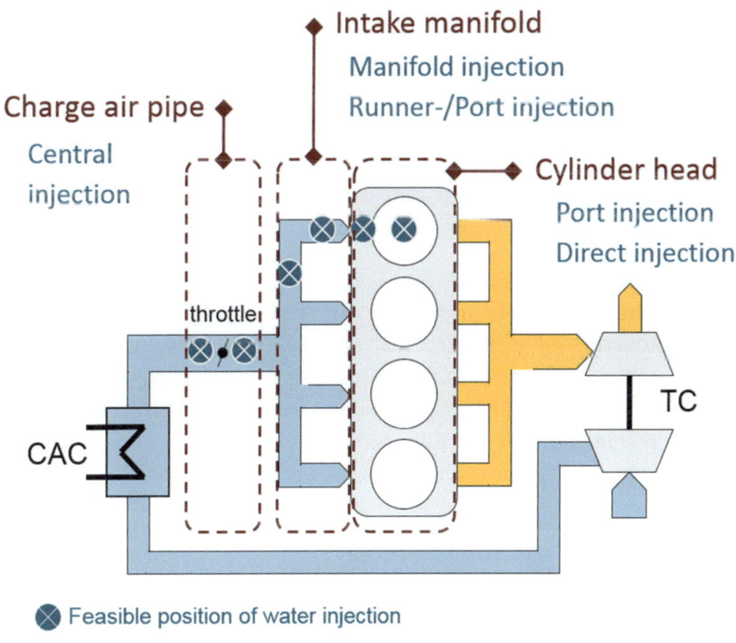

Fig. 3. Potential places for injecting water

The table below compares and summarizes the benefits and drawbacks of the various places for injecting water (Table 1).

Table 1. Evaluation of water injection concepts

Criterion/Position	Central	Intake manifold	Port	Combustion chamber
Cooling of the working gas	-	o	o	+
Equal distribution	-	o	+	++
Response to pre-ignition	-	o	o	+
Droplet size	o	o	o	+
Packaging	+	o	-	--
Pump	o	o	o	--
Costs	+	o	o	--

In principle, there are different system configurations how the water can be introduced. As the above table shows, they come with different benefits and drawbacks. The injected water – initially regardless of where it is introduced – has a cooling effect by extracting heat from the combustion air for:

- evaporation (liquid-gaseous phase change without reaching boiling temperature)
- heating liquid and evaporated water to boiling point
- vaporizing liquid water at boiling point
- heating gaseous water to above boiling point

Heat extraction is rooted in the water's vaporization heat, the higher working gas mass as well as the change in the isentropic exponent of the air / fuel / water mixture.

From the properties of water and fuel presented in Fig. 4, the above-mentioned physical effects show that the vaporization of water has the greatest influence in terms of heat extraction. The huge difference in the vaporization enthalpy between water and fuel shows the potential of the water injection.

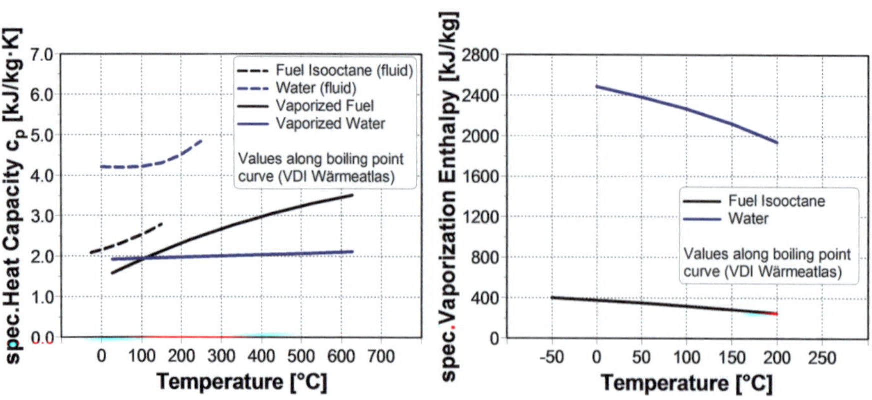

Fig. 4. Specific heat capacity and vaporization enthalpy of water [1, 2]

For instance, at a 30% water component in relation to fuel mass, the cylinder charge's overall thermal capacity only increases somewhere in the region of between 2 and 4%, which is less attributable to the specific thermal capacity of water vapor than it is to the comparatively low, absolute mass of water injected.

The effect of the injected water can be observed particularly well from the temperature curve for the working gas. Figure 5 shows the various temperature curves for one engine operating point.

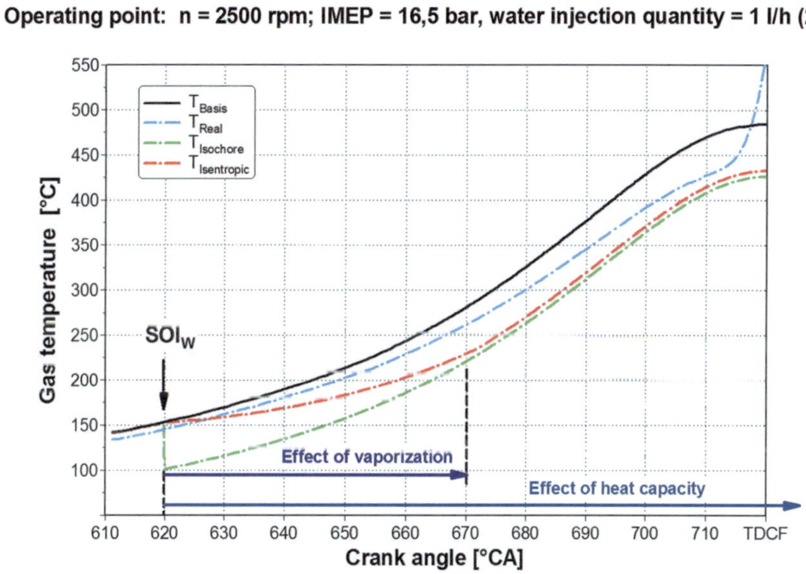

Fig. 5. Influence of water injection on working-gas temperature in the cylinder

The temperature curves for the basis without water injection (black curve) and the basis with water injection (blue curve –25% share in relation to fuel) were plotted using pressure indication measurement data and illustrate the temperature profile in the cylinder with and without water injection. The injected water's cooling effect just before ignition TDC (TDCF) lowers working gas temperature by just under 50 K. Also plotted are the analytically computed temperature curves under the assumption of:

(a) an idealized injection and vaporization curve with once-only isochoric vaporization and subsequent isentropic compression (green curve)
(b) realistic injection and idealized vaporization profile with gradual isochoric vaporization and isentropic compression (red curve)

This provides the basis for identifying two key findings:

Once the water has vaporized [in case (a) at 620 °CA, in case (b) after 670 °CA], hardly any further cooling takes place in relation to the temperature curve without water injection. This means that the water's cooling effect in the order of a temperature reduction of approx. 70 K does not significantly increase between conclusion of vaporization and reaching ignition TDC. In spite of the effect of the cylinder charge's higher overall thermal capacity from the additional water mass in this phase, the effect it does have – at least in comparison to the effect of vaporization heat – is evidently far lower. During combustion too, the higher overall thermal capacity does contribute to reducing working gas temperature and, ultimately, also the temperature of the exhaust gas.

Toward the end of compression (10 °CA before ignition TDC), working gas temperature computed on the basis of water injection is approx. 20 K lower than the temperature calculated from measurement. This being so, it must be assumed that the water's theoretically possible cooling effect is not entirely reached under real-world conditions.

For further analysis, a complex simulation model was generated that reproduces the effects of directly injecting water into the combustion chamber and can reliably predict the thermodynamic effect. Figure 6 shows the simulation results obtained from varying the start of water injection.

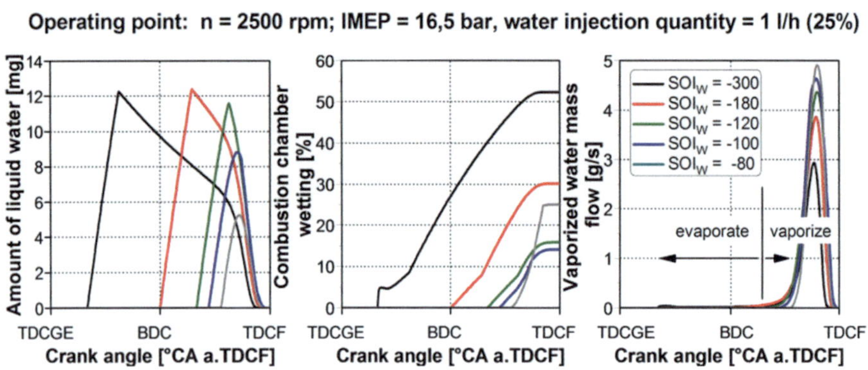

Fig. 6. Influence of the start of water injection on vaporization and wall wetting

This reveals that even if water is injected earlier (300 °CA before ignition TDC), water only starts to vaporize at approx. 120 °CA before TDCF. Some of the water injected until that point is deposited on the combustion chamber walls, evaporates or vaporizes and, in the main, extracts heat not from the working gas but from the combustion chamber walls.

A glance at the pressure and temperature development in the cylinder (Fig. 7) shows that any vaporization condition ($T_{cyl} \geq T_{boiling} = f(p_{cyl})$) is not reached until 125 °CA before ignition TDC in the compression stroke.

Initially, therefore, any earlier injection of water inevitably does nothing more than to evaporate or deposit it on the cylinder wall with subsequent evaporation and vaporization and without extracting any heat directly from the working gas. Figure 7 also clearly shows that the gas temperature in the intake manifold at 45 °C always falls below the boiling point of water. As a result, injecting water into the intake manifold or port does not produce the desired effect in respect of achieving maximum working gas cooling from water vaporization.

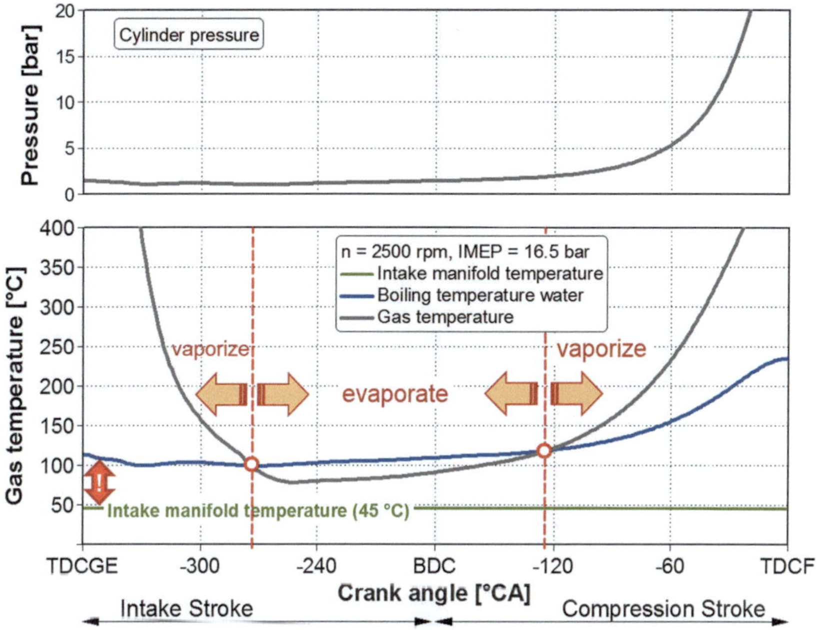

Fig. 7. Pressure and temperature development in the cylinder

Plotted in relation to the start of water injection, the curves shown in Fig. 8 for working-gas temperature at 20 °CA before ignition TDC (left) and at the center of combustion (right – MFB50%) on the one hand demonstrate the positive effect of a lower final compression temperature on gasoline engine knock and, on the other, a direct correlation of this effect with the start of injection (SOI) and, ultimately, also the location of water injection.

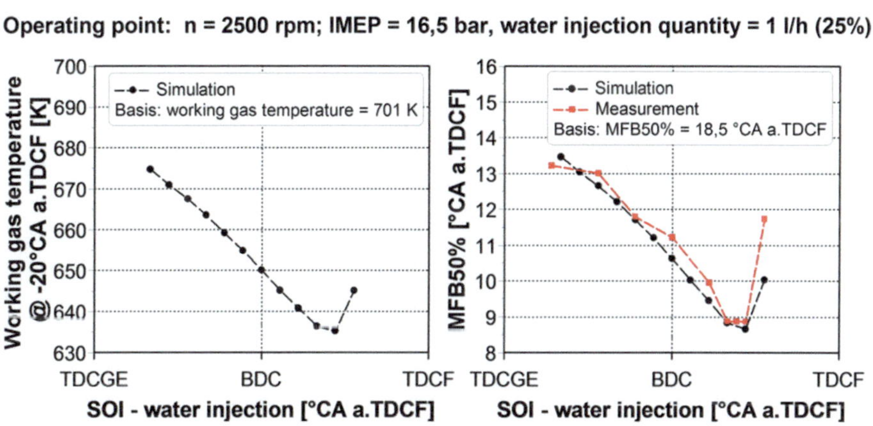

Fig. 8. Influence of the start of water injection on working gas temperature and center of combustion

The thermal efficiency of water was defined as the basis for evaluating and comparing the effect of water injection from an engineering perspective. The thermal efficiency of water describes the relationship between the temperature reduction ΔT_{Real} actually achieved with a specific quantity of water and the theoretically possible reduction in temperature ΔT_{Theor} in conjunction with full vaporization, the full effect of vaporization on working gas as well as the idealized injection and vaporization of the same amount of water at the thermodynamically optimum point at bottom dead center (BDC). This means that the thermal efficiency of water describes the optimum utilization of water in terms of extracting vaporization heat from the working gas while at the same time minimizing wall wetting.

A result of the test bench measurements is shown in Fig. 9. The diagram presents the correlation of the thermal efficiency water injection with the center of combustion (MFB50%) as a function of the variation start of injection (SOI) for the water injection.

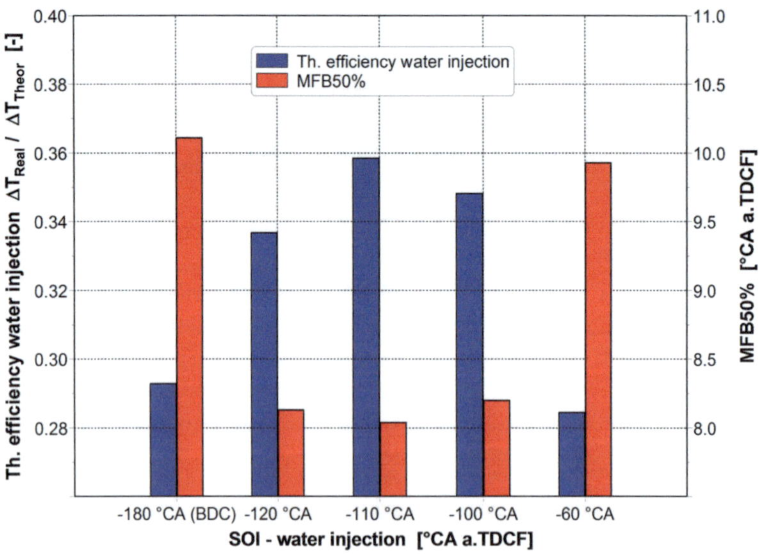

Fig. 9. Water efficiency results

Another result is, that this characteristic clearly demonstrates that even with direct water injection, a significant part of the injected amount of water not only extracts vaporization heat from the working gas but also from the combustion chamber wall. This is confirmed by the simulation results (Fig. 6). The temperature effect cannot be achieved either at the thermodynamically optimum point in time. This means that the thermal efficiency of water also denotes the water component that is unable to contribute to reducing working gas temperature, i.e. this parameter can also provide the basis for the water injection parameters and optimizing injector targeting.

The studies conducted were able to show that thermal water efficiency delivers a simple yet effective parameter for evaluating and optimizing water injection.

3 Design Aspects of Implementing Direct Water Injection

From a thermodynamic perspective, injecting water directly into the combustion chamber provides the greatest potential for reducing knock tendency. In addition to this, implementing such also offers far more degrees of freedom for calibration than injecting water into the engine's intake system since it is not subject to any constraints from intake valve timing. This contrasts with a greater level of work involved in integrating the system – particularly for package reasons – as well as complex components resulting from higher system pressure, and higher costs.

Probably the greatest challenge in implementing direct water injection lies in designing the combustion chamber. Today, modern downsized engines have a relatively small cylinder bore and a combustion chamber roof which, with four gas exchange valves, the fuel injector and the spark plug, provide only limited options for integrating a further injector. On top of this, the valve train and design structure on the cold side of the engine may also make it necessary to create access points. Obvious places for locating the water injector in the cylinder head is a central position in the combustion chamber roof as well as at the side below the intake valves. A further conceivable option is to position the water injector in the cylinder housing, whereby piston and ring set would have to be adapted accordingly. Regardless of water injector positioning, the aim must be to ensure maximum interaction with the charge motion in the combustion chamber and minimal wetting of components and combustion chamber walls. Vaporization characteristics are the key criterion in configuring the injector.

As part of preliminary development activities, single-cylinder test engines with direct water injection were set up to conduct fundamental experimental studies and analyze the thermodynamic effects of injecting water with regard to avoiding knock, improving efficiency and reducing exhaust gas temperature. The engines with a displacement of 500 cc have a compression ratio of 10:1 and a fuel injector centrally positioned in the combustion chamber roof. The water injector is positioned on the side of the cylinder head at right angles to the inflowing combustion air and can be turned about the injector's longitudinal axis in the course of testing. Figure 10 shows where the water injector is positioned in the cylinder head.

In addition to positioning in the cylinder head, one configuration also enables the water injector to rotate in the test engine's cylinder housing so that the water can be injected into the combustion chamber towards the combustion chamber roof.

For the injected amount of water to achieve the full potential in terms of cooling the charge mass, the aim must be to maximize interaction with the working-gas and, wherever possible, avoid any wall and component wetting. For this, it is essential to optimize injector spray and targeting. For investigation purposes, various injectors of the same type were configured in the light of previous experience from laying out spray pattern for direct gasoline injection whereupon they underwent optical and hydraulic measurement

Here, allowance must be made for the particular vaporization characteristics of water as compared to gasoline. Key configuration criteria in achieving maximum water efficiency is high spray atomization quality with small droplet diameter – governed by hole diameter and system pressure – as well as minimal wall and component wetting.

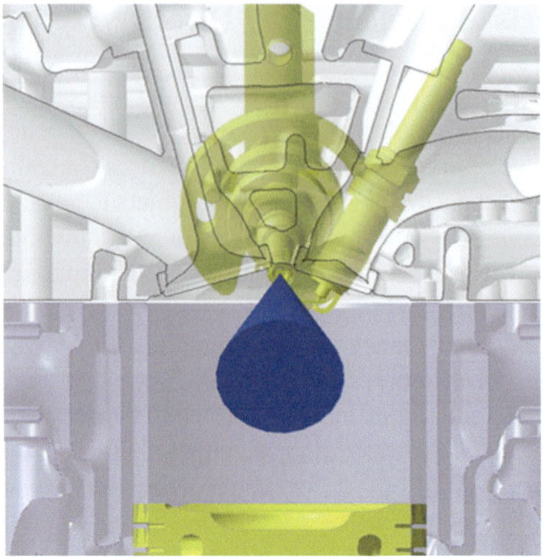

Fig. 10. Cross section of a test engine with water injector in the cylinder head

4 Selected Results of Influencing the Gasoline Engine Working Cycle by Direct Water Injection

The studies on direct water injection were conducted on a single-cylinder test engine with external supercharging. The boundary test conditions, such as valve timing, boost pressure and exhaust gas backpressure, charge air temperature as well as fuel injection, were parameterized in the same way as for the mass-produced full-scale engine.

Besides the standard measurement systems (pressure and temperature measurement, air mass, fuel mass, lambda sensor, exhaust gas and particulate measurement), the test engine also features pressure indication in the combustion chamber as well as in the intake and exhaust port. These measurement data were used in post-processing for thermodynamic analyzing of the working cycle and for constructing a model for simulating water injection. IAV's KIS system has been used for determining and evaluating knock-specific parameters.

Below, attention turns to discussing the results of optimizing injection parameters as well as the influence of direct water injection on thermodynamic characteristics at selected operating points. All of the results shown exhibit a stoichiometric air-fuel ratio.

Optimization of Injection Parameters for Direct Water Injection

To maximize water efficiency, attention first focused on adjusting the parameters for optimizing water injection, such as injector positioning and targeting, injection pressure as well as start of injection (SOI). The results presented reflect the water injector position shown in Fig. 10. Figure 11 shows the center of combustion (MFB50%) as a function of variation in water injection pressure as well as the start of injection at operating point n = 2500 rpm and IMEP = 16.5 bar.

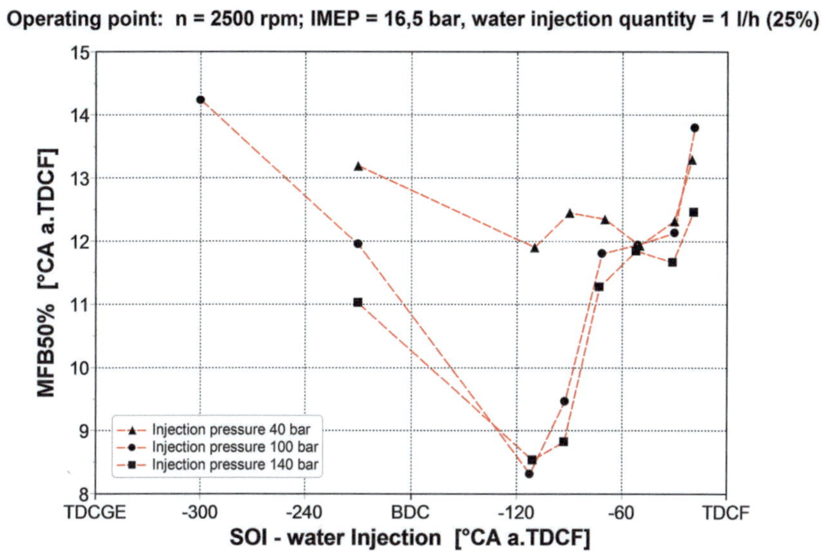

Fig. 11. Influence of injection pressure and injection timing for direct water injection

The curves showing the center of combustion (MFB50%) for the three different water injection pressures demonstrate that the water's poor vaporization characteristics require a minimum injection pressure of approx. 100 bar to assist water vaporization as a way of reducing knock tendency and advancing the center of combustion to efficiency optimized values.

This means it is necessary to provide the best compromise between droplet size and penetration length (wall wetting) so that injection can generate optimum vaporization conditions. In other words, insufficient injection pressure (40 bar) produces hardly any influence on knock tendency. For the configuration shown, any higher injection pressure (140 bar) fails to improve the above-mentioned compromise between droplet size and the need to avoid wall wetting.

Among the variants with an injection pressure in excess of 100 bar, the potential to reduce knock tendency is shown to correlate closely with the start of injection. This is essentially attributable to the vaporization characteristic plotted in Fig. 7.

Figure 12 below shows this correlation between knock tendency and start of injection in detail for operating point n = 2500 rpm and IMEP = 20 bar. The graph on the left presents the center of combustion as a function of start of injection and, in the right-hand diagram, the CA-related vaporization curve for this operating point as well as the working gas temperature for the SOI measuring point = −120 °CA after ignition TDC (TDCF) across the same crank-angle window. The basis measurement point without water injection has a center of combustion of 25 °CA after TDCF at knock limit.

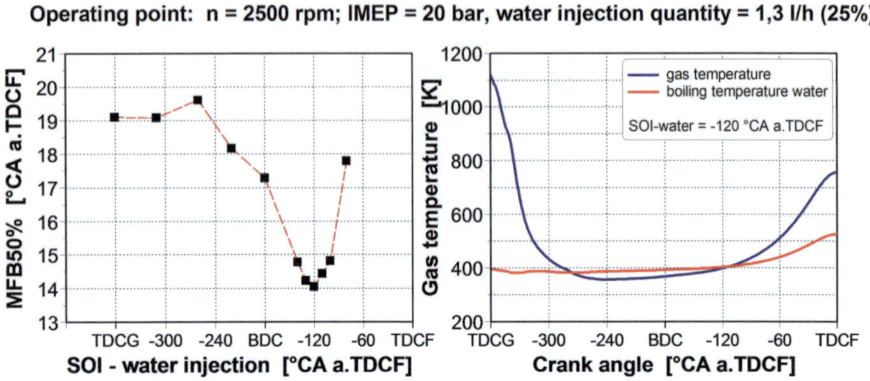

Fig. 12. Influence of water injection on the center of combustion

The characteristic curve for the center of combustion can be seen at this operating point too. The optimum injection window is approx. −120 °CA after TDCF and advances the center of combustion by 11 °CA over the basis. This can be explained by looking at the temperature curves in the right-hand diagram. Vaporization conditions only prevail in the combustion chamber from −120 °CA after TDCF, resulting in optimum water utilization in terms of extracting vaporization heat from the working gas while at the same time minimizing wall wetting. This produces a reduction in knock tendency from a lower final compression temperature, making it possible to select an efficiency-optimized ignition point. This is also confirmed by the results of the 1-D simulations carried out in Sect. 2 (Fig. 6).

Later starts of injection diminish the potential for reducing knock tendency since the necessary vaporization time is no longer available and any start of injection closer to TDC at the same time comes with a greater degree of wall wetting.

Starts of injection in the intake stroke show very little influence on the center of combustion as vaporization conditions still remain absent. In this period, the water can merely evaporate, with the simulation results in Sect. 2 (Fig. 6) verifying a high degree of wall wetting. As a result, this water component can only make a minor contribution to reducing the working-gas temperature.

In addition, at this operating point fuel injection takes place in the intake stroke. This means that the vaporization processes for the two media of fuel and water take place side by side at the time of intake stroke injection. Despite the more favorable vaporization characteristics of fuel, homogenization of the air-fuel mixture must under no circumstances be allowed to deteriorate as this would reduce efficiency and increase untreated emissions (HC, particulate matter).

At the same time, the results of injecting water in the intake stroke can, by way of example, be used for evaluating a manifold injection or mixture injection in the combustion chamber (fuel-water). This means that for the same amount of water, directly injecting water with an optimum start of injection increases the potential for advancing the main center of combustion by about 50% over manifold or mixture injection.

The combustion engine's lack of smoothness (COV) correlates with the center of combustion. Advancing the center of combustion at the same time reduces irregularities in operation.

Influence of Water Quantity on the Center of combustion

For the same operating point, Fig. 13 shows the result of varying the injected water quantity in relation to the center of combustion for optimized start of injection. The working gas temperatures determined from thermodynamic process analysis at the ignition point are plotted in the right-hand diagram.

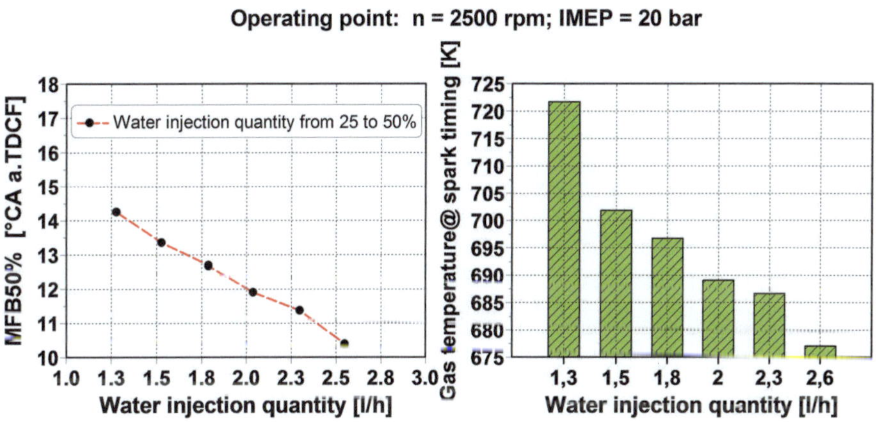

Fig. 13. Influence of injected water quantity on the center of combustion

The basis measurement point without water injection has a center of combustion of 25 °CA after TDCF at the knock limit.

In the right-hand diagram, it can be seen that increasing the amount of injected water results in a constant reduction in working gas temperature at the ignition point, making it possible to reduce knock tendency which ultimately leads to the earlier center of combustion (left-hand diagram). As the injected water quantity increases, the optimum start of injection must be continuously advanced in order to provide the necessary vaporization time. However, this also reduces water efficiency as the optimum start of injection is essentially governed by the vaporization characteristic or wall wetting.

Furthermore, a correlation can be identified between injected water quantity and advancing the center of combustion. The center of combustion can be advanced by 1 °CA with approx. 5% of the amount of directly injected water.

Implementing these potentials of direct water injection may take the permissible peak pressure beyond the map range close to full load. This must be taken into consideration at concept layout level.

As expected, advancing the center of combustion reduces exhaust gas temperature.

In this specific case, directly injecting water at a rate of 2.6 l/h (50% share in relation to fuel) – while advancing the center of combustion – is capable of achieving a 125 K reduction in working gas temperature compared with the basis without water injection.

This means that direct water injection provides a concept for increasing power output at a given maximum exhaust gas temperature and stoichiometric air-fuel ratio.

Direct water injection also provides the capability of multiple injection throughout the working cycle. By way of example, this was examined at operating point n = 2500 rpm and IMEP = 16.5 bar, with the results for the center of combustion or exhaust gas temperature being shown in Fig. 14. At the basis measuring point without water injection, the center of combustion is 16.5 °CA after TDCF at knock limit.

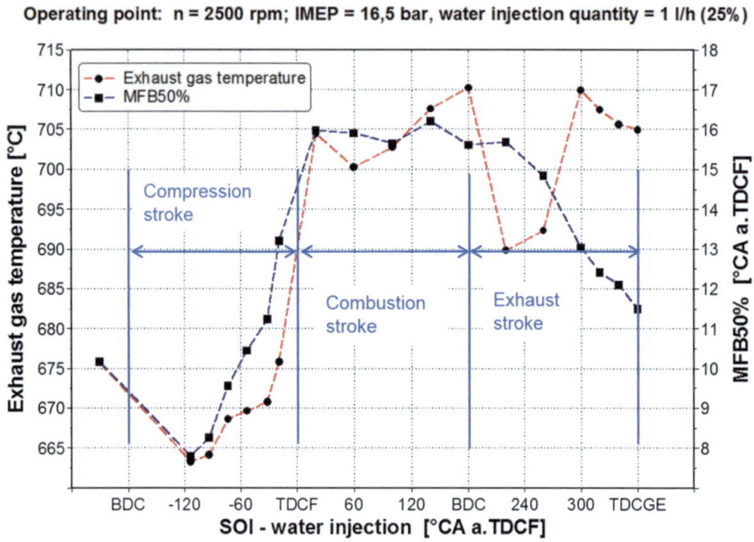

Fig. 14. Influence of water injection on exhaust gas temperature

At the selected operating point, water injection in the compression stroke advances the center of combustion by 8 °CA over the basis, equating to a reduction in exhaust gas temperature of 60 K. Direct water injection during the exhaust phase brings about direct exhaust gas cooling. This is most effective when injection takes place during the pre-exhaust phase. In the example shown, this leads to a reduction in working gas temperature of 50 K that does not result from shifting the center of combustion. The above-mentioned effects can be combined by applying a double injection consisting of an injection event into the compression phase and a further one into the exhaust phase. This is of particular interest if the quantity of water injected into the compression phase is limited.

Influence of Water Injection on Gasoline Engine Combustion

For operating point n = 2500 rpm and IMEP = 20 bar, Fig. 15 shows the influence of water injection on gasoline engine combustion using the key parameters of center of combustion, burn delay and burn duration. Measuring point 1 represents the basis without water injection. By comparison, water injection at optimum start of injection, but with the same center of combustion and cylinder charge as the basis, is shown at measuring point 3. It can be seen that the inert effect of water vapor and higher charge mass in the combustion chamber adversely affects gasoline flame front combustion. Burn delay is slightly increased by 1.5 °CA, burn duration, in contrast, by 5.5 °CA.

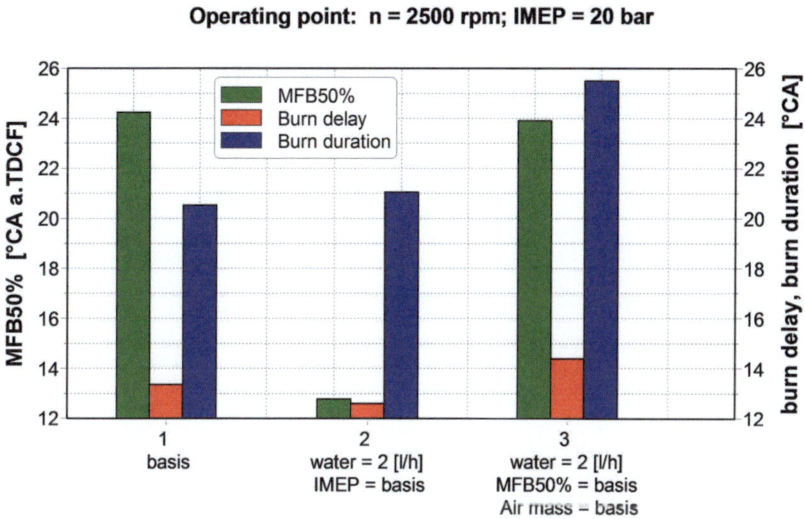

Fig. 15. Influence of water injection on gasoline engine combustion

Although measuring point 2 represents the same injected water quantity, the ignition point is at the knock limit. As a result of the higher pressure and temperature level (compared to measuring point 3) on releasing heat, advancing the center of combustion leads to an overcompensation of the inert effect of the water vapor so that the parameters of center of combustion, burn delay and burn duration again reach the level of the basis without water injection.

Influence of Water Injection on Untreated Emission

For operating point n = 2500 rpm and IMEP = 20 bar, Fig. 16 shows the influence of water injection on untreated emissions in relation to hydrocarbon as well as nitrogen-oxide emission. All shown emissions are normalized to the basis. Measuring point 1, in turn, represents the basis without water injection. By comparison, water injection at optimum start of injection, but with the same center of combustion and cylinder charge as the basis, is shown at measuring point 3. The inert effect of the water vapor and the higher charge mass in the combustion chamber give rise to a reduction in working-gas temperature in the burned and unburned zone, causing nitrogen oxide and HC emissions to rise.

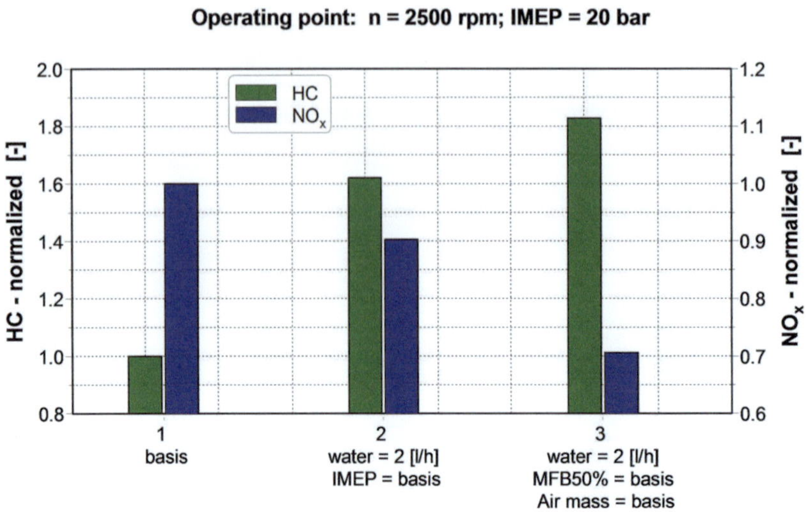

Fig. 16. Influence of water injection on untreated emissions

Although measuring point 2 represents the same injected water quantity, the ignition point is at knock limit. As a result of the higher pressure and temperature level (compared to measuring point 3) on releasing heat, advancing the center of combustion leads to an increase in nitrogen oxide emission, but also to a slight reduction in hydrocarbon emission.

However, the level of hydrocarbon emission from the basis is not reached with any injected water quantity.

After evaluating all efficiency-relevant parameters, an injected water quantity of 2.6 l/h, for example, reduces indicated fuel consumption ISFC for the single-cylinder test engine by 10.5% at operating point n = 2500 rpm and IMEP = 20 bar through advancing the center of combustion.

Summarizing and in simplified form, Fig. 17 shows the relationship between optimum start of injection, injection concept and potential for reducing knock tendency (center of combustion).

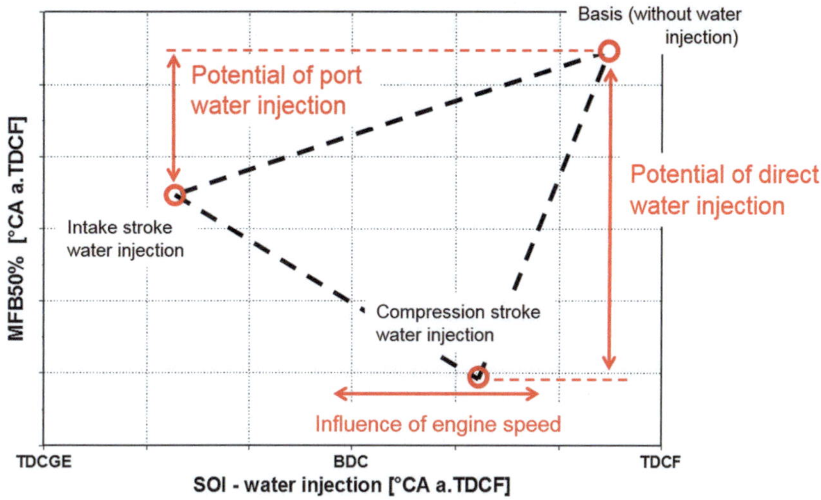

Fig. 17. Schematic diagram showing the potential of different water injection concepts

Compared to intake stroke injection, direct water injection in the compression stroke provides more than twice the potential for advancing the center of combustion. This results from a higher water efficiency, i.e. through a start of injection optimized for the vaporization conditions, minimizing wall wetting while at the same time maximizing the reduction in working gas temperature.

Here, the far lower potential of intake stroke injection can be interpreted as the maximum achievable potential of manifold injection (e.g. from injection through the opened intake valves). This results from the lower working gas temperature (below the boiling curve) during the intake stroke, with the consequence that a large component of the liquid water mass is precipitated on the walls of the combustion chamber and piston, preventing it from making any contribution to reducing charge temperature. This is illustrated by the simulation results (Fig. 6).

The potential of direct water injection in relation to engine speed is evaluated in the lower part of the diagram. As a result of the shorter time available for vaporization, the optimum start of injection steadily shifts from the compression stroke towards bottom dead center as engine speed increases. However – at the same time – also into crank-angle windows in which no vaporization conditions yet prevail, leading to steadily increasing water wall wetting. This means that as engine speed increases, the major advantage (approx. 50%) of direct water injection continuously decreases in comparison to manifold injection (comparable with the result of intake stroke injection).

5 Challenges in Implementing Water Injection

To implement a system of water injection in a vehicle, answers must be found to a many different issues ranging from storing or supplying water to operating strategy and diagnosing the overall system. Ultimately, for the resultant trade-off, the best possible overall compromise needs to be found in relation to the vehicle/engine combination.

Besides the activities on the single-cylinder test engine, experiences were gathered with full-scale engines on engine test benches and in the vehicle with various concepts for water injection. These provide the basis for identifying pending challenges, some of which are separate from defining the place at which the water is injected.

Water Quality and Water Availability

Constant quality of water must be ensured over service life. Mains water, demineralized or distilled water and a mixture of water and alcohol may be an option where water is stored in a tank system with replenishment intervals. Although availability speaks in favor of mains water, using it involves considerable drawbacks as a result of water quality, which differs widely from region to region, and from the effects of chemical constituents on emission behavior – such as particle number – also on the engine's mechanical parts. This makes demineralized or distilled water the preferred option since water quality remains constant. Using a mixture of water and alcohol has advantages in terms of cooling medium solidifying at low temperatures. Depending on the alcohol used and its percentage component, it can play a part in influencing knocking resistance. Using a water-alcohol mixture comes with drawbacks to the extent that its alcohol content affects energy conversion, demanding intervening control action in the supply of primary fuel and possibly making it necessary to extend the On-Board Diagnostics.

Besides storing and replenishing water, concepts are also being discussed that involve the sourcing of water in the vehicle. For example, this includes surface water (precipitation), using water condensed from the air by the passenger compartment air-conditioning system, and condensing water from the exhaust gas. Besides the huge effort involved, it will be difficult with any of these concepts to guarantee a constant water quality that remains stable over time.

Water Demand

One central question in configuring the overall system of water injection lies in the quantity of water needed. The aim must be to keep the amount as low as possible. For logical reasons, water injection will only be activated when it becomes necessary for reasons of severe knock or high exhaust gas temperature. In respect of water mass flow, a water component of approx. 30% in relation to the injected fuel mass seems appropriate. Integral water demand in the vehicle largely depends on the vehicle/engine combination, load population and, consequently, also on driving behavior. Aimed at maximizing the replenishment interval, the water reservoir selected should also be as large as possible. In terms of dimensioning, however, consideration must also be given to the situation in the vehicle and to operating strategy at low ambient air temperatures.

Water Injection System

To avoid damaging the system and ensure proper working order at all times, using water makes it advisable to evacuate the carburetion system before or after switching off the engine. This involves the use of a feed pump with a delivery direction that can be varied and selected. To achieve the highest possible quality of atomization, injecting water directly into the combustion chamber demands a system pressure that is far higher than for manifold injection. Injection pressures about 150 bar are recommendable, which poses a major challenge and results in the need for new pump concepts. An injector suitable for direct injection can be derived from existing designs for the direct injection of gasoline.

Basis Engine Mechanical System

The water added to combustion air naturally has an influence on the engine's mechanical functions. On the one hand, the more favorable centers of combustion increase peak pressure in the combustion chamber, and lead to a higher level of strain which must be taken into account in configuring the system – particularly the engine. On the other hand, injected water can gather on the cylinder liner and, from there, penetrate the cylinder crankcase, affecting the friction and wear of various components. Water entering the engine oil also poses a potential risk. Water evaporated from engine oil or present in a gaseous state in the crankcase increases crankcase pressure, this being dissipated via the crankcase valve. Hence, the venting system must be adapted on account of the higher throughput and to ensure proper working order at low ambient air temperatures. In providing the right answer to implementing water injection, there will be no way around extensive development work on the base engine side.

Behavior at Low Ambient Air Temperatures

The material properties of water – particularly high vaporization heat – makes this medium ideally suited to reducing the knock tendency of gasoline engines. Opposing this is the major disadvantage that water freezes at a temperature of below 0 °C. This means that components in which water would be able to freeze must be of frost-proof design. Once water is in a solid aggregate state, re-liquefying it involves supplying a very high amount of energy because of the high level of heat that is needed to melt water. Various concepts, such as the use of heated engine coolant or the integration of an electric heating device, are being discussed in this regard. When using water-alcohol mixtures to reduce the freezing point, they must not segregate over time. This, too, demands parallel development activities.

6 Concept Approaches for Water Injection

The engine concept must be designed to make optimum use of the thermodynamic potential provided by water injection. Although applying water injection to an existing engine concept as an "add-on" solution is fundamentally possible, it nowhere near utilizes the maximum possible potential provided by water injection. In developing a gasoline engine with water injection, a distinction can be drawn between two general concept approaches (Fig. 18).

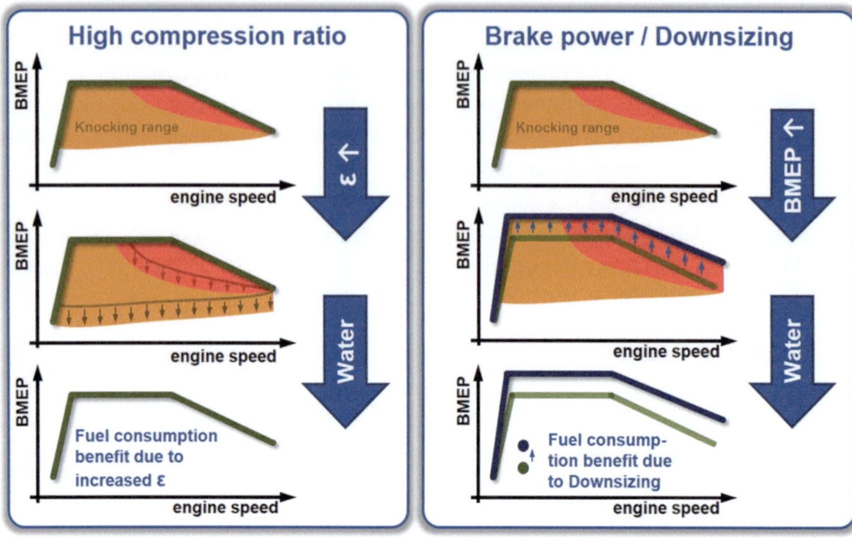

Fig. 18. Concept approaches for gasoline engines with water injection

One concept approach is based on raising the engine's geometric compression. Under otherwise identical boundary conditions, this would lead to a significant increase in knock tendency – however also to poorer efficiency and a higher exhaust gas temperature. Ultimately, to higher fuel consumption and lower rated power output. As a result of its knock-reducing effect, however, water injection can compensate for these drawbacks or reduce them to such an extent that a higher level of efficiency dominates as a result of higher compression, significantly improving fuel consumption across the entire map range, particularly in the range of relevance to the customer.

Alternatively, water injection can be used to provide a power or downsizing concept which, however, will make it necessary to adapt the turbocharging system. Knock tendency, which increases in line with mean effective pressure, is then reduced by water injection to such an extent which, in spite of higher full load, not only produces an efficiency-promoting high-pressure cycle but also remains within the permissible exhaust gas temperature. Depending on concept layout, the customer either benefits from a higher driving performance or, in the case of the downsizing option with reduced displacement, from lower consumption through the shift in operating point.

7 Summary/Outlook

The results presented in this paper show that direct water injection provides immense thermodynamic potential in terms of avoiding gasoline engine knock and reducing exhaust gas temperature. They are based on experimental studies carried out on single-cylinder test engines as well as on computational analyses using working-cycle computation.

The cooling, and therefore knock-reducing effect of injecting water is largely governed by the start of injection and by the pressure and temperature conditions prevailing in the cylinder at the time. The maximum effect is achieved when vaporization temperature in the cylinder is exceeded at the start of injection, the water can immediately vaporize and therefore extensively avoid any wall wetting. The vaporization heat directly extracted from the working gas leads to a reduction in the final compression temperature, significantly lowering knock tendency.

At operating point n = 2500 rpm and IMEP = 20 bar, which was the focus of detailed analysis, a water component of 25% (measured against fuel mass) significantly advances the center of combustion by 11 °CA. Doubling the quantity of water (producing a volumetric flow of 2.6 l/h) advances the center of combustion by an additional 5 °CA. This achieves a 10,5% reduction in indicated fuel consumption ISFC and reduces the exhaust gas temperature by 125 K.

Directly injecting water also provides the capability of targeted, multiple water injection in the work cycle, and also of reducing exhaust gas temperature even further. For this, the period on opening the exhaust valves is shown to be a favorable point for starting injection.

The engine concept must be designed to optimize the thermodynamic potential provided by water injection in terms of enhancing efficiency or lowering fuel consumption for the same driving performance. Basically, the options available here involve raising the geometric compression ratio or increasing maximum mean effective pressure while reducing displacement (downsizing).

Today, water injection is an interesting technology for enhancing the efficiency of gasoline engines. Implementing it is shown to be a complex matter and has major impact on many sub-systems of today's familiar gasoline engine. Only an integrated approach will make it possible to implement a water injection system while meeting today's applicable quality criterion on the basis of a reasonable cost-benefit ratio.

References

1. VDI-Wärmeatlas, VDI e.V., 10. Auflage (2006)
2. List, H., Pischinger, A.: Thermodynamik der Verbrennungskraftmaschine. Springer, New York (1989). Band 5

Combustion Process

Suppressing Knocking by Using CleanEGR – Better Fuel Economy and Lower Raw Emissions Simultaneously

Michael Fischer[1(✉)], Michael Günther[2], Carsten Berger[2], Ralf Troeger[2], Michal Pasternak[3], and Fabian Mauss[3]

[1] Tenneco GmbH, Edenkoben, Germany
Mfischerl@tenneco.com
[2] IAV GmbH, Chemnitz, Germany
{michael.guenther, carsten.berger, ralf.troeger}@iav.de
[3] BTU Cottbus, Cottbus, Germany
{Michal.pasternak, maussf}@b-tu.de

Abstract. The use of external Exhaust Gas Recirculation (EGR) is a common technology available for the reduction of nitrogen oxides (NOx) emitted by internal combustion engines. With regard to gasoline engines, the addition of EGR at higher loads reduces knock tendency and improves fuel economy by reducing the necessity for fuel enrichment. To further maximize these benefits, the recirculated exhaust gases are cooled down that improves engine efficiency by enabling an advanced center of combustion (MFB50). Hereby gasoline engines can be operated at EGR rates up to 20%, which is enabling stoichiometric operation in the entire engine map. On the other hand, cooled EGR is leading to well-known low-temperature issues such as fouling, corrosion and condensation. In response to that challenge, in this work the use of cooled LP- and HP-EGR is analyzed by engine testing for different engine intake temperatures. For dedicated tests a higher compression ratio for improved fuel economy and a coated Gasoline Particulate Filter (CleanEGR™) for cleaning the EGR gas is investigated as well. As a conclusion, external cooled and cleaned EGR is measure to meet future RDE requirements (stoichiometric operation in entire map) by achieving improvements in engine efficiency and engine out emissions (PN, NOx) simultaneously.

1 Introduction

The benefits of external Exhaust Gas Recirculation (EGR) on gasoline engine efficiency are well known [1], and advantages for low-pressure EGR (LP-EGR) [1] and high-pressure EGR (HP-EGR) [2] have been presented. The usage of intensive EGR for normal aspirated or charged gasoline engine were announced by several carmakers [3–5] to reach engine efficiency of more than 40%. In addition to fuel economy improvements, future emission legislation in Europe and China limits the particulate numbers (PN) of tail pipe emissions under Real Driving Conditions (RDE) from 2017 onwards. This will be an added challenge, especially for downsized engines because they often operate at higher loads resulting in higher exhaust temperatures, Fig. 1.

M. Günther and M. Sens (eds.), *Knocking in Gasoline Engines*,
https://doi.org/10.1007/978-3-319-69760-4_21

"Fuel enrichment" is typically applied under such conditions for component thermal protection, but it increases fuel consumption and emissions, including both particulate matter (PM) and HC [6]. For improved low-end torque (LET) characteristic "Scavenging" is applied allowing better turbo efficiency and knocking occurs at higher loads for all engine speeds. Further engine downsizing is leading to even higher share of these critical load points in the legislation test procedures.

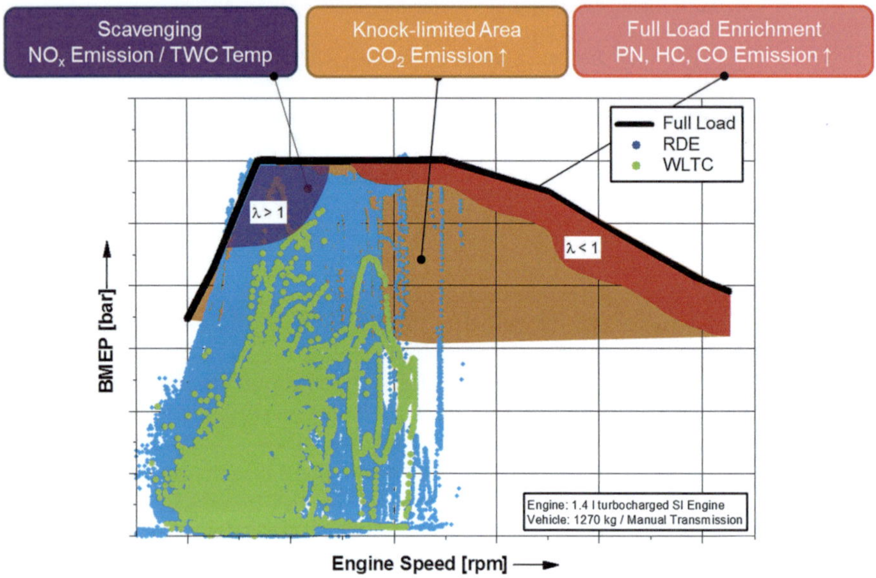

Fig. 1. Gasoline operation map for WLTC/RDE conditions: efficiency and emissions challenges at higher engine loads for a downsized turbocharged gasoline engine in a midsize car

Also known is that gasoline engines with direct injection (GDI) of the fuel produce more soot particulates during combustion than Port Fuel Injection engines (PFI) [7]. However, external EGR significantly reduces engine out soot particulates [8].

External EGR is also well known for almost 25 years to produce fouling layers in EGR coolers of internal combustion engines [9], which reduces EGR cooler efficiency up to 30% [10]. For explaining the effect which leads to deposit formation it is stated in [10]: "The primary drive for soot deposition is thermophoresis, proportional to the difference between gas temperature and wall temperature. Thus, lower coolant temperature and hotter gas temperature increase soot collection efficiency. HC and acid condensation, on the other hand, depend primarily on wall temperature. Condensation occurs on cold surfaces and depends on the local partial pressure, not on the gas temperature in a direct way." And even though Diesel engines produce significantly more particulates than gasoline engines, the effect of deposit formation is known for gasoline engines as well [11]. Lower wall temperatures significantly increase the soot deposit mass [12] or fouling, and turbulence induced by surface dents or fins on the cooler wall does not help to reduce this fouling risk [10, 11].

External EGR suppresses engine knocking, which improves the engine efficiency [1–5]. But conversely, increased inlet gas temperatures with external EGR reduce engine efficiency by increasing gas exchange losses and knock tendency [13]. To maintain the positive effect of EGR on knocking of gasoline engines the cooling of EGR or EGR/Air mixtures close to 50 °C or less needs to be considered and is typically avoided due to the risk of fouling and corrosion with such low EGR gas temperatures. But for this investigation this method is selected, and fouling and corrosion risks are reduced as the EGR gas is cleaned with a purification device before it enters the EGR cooler [14].

In this paper, the effect of cooled and cleaned EGR on knocking is analyzed in various LP- and HP-EGR application for different engine load points.

2 Thermodynamic Aspects of High-Load EGR for Gasoline Engines

Knocking is influenced significantly by the temperature within the unburned zone and thus can be prevented by reducing the in-cylinder temperature at the end of compression and during combustion.

Table 1 shows three possible technology approaches which are capable of decreasing in-cylinder temperature during the combustion phase and thereby mitigate knocking.

Table 1. Technology approaches for knock mitigation

Water Injection	Miller-Cycle	Cooled EGR
Increasing Efficiency		
$MFB_{50\%}$ ↓ Gas Properties (κ ↑↑) Wall Heat Losses ↓	$MFB_{50\%}$ ↓ Gas Properties (κ ↑) Wall Heat Losses ↓	$MFB_{50\%}$ ↓ Gas Properties (κ ↑↑) Wall Heat Losses ↓
Reducing NO_x Emission		
Evaporation + Heat Capacity	Miller Effect	Total Heat Capacity
Reducing Exhaust Gas Temperature		
$MFB_{50\%}$ ↓ Evaporation + Heat Capacity	$MFB_{50\%}$ ↓ Reduced Effective Compression	$MFB_{50\%}$ ↓ Total Heat Capacity

As a consequence of the reduced knocking tendency, an earlier center of combustion (MFB50) can be realized leading to a higher combustion efficiency as well as a lower exhaust gas temperature. In addition, the lower in-cylinder temperature results in decreased wall heat losses, reduced NOx emission and an increased isentropic coefficient are leading to a higher thermal efficiency:

$$\eta_{th} = 1 - \frac{1}{CR^{\kappa - 1}} \tag{1}$$

In contrast to the Miller Cycle, Water Injection and cooled EGR benefit from an increased mass within the cylinder resulting in a higher total heat capacity, even if the cooling effect in case of Water Injection is realized primarily by water evaporation. Due to the gas properties of water vapor and exhaust gas, respectively, Water Injection and cooled EGR have a bigger influence on the isentropic coefficient than just by lower temperature.

To understand the thermodynamic effects of cooled EGR, which is subject of this publication, it is reasonable to consider the influence of exhaust gas on the gas properties of the cylinder mixture as shown in Fig. 2.

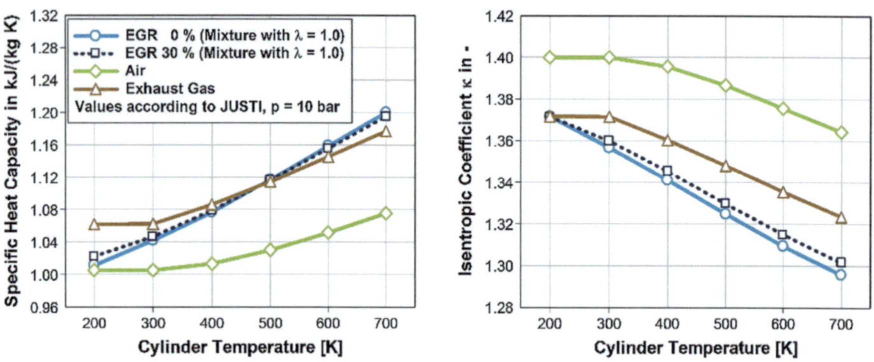

Fig. 2. Gas properties of air, exhaust gas and mixture in dependence on temperature

In general, both specific heat capacity and isentropic coefficient are clearly dependent on temperature. While specific heat capacity is increasing with temperature, isentropic coefficient is dropping. Although the differences between clean air (green line) and exhaust gas (brown line) are obvious, the influence of exhaust gas on the gas properties of a cylinder mixture (air, fuel and exhaust gas) is relatively low. In case of specific heat capacity, the influence can even be considered as negligible. So, the effect of EGR on heat capacity is dominated not by the specific value, but by the additional charge mass. The impact on the isentropic coefficient can be estimated at 0.5% (in case of 30% EGR compared to 0% EGR). This value also seems to be very low, but it has to be taken into account, that the isentropic coefficient has an exponential influence. Nevertheless, the temperature dependence means the major impact.

A high isentropic coefficient during compression phase leads to a better thermal efficiency, but also to a higher in-cylinder temperature at end of compression, see Fig. 3, which would be disadvantageously with regard to knocking.

But considered for a high engine operating point with different external EGR ratios from 0 to 30% that temperature increase is very low, see Fig. 4 – as long as there is no additional heat input by adding exhaust gas.

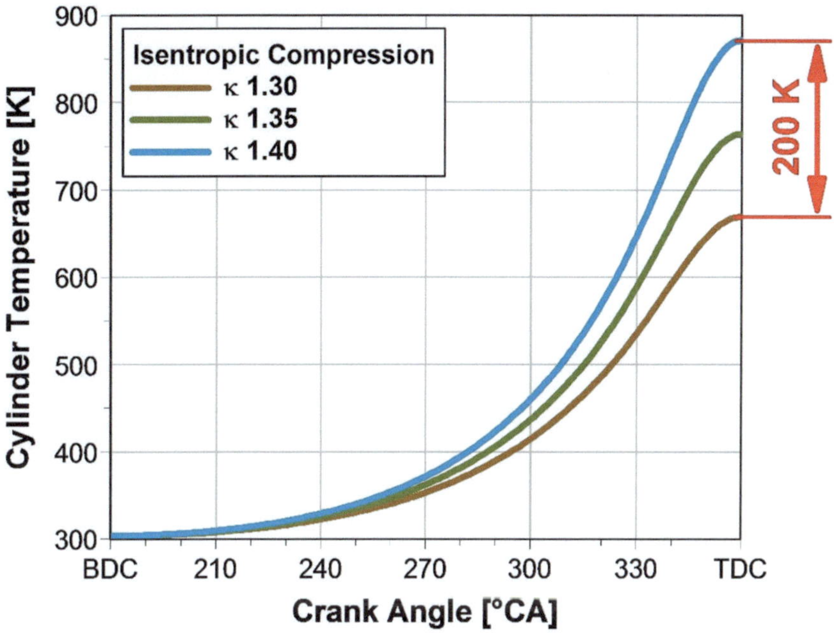

Fig. 3. Influence of isentropic coefficient on in-cylinder temperature in case of a isentropic compression

That means, cooling the recirculated exhaust gas is essential for two reasons: keeping the in-cylinder temperature low for reaching a high isentropic coefficient during compression (high thermal efficiency) and even though reaching a low temperature at the end of compression (knock mitigation).

Regarding the phase of the heat release, the thermodynamic effects are similar. The heat release results in a temperature rise within the cylinder, which leads to a decrease in isentropic coefficient. As a consequence, the temperature and pressure rise would also decrease, what is positive with regard to knocking, but negative with regard to thermal efficiency. But it has also to be taken into account, that the heat input is made to a higher charge mass and total heat capacity, respectively. As it can be seen in Fig. 5, the in-cylinder temperature during combustion is reduced by EGR.

Although the isentropic coefficient is increased by the recirculated exhaust gas and the lower in-cylinder temperature (compared to a heat release without recirculated gas) which would lead to a higher temperature rise during heat release, this effect is clearly overcompensated by the influence of the increased total heat capacity. The ratio of released heat amount (fuel energy) referred to the total mass trapped in the cylinder, drops by about 30%. In sum, the reduced peak temperature will not only result in knock mitigation, but also in reducing wall heat losses (as shown in Fig. 5) and NOx emission.

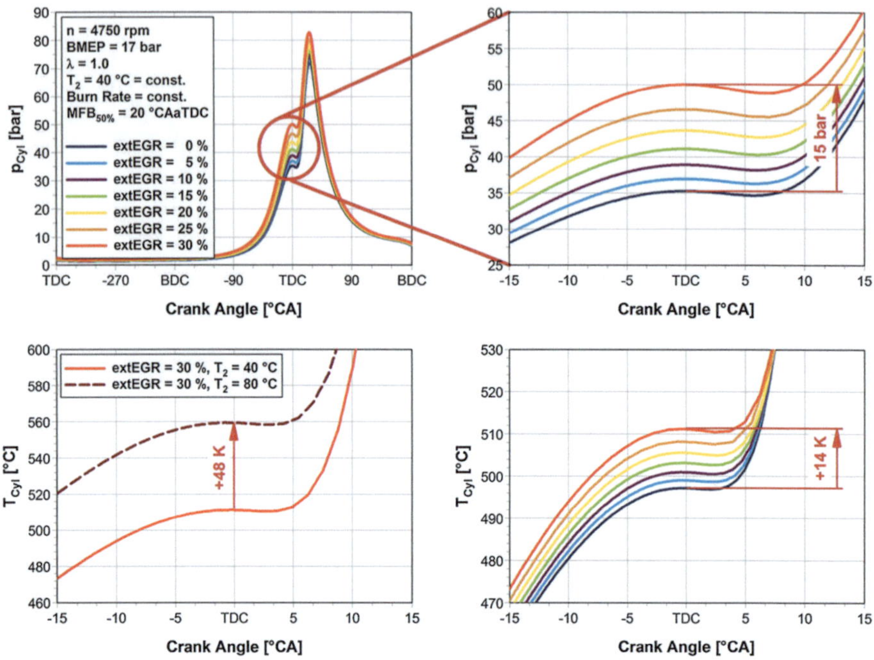

Fig. 4. Influence of external EGR ratio on cylinder pressure and temperature during compression

As already mentioned for the compression stroke, the targets for the phase of the heat release are the same: high isentropic coefficient for high thermal efficiency and even though a low peak temperature for the sake of knock mitigation. To achieve both targets, a high recirculated exhaust gas mass in combination with an intense cooling is required.

As a consequence of the reduced in-cylinder temperature, a high thermal efficiency and an earlier center of combustion that can be realized because of knock mitigation, cooled EGR also leads to a remarkable drop in exhaust gas temperature. Figure 6 illustrates that effect based on a simulation study with constant burn rate and preset MFB50.

It can be seen, that even without changing the center of combustion, exhaust temperature decreases by 100 K if cooled external EGR is applied with a rate of 15%. Another drop of 25 K can be achieved by an earlier center of combustion due to reduced knocking. The downside of filling the cylinder with recirculated exhaust gas is a higher boost pressure demand that is necessary to keep the trapped fresh air mass constant. To reach the same mean effective pressure, Fig. 6 indicates, that an external EGR rate of 15% requires an additional boost pressure 0.2 bar – an amount, which implies the adaptation of the charging system including the enhancement of the charge air intercooler. This also leads to the requirement for cooling the recirculated EGR, since the boost pressure demand increases further with higher EGR temperature.

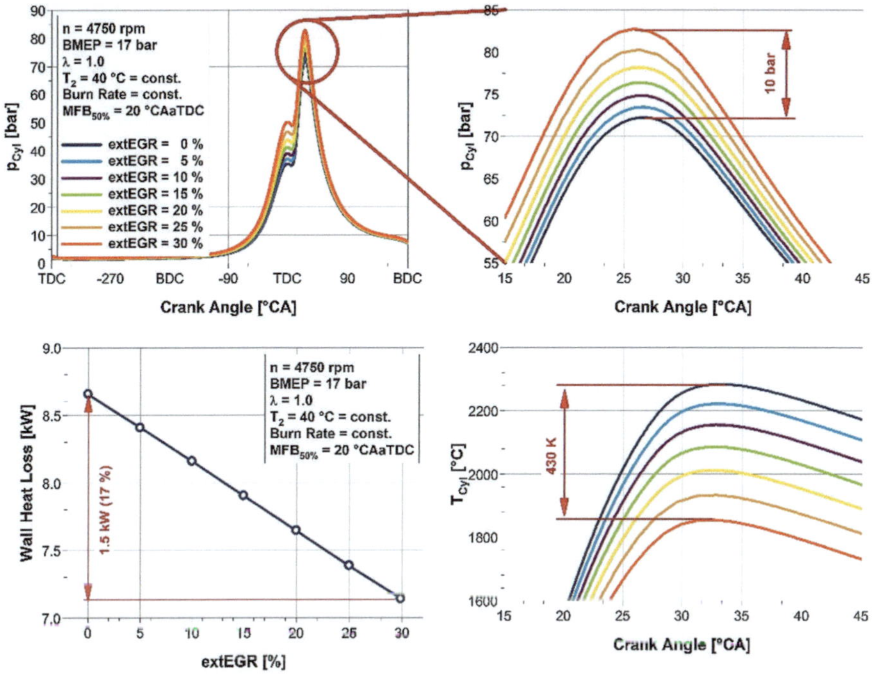

Fig. 5. Influence of external EGR ratio on cylinder pressure and temperature during heat release

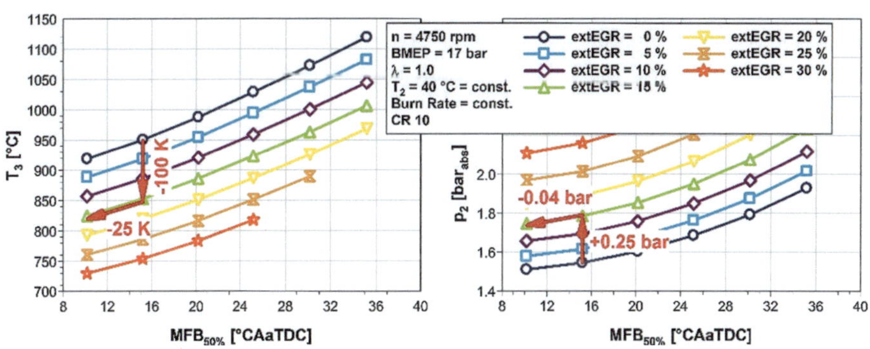

Fig. 6. Influence of external EGR ratio on exhaust gas temperature and boost pressure demand

To gain the full thermodynamic potential of an engine equipped with an cooled EGR system, it can be recommended to increase the compression ratio. By that measure, part load operating points which are not limited by knocking also benefit from the higher thermal efficiency coming along with an increased compression ratio. Under

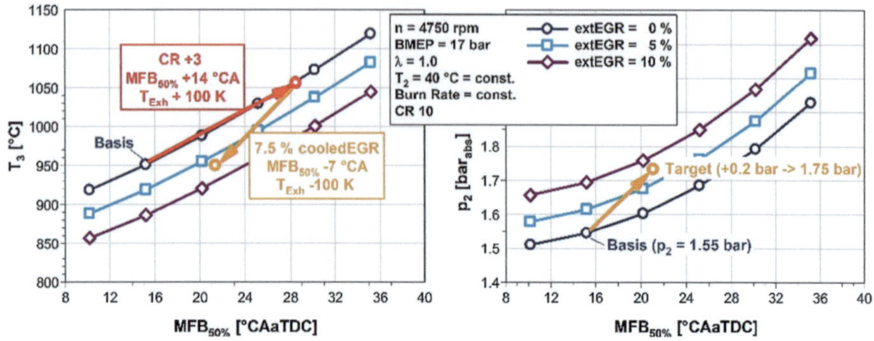

Fig. 7. Concept approach with cooled EGR and increased compression ratio

the assumption, that an increase in compression ratio by +3 leads to a shift in center of combustion by +14 °CA, the engine would exceed its permitted exhaust gas temperature by 100 K, see Fig. 7.

Then, applying a cooled EGR system on that engine with an EGR ratio of 7.5%, a center of combustion of 21 °CA can be realized. The exhaust temperature decreases by 100 K back to the permitted value of the base engine. It's important to mention, that such a concept approach does only work, if the additional boost pressure and cooling power demand can be provided in the whole engine map. By increasing the compression ratio by +3, the fuel consumption in part load operating points can be reduced by up to 3.5%

In principle, there are different EGR system configurations that are usually distinguished by their exhaust sampling point and the point where the gas is fed back to the intake air. The two most common systems, Low Pressure EGR and High Pressure EGR, are compared schematically in Fig. 8. Amongst others, one important criterion for an EGR System is the pressure difference between sampling and feeding point as a measure for the maximum external EGR rate that can be achieved. Concerning this matter, Fig. 8 indicates that LP EGR has an advantage in low engine speed area while HP EGR can obtain higher EGR rates at high engine speeds, e.g. rated power.

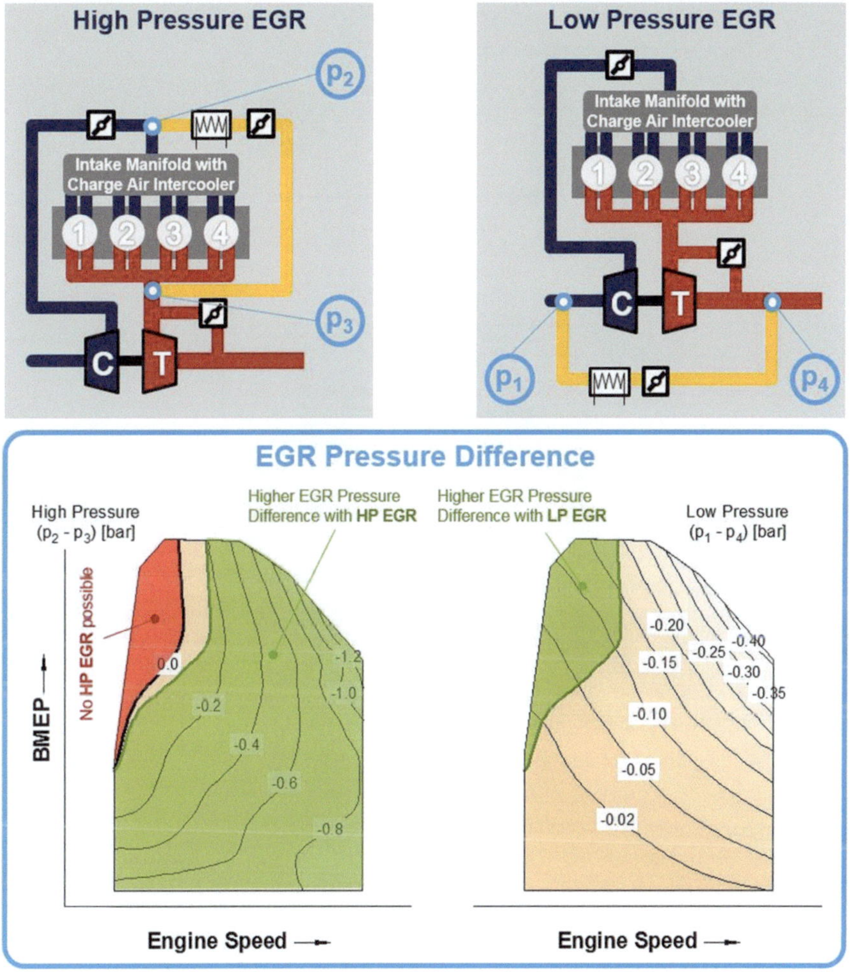

Fig. 8. Comparison of low and high pressure EGR

3 CleanEGR– Principal, Experimental Setup and Effects

3.1 CleanEGR: Principal - an Innovative Purification Device

The EGR purification device is a catalytically coated Gasoline Particulate Filter (cGPF), which consists of a ceramic substrate with alternatingly plugged channels to force gas flow through porous walls, filtering the particulates. And, if catalytically coated, it also eliminates pollutants (HC, CO, NO_x) to yield cleaned EGR gas. It is integrated in the hot section of the LP- or HP-EGR system before the valve and cooler, as described in Fig. 9. It is named CleanEGR™[1]. The particulate filter applied is a

[1] CleanEGR™ is a registered trademark by Tenneco Inc.

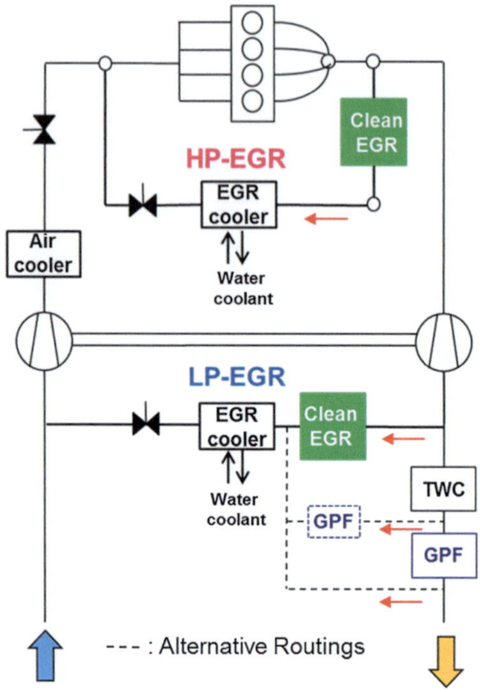

Fig. 9. CleanEGR application and alternative routing for LP- and HP-EGR system for a gasoline engine

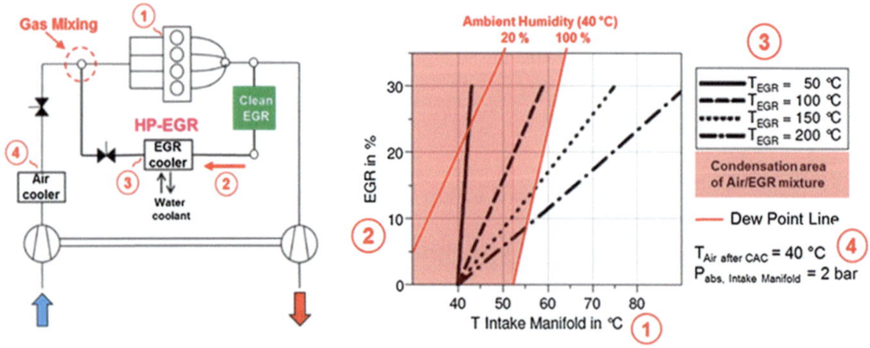

$$T_{Intake} = \frac{(1-x_{EGR})}{1-x_{EGR}*(1-R_C)} * T_{Air} + \frac{x_{EGR}*R_C}{1-x_{EGR}*(1-R_C)} * T_{EGR}$$

with heat capacity ratio: $R_C = \frac{c_{p,EGR}}{c_{p,Air}}$

Fig. 10. Calculation of Air/EGR mixture temperature and water condensation area in the engine intake system wall-flow cordierite filter with a high porosity to enable a catalytic coating while minimizing its pressure drop. This coating is an Pd/Pt-coating which is often used in Three-Way-Catalyst (TWC) to reduce the pollutants CO, NO_x, and HC during stoichiometric operation of the combustion engine. For comparison, an uncoated particulate filter (GPF) is used in the test program as well. The amount of cleaned EGR (N_2, CO_2, H_2O) is controlled by an EGR valve before it enters the EGR Cooler. Although the EGR gas is not released to the atmosphere, such cleanup of pollutants offers benefits in engine performance, as described within this study.

In a HP-EGR application, the manifold intake temperature can calculated as illustrated in Fig. 10. More details of the calculation approach are described in [14]. As red marked in Fig. 10 there is a wide area of condensation risk, which is influenced strongly by the fluctuation of the ambient humidity. As a consequence, for almost all real world situation a condensation risk in the intake manifold needs to be taken into account and by applying EGR, a fall-out of EGR ingredients in the engine intake system might be occur.

3.2 CleanEGR: Experimental Setup

A series production gasoline turbocharged engine with direct fuel injection is used to evaluate the concept, as described in Table 2.

Table 2. Test engine specifications

Engine type and displaced Volume	I4-1395 cc
Stroke	80.0 mm
Bore	74.5 mm
Max power and power density	103 kW and 73.5 kW/L
Max torque and max BMEP	250 Nm and 20 bar
Compression ratio	10.5:1
Number of valves and timing	4 with Inlet-/Outlet –VVT
Exhaust manifold	4-in-1, integrated in Cylinder head
Charging system	Single Scroll WG-Turbocharger
Air cooling	Air/water cooler
Fuel injection system	Direct injection, max. 200 bar
Fuel type	RON 91

A high-pressure (HP)- or low-pressure (LP)-EGR system is added to this engine, see Fig. 9. Within this EGR loop a purification device to clean EGR is integrated before the cooler, and between them is an EGR valve to control the EGR mass flow rate. The EGR cooler is cooled by an external coolant system, characterized with flexible coolant temperatures and coolant mass flow rates enabling EGR cooling down to 50 °C for all required EGR mass flows.

The engine tests are performed at steady state conditions on an engine test bed for various engine load points. Maximum EGR rates of 15–20% are achieved, depending on engine load point and respecting engine combustion instability (COV) of max 3%. The focus of the investigation is to identify benefits and challenges of clean EGR specifically at high engine loads, where cooled EGR ($\leq$ 100 °C) suppresses knocking. Knocking in combustion occurs at higher engine loads as the exhaust gas temperature is often above 500 °C. In the steady state load points EGR rate and inlet gas temperature are varied. Based on these values, the ignition set point, Air/Fuel Ratio (AFR) and Turbo-Wastegate valve position are tuned to meet required engine load by keeping limits for exhaust gas temperature, COV and knock border. Knocking combustion is detected by cylinder individual pressure sensors and the knock limit was set be the criteria limit Eq. (2)

$$dp_{98.5,max} = \frac{n}{1000} \, in \, bar \tag{2}$$

Therein $dp_{98.5,\ max}$ is a statistic value, where 98.5% of all high frequency peak pressure values of the measured combustion cycles are below this amount [16]. Under such operating conditions, as well as across much of the engine map, the EGR gas temperature in the cGPF is hot enough for high catalytic conversion efficiency and passive filter regeneration, avoiding the accumulation of soot within the filter.

Added to the steady state testing, rapid fouling tests are performed to evaluate the efficiency of the EGR purification, applying a 12-hour test with clogging and solidifying cycles including engine shut down and restart periods. And, in order to assess the effects on the engine combustion efficiency of sub-cooled gas mixtures in the engine intake system, the intake gas is cooled by a water-cooled air cooler using the cold water circuit of the engine test bench. With this the gas mixture temperature in the engine intake system is varied between 20 and 70 °C. All engine tests are performed with gasoline fuel with RON 91.

3.3 CleanEGR: Effects

Based on the findings in Sect. 3.1 the EGR cleaning effect of CleanEGR and the minimizing of fouling risk was investigated and analyzed in rapid EGR fouling tests, Fig. 11. By using CleanEGR technology, the fouling mass was reduced by 73% by using an uncoated Filter and by 82% by using a coated Filter. The mass difference indicated that the reduction of HC, NOx and CO by the coated filter have a positive effect on minimizing the fouling mass as well. As presented in [14], the condensed water have a clear appearance with nearly no combustion products and the pH value changed from acidic to neutral by applying a coated filter in the EGR system.

While soot and HC are the main drivers for creating the fouling mass at cold surfaces in the EGR systems. NOx emission of EGR system has an relevant influence on knocking combustions [1]. As presented in [14, 15] knocking is suppressed by the reduction of NOx in EGR system. The simulation in [14] is indicating a potential of 6 CA advanced MFB50, while [15] measured 2 CA benefit by engine testing.

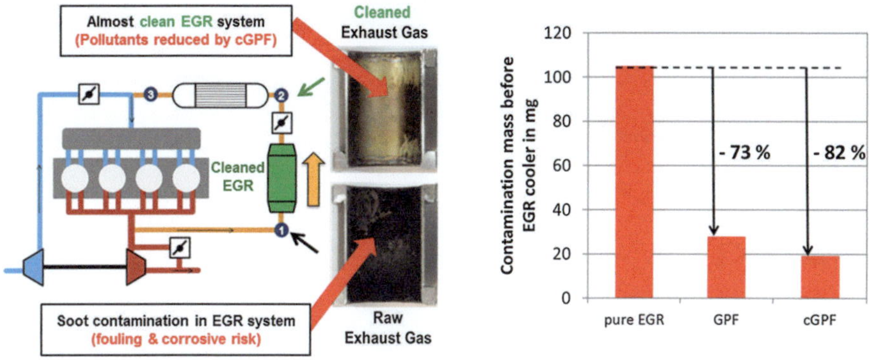

Fig. 11. Experimental setup and results of EGR fouling tests by using CleanEGR (fouling mass at position 2)

4 Selected Results of the Impact of Cooled EGR and CleanEGR on Knocking

4.1 Effects of Cooled EGR in LP-EGR Application w High Compression Ratio of 12.7 at n = 2500 rpm, 14 bar BMEP

The following results show the thermodynamic potential of a cooled Low Pressure EGR system in an exemplary engine operating point (2500 rpm, BMEP 14 bar). For these measurements, the engine has been equipped with a higher compression ratio (CR = 12.7) and an exhaust flap after the Three Way Catalyst (TWC). The exhaust flap has been used to increase the pressure at the exhaust sampling point (before TWC) in order to achieve a higher EGR ratio. Figure 12 shows that in this operating point, the exhaust flap had to be used to reach EGR ratios higher than 20%, because beyond that ratio the EGR pressure difference (p1 – p4) became too low to surpass the pressure drop in the EGR path.

The diagrams also show an increase in boost pressure demand (p Intake Manifold) of about 350 mbar, which is necessary to obtain the same fresh air charge in the cylinder. The increase in boost pressure is realized by closing the wastegate of the turbocharger resulting in a higher exhaust back pressure (p Exhaust Gas before TC) of 450 mbar which leads to an increased internal residual gas ratio and higher pumping losses (+60 mbar).

The influence of external EGR on the combustion is shown in Fig. 13. It can be seen that the recirculated exhaust gas extends the ignition delay as well as the burning duration due to its inert gas effect. The effect on the burning duration would even be higher, if the ignition timing was constant.

But for this investigation, the ignition timing has been modified to operate the engine at the knock limit. On the basis of MFB50% it can be summarized, that cooled

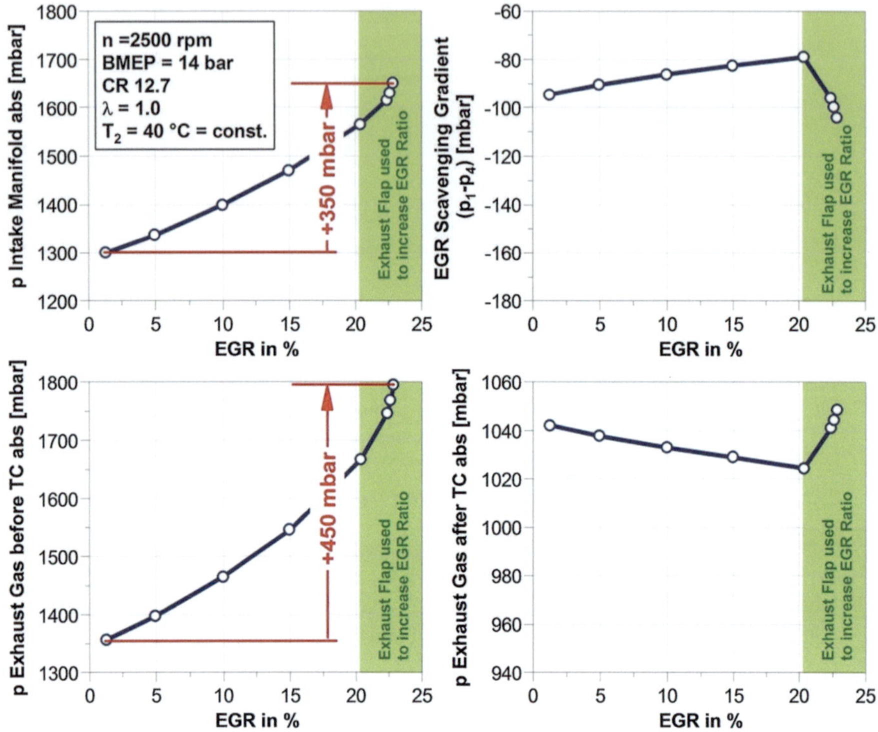

Fig. 12. Influence of cooled LP EGR on boost and exhaust pressure

EGR has a significant effect on knocking. The earliest possible MFB50% is achieved with an EGR ratio of 20%. Along with the positive influence of EGR on isentropic coefficient (high thermal efficiency) and reduced wall heat losses, this means a benefit in specific fuel consumption of 9 g/kWh (−3.8%), see Fig. 14.

Here, also the thermodynamic advantage of the increased compression ratio is plain to see: The operating point driven with base compression ratio has a specific fuel consumption of 238 g/kWh with an MFB50% of 9.1 °CAaTDC. That means, from the view of center of combustion there would have been only little room for improvement. By increasing the compression ratio, the center of combustion has to be shifted to 22.5 °CAaTDC due to increased knocking tendency. But with the thermodynamic effects of cooled EGR and the benefit of a higher thermal efficiency due to the higher compression ratio, a BSFC of 230 g/kWh can be realized – a value that would not have been possible with cooled EGR applied on the engine with base compression ratio. Furthermore as Fig. 14 shows the impressive influence on exhaust gas temperature and NOx emission, that can be significantly reduced.

As mentioned in Sect. 2, it is absolutely necessary to cool the recirculated exhaust gas in order to achieve the maximum thermodynamic potential. Since cooling power is

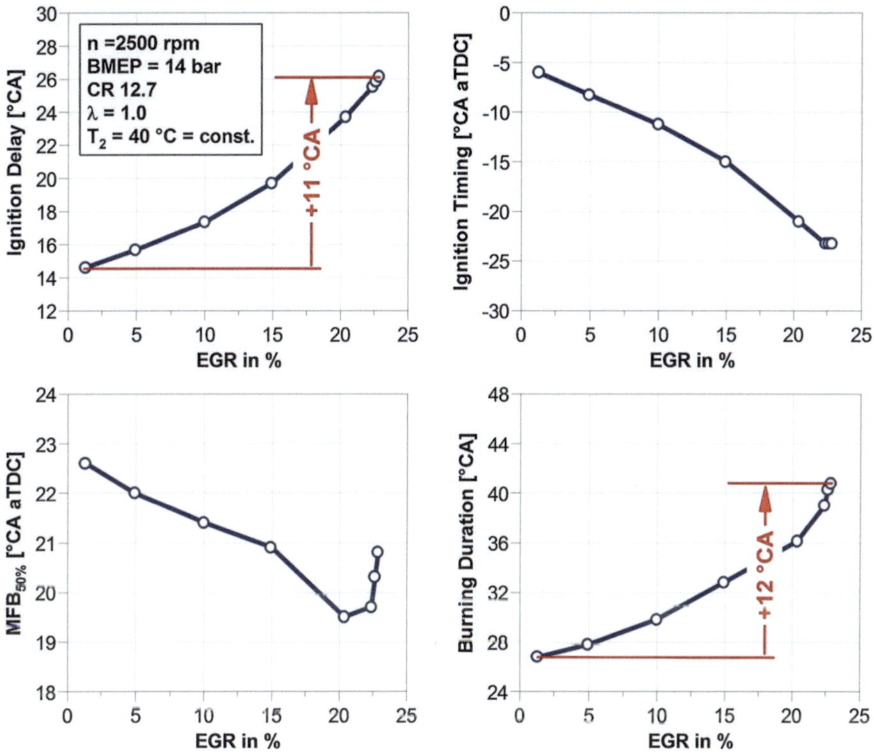

Fig. 13. Influence of cooled LP EGR on combustion

limited in vehicle application, it is important also to disclose the additional cooling power demand that consists of the cooling power of the EGR cooler and of the charge air intercooler. As shown in Fig. 15 the additional cooling power demand for an EGR ratio of 20% and a target intake temperature of 40 °C can be estimated at +5.1 kW.

4.2 Effects of Cleaned EGR in a HP-EGR Application with Reference Compression Ratio of 10.5 at n = 4750 rpm, 17 bar BMEP

Figure 16, left side shows the engine BSFC as a function of EGR percentage and intake manifold temperature (IMT), and the red dots illustrate the testing points. As shown in the figure, engine BSFC decreases with increasing amount of EGR and more EGR cooling which reduces IMT. At IMT of 70 °C without EGR, the engine BSFC is higher than 300 g/kWh due to engine knock limit and enrichment, see Fig. 16, left side. When reducing the IMT to 40 °C without EGR, 253 g/kWh BSFC is achieved, and adding 15% EGR further reduces BSFC to 243 g/kWh, yielding an additional 4% fuel economy improvement. Based on this, combining a reasonable amount of EGR with extreme cooling under high speed and high load conditions can eliminate fuel

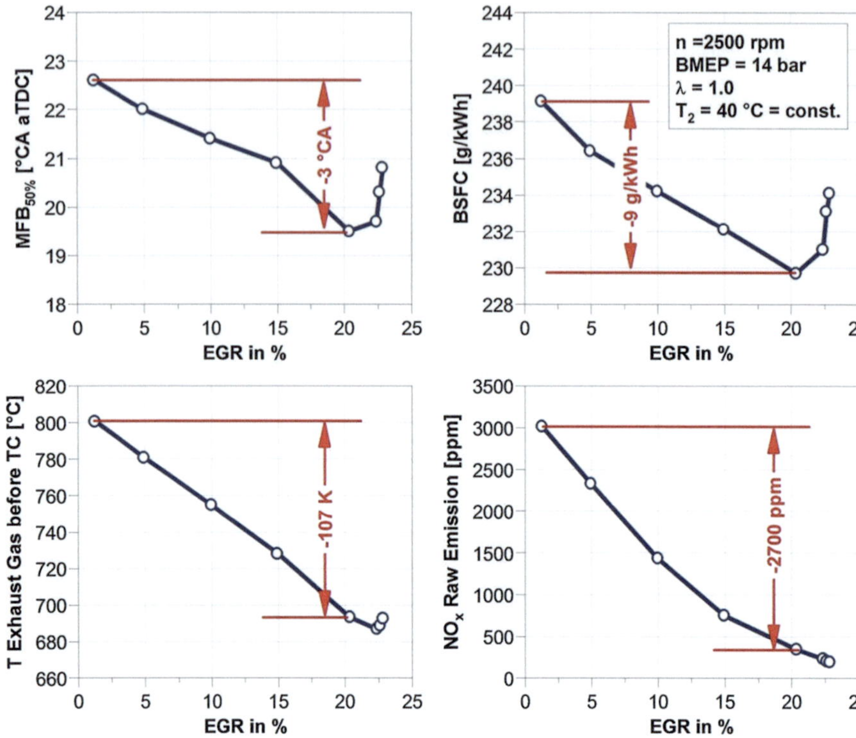

Fig. 14. Influence of cooled LP-EGR on MFB50, BSFC, $T_{Exhaust}$ and NOx emission

enrichment, mitigate knock, and therefore significantly reduce engine fuel consumption. It can also be seen, that in the cases with the EGR cooler outlet temperature over 100 °C, the IMT increases significantly and the fuel economy benefit of external EGR is reduced accordingly. When the EGR cooler outlet temperature is constant at 200 °C, there is no notable fuel economy improvement by increasing external EGR from 0 to 15%. The lower the EGR outlet temperature, the more BSFC benefit by adding 15% EGR. And, although sub-cooling of the gas in the engine intake system (i.e. to 20° C) enables benefits in BSFC while applying no EGR, as EGR rates increase more than 10%, the sub-cooling offers no added benefit.

Figure 16, right side illustrates the phasing of combustion center (MFB50) as a function of EGR rate and IMT, demonstrating combustion center phasing can be advanced with increased EGR rates and lower IMT. With 15% EGR the IMT is reduced from 65 °C to 42 °C as the EGR cooler outlet temperature is reduced from 200 °C to 50 °C, respectively, and the MFB50 is advanced by about 7 °CA. Combustion phasing advance is due to the spark timing advance. Adding EGR and cooling down the intake temperature reduces the combustion flame speed. Due to the advance in spark timing and combustion, phasing is without knocking, the engine is able to achieve better fuel efficiency, as illustrated in Fig. 16, left side. It can also be seen in Fig. 16, right side that with a 15% EGR rate, cooling down the intake gas lower than

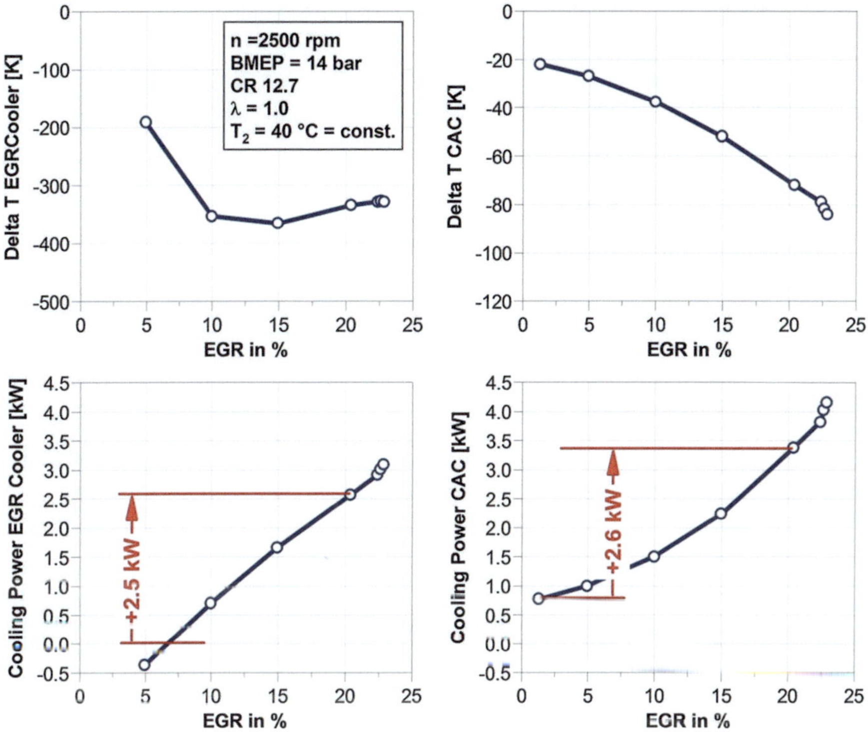

Fig. 15. Influence of cooled LP EGR on cooling power demand

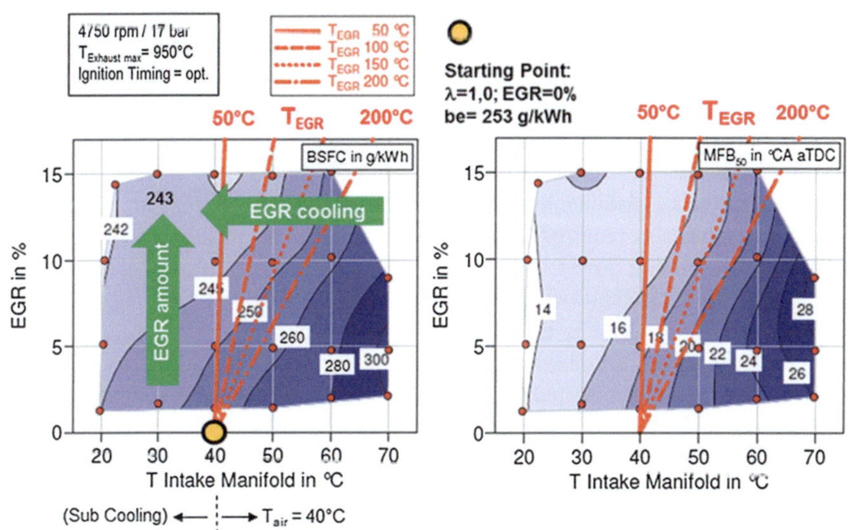

Fig. 16. Brake specific fuel consumption (BSFC, left side) and center of combustion (MFB50, right side) as a function of EGR rate and intake manifold temperature (IMT) at 4750 rpm, 17 bar BMEP

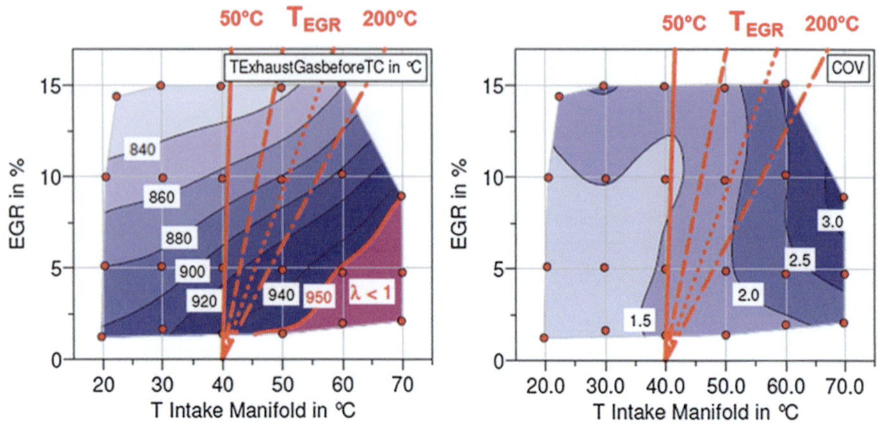

Fig. 17. Turbine inlet temperature (left side) and Combustion stability criterion (COV, right side) as a function of EGR rate and intake manifold temperature (IMT) at 4750 rpm, 17 bar BMEP

40 °C will not further improve the MFB50 position, while its improvement with further cooling is still possible without applying EGR. In this case, cooling is needed to reach 20 °C IMT to get the same fuel consumption and MFB50 position, suggesting a 15% EGR rate will gain the same fuel consumption benefit as additional sub-cooling down to 20 °C IMT without EGR.

Figure 17, left side illustrates the turbine inlet temperature as a function of EGR rate and IMT. At high IMT and little or no EGR, the turbine inlet temperature reaches over 950 °C and fuel enrichment has to be applied to control the exhaust temperature to protect engine hardware, such as the exhaust manifold, turbocharger, and downstream catalyst. Clean EGR enables the avoidance of fuel enrichment, reducing exhaust temperatures below 840 °C with extreme cooling of the EGR to 50 °C and a rate of 15%. Not only does this further improve fuel efficiency, it also enables the use of lower grade materials for the exhaust manifolds and turbocharger, or at least it reduces material failure risks. Admittedly, to achieve this low EGR temperature an effective EGR cooling system is required, and at this operation point near engine peak power, 11 kW of EGR cooling power is required.

Figure 17, right side shows the combustion stability criterion COV, as a function of EGR rate and IMT. COV is the coefficient of variation analyzing mean indicated pressure values, which are measured by individual cylinder pressure sensors. Within the range of the parameters tested (EGR 0–15%, IMT 20–70 °C), the combustion COV is kept under 3%. With EGR temperature below 100 °C, the combustion stability is not sensitive to EGR rate. When an upper combustion stability limit is reached, such as at 2.5% while the EGR rate is at 15% and EGR temperature is nearly 200 °C, a further increase of the EGR rate is possible by further lowering the EGR temperature. Such benefits demonstrate the expanding capability and benefits of clean EGR in the combustion system.

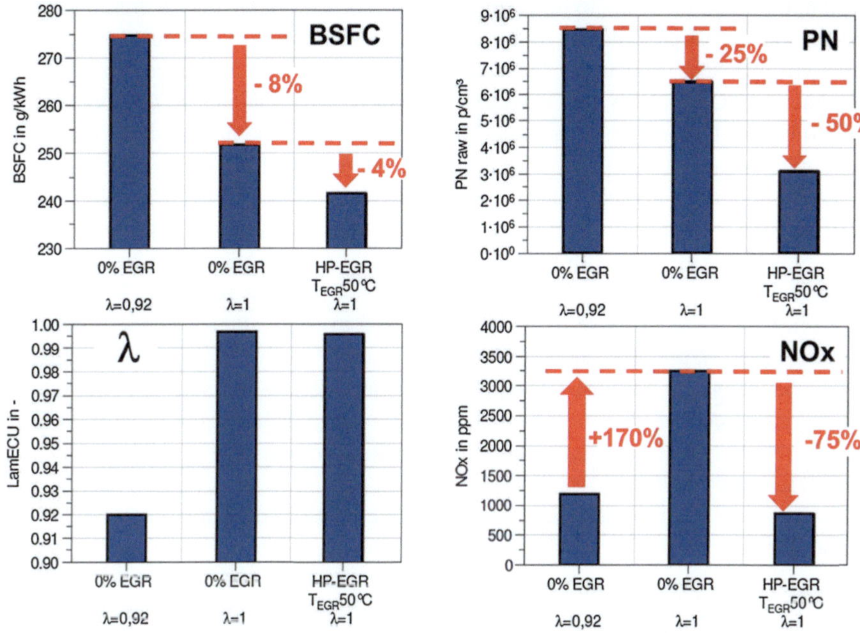

Fig. 18. Fuel efficiency (BSFC), Air/Fuel-Ratio ((λ) NOx and PN emissions at 4750 rpm, 17 bar BMEP with optimal Ignition Timing (knocking threshold)

Figure 18 illustrates the benefit of cleaned and cooled EGR and extreme EGR cooling on engine raw PN- and NOx- emissions. The stock engine uses fuel enrichment (λ = 0.92) under the testing condition (4750 rpm, 17 bar BMEP). By eliminating fuel enrichment the BSFC is reduced by 8%. Hereby the PN emission is reduced by 25%, but NOx emission is increased by 170%. By applying cooled and cleaned EGR the BSFC is improved by further 4% (caused by earlier MFB50 position and less heat losses). In-cylinder temperature is reduced significantly by EGR, see Sect. 2. Hereby the PN and NOx emission are reduced further more. Compare to rich operation point the PN emissions are reduced by ~75% and the NOx emission reduced by ~25% as well.

5 Summary and Conclusion

Cooled external EGR is an effective measure to suppress knocking and to increase the efficiency of a gasoline engine due to an earlier center of combustion, higher isentropic coefficient and reduced wall heat losses. By higher in-cylinder charge mass, the combustion temperature is reduced as PN/NOx emission and exhaust gas temperature as well.

The thermodynamic potential of a cooled Low Pressure EGR system has been investigated on a 1.4 l gasoline engine with CR 12.7 on an engine test bench in different WLTC-relevant operating points. With an EGR rate of up to 20%, a fuel consumption benefit of up to 7% can be achieved. The main reason for this is the advanced center of combustion by up to 6 °CA which leads in combination with the increased total heat capacity of the cylinder charge to a 100 K reduced exhaust gas temperature. NOx and PN raw emission is reduced by 93% and 30%, respectively. Therefore, an additional boost pressure of 175 mbar has to be provided. To ensure an intake temperature of 40 °C, an additional cooling power of 5 kW is required.

The benefit of a cooled + cleaned High Pressure EGR has been investigated as RDE measure on the same engine in a relevant operating point at high engine load (4750 rpm/17 bar BMEP) with basis CR. With this, fuel enrichment can be completely eliminated. FEI of 6–12% were demonstrated by significant reduced PN and NOx emission. Cooling EGR close to 50 °C is required to enable these benefits. As drawback, higher boost pressure, higher in-cylinder pressure and high EGR cooling demand have to be taken into account. Cleaning EGR is an effective measure to minimize the fouling risk of cooled EGR in the engine intake system. Minimizing of NOx in EGR owns furthermore a portion for suppressing knocking combustion, up to 2 °CA benefit in MFB50 were observed.

By means of technologies such as cooled and cleaned EGR, future gasoline engines will be able to comply with RDE legislation and to reduce fuel consumption in relevant drive cycle operation.

Notation

BDC	Bottom dead center
BMEP	Brake mean effective pressure
BTE	Brake thermal efficiency
BSFC	Brake specific fuel consumption
CA	Crank angle
CleanEGR™	Trademark by Tenneco
cGPF	coated Gasoline Particulate Filter
CO	Carbon monoxide
CO_2	Carbon dioxide
COV	Coefficient of Variation
CR	Compression ratio
Cyl	Cylinder
EGR	Exhaust Gas Recirculation
FEI	Fuel efficiency improvement
GPF	Gasoline particulate filter
HC	Hydrocarbon, exhaust gas emission
HP	High Pressure
IMT	Intake manifold temperature
LET	Low end torque
LP	Low Pressure
MFB50	Crank angle, where 50% of mass fraction are burned
N	engine speed
NOx	Oxides of nitrogen, exhaust gas emission

P, p	Pressure
PN	Particulate number
PM	Particulate mass
RDE	Real driving emission, test procedure in EU6dff
T	Temperature
TDC	Top dead center
TWC	Three-way-catalyst
WLTC	World Light-duty Test Cycle
η	Efficiency
κ	Isentropic coefficient
λ	Air/fuel-ratio
c_p	Specific heat capacity
x_{EGR}	Exhaust recirculation gas rate

References

1. Takaki, D., Tsuchida, H., Kobara, T., Akagi, M., et al.: Study of an EGR system for downsizing turbocharged gasoline engine to improve fuel economy. SAE Technical Paper 2014-01-1199 (2014). https://doi.org/10.4271/2014-01-1199
2. Miller, J., Taylor, J., Freeland, P., Warth, M., et al.: Future gasoline engine technology and the effect on thermal management and real world fuel consumption. SAE Technical Paper 2013-01-0271 (2013). https://doi.org/10.4271/2013-01-0271
3. Ohta, Y., Fushiki, S., Matsuo, S.: The New PRIUS Powertrain:The New 1.8L ESTEC 2ZR-FXE Engine with the New Generation Hybrid System. Vienna Motor Colloquium (2016)
4. Hwang, I., Lee, H., Park, H., et al.: Hyundai-Kia's Highly Innovative 1.6L GDI Engine for Hybrid Vehicle. Vienna Motor Colloquium (2016)
5. Hirose, I.: Mazda 2.5L SKYACTIV-G Engine with New Boosting Technology. Vienna Motor Colloquium (2016)
6. Macaudiere, P.: Remaining challenges to face forthcoming RDE and WLTP constraints. In: 10th Integer Emission Summit, Düsseldorf, June 2014
7. Gladstein, Neandross & Associates: Ultrafine particulate matter and the benefits of reducing particle numbers in the United States. MECA-Report, July 2013
8. Alger, T., Gingrich, J., Khalek, I.A., Mangold, B.: The role of EGR in PM emissions from gasoline engines. SAE Technical Paper 2010-01-0353
9. Lepperhoff, G., Houben, M.: Mechanisms of deposit formation in IC engines and heat exchangers. SAE paper 931032
10. Hoard, J., Abarham, M., Styles, D., Giuliano, J.M., et al.: Diesel EGR cooler fouling. SAE Technical Paper 2008-01-2475
11. Yoo, K., Hoard, J., Boehman, A., Gegich, M.: Experimental studies of EGR cooler fouling on a GDI engine. SAE Technical Paper 2016-01-1090 (2016). https://doi.org/10.4271/2016-01-1090
12. Warey, A., Bika, A.S., Vassallo, A., Balestrino, S., Szymkowicz, P.: Combination of Pre-EGR cooler oxidation catalyst and water vapor condensation to mitigate fouling. SAE Technical Paper 2014-01-0636. https://doi.org/10.4271/2014-01-0636

13. Guhr, C., Zellbeck, H.: Turbocharging with low temperature charge air cooling and EGR. MTZ 10/2012
14. Fischer, M., Kreutziger, P., Sun, Y., Kotrba, A.: Clean EGR for gasoline engines – innovative approach to efficiency improvement and emissions reduction simultaneously. SAE Technical Paper 2017-01-0683 (2017). https://doi.org/10.4271/2017-01-0683
15. Pasternak, M., Netzer, C., Mauss, F., Fischer, M., Sens, M., Riess, M.: Simulation of the effects of ignition timing and external EGR on gasoline combustion under knock-limited operation and high speed and load. In: 5th International Conference on Knocking in Gasoline Engines, Berlin (2017)
16. Fischer, M., Günther, M., Röpke, K., Lindemann, M., Placzek, R.: Knock detection in spark-ignition engines. MTZ Worldwide 3/2003 Volume 64 (2003)

MIX
Papier aus verantwortungsvollen Quellen
Paper from responsible sources
FSC® C105338

If you have any concerns about our products,
you can contact us on
ProductSafety@springernature.com

In case Publisher is established outside the EU,
the EU authorized representative is:
Springer Nature Customer Service Center GmbH
Europaplatz 3, 69115 Heidelberg, Germany

Printed by Libri Plureos GmbH
in Hamburg, Germany